QH506 .S28 2005
Sarkar, Sahotra.
Molecular models of
 life :
Northeast Lakeview Colleg
33784000114579

Molecular Models of Life

Life and Mind: Philosophical Issues in Biology and Psychology
Kim Sterelny and Robert A. Wilson, editors

Cycles of Contingency: Developmental Systems and Evolution, Susan Oyama, Paul E. Griffiths, and Russell D. Gray, editors, 2000

Coherence in Thought and Action, Paul Thagard, 2000

Evolution and Learning: The Baldwin Effect Reconsidered, Bruce H. Weber and David J. Depew, 2003

Seeing and Visualizing: It's Not What You Think, Zenon Pylyshyn, 2003

Organisms and Artifacts: Design in Nature and Elsewhere, Tim Lewens, 2004

Molecular Models of Life: Philosophical Papers on Molecular Biology, Sahotra Sarkar, 2005

Molecular Models of Life

Philosophical Papers on Molecular Biology

Sahotra Sarkar

A Bradford Book
The MIT Press
Cambridge, Massachusetts
London, England

© 2005 Massachusetts Institute of Technology

All rights reserved. No part of this book may be reproduced in any form by any electronic or mechanical means (including photocopying, recording, or information storage and retrieval) without permission in writing from the publisher.

MIT Press books may be purchased at special quantity discounts for business or sales promotional use. For information, please email special_sales@mitpress.mit.edu or write to Special Sales Department, The MIT Press, 5 Cambridge Center, Cambridge, MA 02142.

This book was set in Stone serif and Stone sans on 3B2 by Asco Typesetters, Hong Kong, and was printed and bound in the United States of America.

Library of Congress Cataloging-in-Publication Data

Sarkar, Sahotra.
 Molecular models of life : philosophical papers on molecular biology / Sahotra Sarkar.
  p. cm. — (Life and mind)
 "A Bradford book."
 ISBN 0-262-19512-7 (hc : alk. paper)
 1. Molecular biology—Philosophy. 2. Biology—Philosophy. I. Title. II. Series.
QH506.S28 2004
572.8′01—dc22   2004042614

10 9 8 7 6 5 4 3 2 1

To the memory of John Maynard Smith

Contents

Preface   ix
Acknowledgments   xiii
Sources   xv

1   Introduction   1

Part I   Reduction   53

2   Models of Reduction and Categories of Reductionism (1992)   55

3   Genes versus Molecules: How To, and How Not To, Be a Reductionist (2002)   85

4   Reduction: A Philosophical Analysis (2001)   105

Part II   Function   115

5   Reductionism and Functional Explanation in Molecular Biology (1991)   117

6   Natural Selection, Hypercycles, and the Origin of Life (1988)   145

7   Form and Function in the Molecularization of Biology (1996)   161

Part III   Information   181

8   Decoding "Coding": Information and DNA (1996)   183

9   Biological Information: A Skeptical Look at Some Central Dogmas of Molecular Biology (1996)   205

10   How Genes Encode Information for Phenotypic Traits (2004)   261

Part IV Evolution 285

11 Neo-Darwinism and the Problem of Directed Mutations (1992) 287

12 Lamarck *contre* Darwin, Reduction versus Statistics: Conceptual Issues in the Controversy over Directed Mutagenesis in Bacteria (1991) 303

13 Directional Mutations: Fifteen Years Afterward (2004) 347

14 From Genes as Determinants to DNA as Resource: Historical Notes on Development, Genetics, and Evolution (2004) 365

Index 389

Preface

The papers collected in this volume span a period of over fifteen years, from 1988 to 2003. When the first few of them were being written philosophical interest in molecular biology had become negligible. This was strange for at least two reasons: (i) starting in the late 1950s and 1960s, molecular biology had been carrying out a transformation of biological practice and theory that was arguably as radical as that wrought by Darwin and Wallace in the nineteenth century. The extent of this transformation had become fully obvious by the 1980s; historians of biology, in particular, had begun to shift much of their attention to the molecular era.[1] And, (ii) in the late 1960s and early 1970s, philosophers had paid ample attention to molecular biology, debating the virtues of reductionism[2] and exploring the nature of conceptual change in molecular biology.[3]

Yet, philosophical interest in molecular biology declined in the late 1970s and 1980s when, with few exceptions,[4] philosophers of biology focused only on evolutionary biology and, within evolutionary biology, on the problem of identifying units of selection. There should be little doubt that this philosophical scrutiny served evolutionary biology well,[5] particularly because many of the most noteworthy evolutionary biologists of that generation also became engaged in the task of clarifying the foundations of evolutionary theory.[6] However, it also had the unfortunate unintended consequence that the philosophy of biology became out of step with most of contemporary biology. This was the era when ubiquitous molecular variation was believed to be challenging the neo-Darwinian interpretation of evolution, when the unexpected complexity of eukaryotic genetics was tantalizing a new generation of molecular geneticists, and the complexity and apparent universality of genetic regulatory mechanisms was beginning to offer hope for a theory of morphogenesis based on developmental genetics. With the exception of some of the controversies surrounding neo-Darwinism, in the 1980s, philosophers paid little attention to these developments. When the Human Genome Project was being initiated in the late 1980s and early 1990s, philosophers of biology, unlike historians and social scientists, contributed little to the ongoing debates.

The last decade has seen a welcome broadening of scope of the philosophy of biology. Whereas a textbook published in 1993 was written as if evolution were all there is to biology (as if molecular biology does not exist),[7] the most recent one, published in 1999, is catholic in scope, encompassing everything from molecular genetics to ecology.[8] The papers collected in this book are also symptomatic of this broadening. Their immediate focus is molecular biology, but they are all ultimately motivated by the question of what molecular biology contributes to the traditional areas of biological research. A theme that pervades much of this book is that molecular biology provides a unifying perspective on life that complements the one provided by evolutionary biology. Few of the questions discussed in this book are resolved satisfactorily, even from the perspective of the author. However, the main purpose of this book is to encourage further philosophical work on molecular biology because—with the obvious exception of ecology—that is the biology that there is.

Notes

1. Judson (1979) was the pioneering historical work that helped establish this shift.

2. See, for example, Schaffner (1967), Hull (1972), Ruse (1976), and Wimsatt (1976). Sarkar (1989) provides a detailed history.

3. See, for example, Schaffner (1974a,b), Wimsatt (1976), and Maull (1977).

4. Kitcher (1982, 1984) and Rosenberg (1985) are notable as such exceptions.

5. See, for example, Wimsatt (1980), Brandon (1982), and Sober (1984).

6. See, for instance, Lewontin (1970), Mayr (1975), and Maynard Smith (1976).

7. See Sober (1993), which, nevertheless, is admirable within its limited scope.

8. See Sterelny and Griffiths (1999).

References

Brandon, R. N. 1982. "The Levels of Selection." In P. Asquith and T. Nickles, eds., *PSA 1982:* 1. East Lansing: Philosophy of Science Association, pp. 315–322.

Hull, D. 1972. "Reduction in Genetics—Biology or Philosophy?" *Philosophy of Science* 39: 491–499.

Judson, H. F. 1979. *The Eighth Day of Creation*. New York: Simon and Schuster.

Kitcher, P. 1982. "Genes." *British Journal for the Philosophy of Science* 33: 337–359.

Kitcher, P. 1984. "1953 and All That: A Tale of Two Sciences." *Philosophical Review* 93: 335–373.

Lewontin, R. C. 1970. "The Units of Selection." *Annual Review of Ecology and Systematics* 1: 1–18.

Maull, N. 1977. "Unifying Science without Reduction." *Studies in the History and Philosophy of Science* 8: 143–171.

Maynard Smith, J. 1976. "Group Selection." *Quarterly Review of Biology* 51: 277–283.

Mayr, E. 1975. "The Unity of the Genotype." *Biologisches Zentralblatt* 94: 377–388.

Rosenberg, A. 1985. *The Structure of Biological Science*. Cambridge: Cambridge University Press.

Ruse, M. 1976. "Reduction in Genetics." *Boston Studies in the Philosophy of Science* 32: 631–651.

Sarkar, S. 1989. "Reductionism and Molecular Biology: A Reappraisal." Ph.D. Dissertation, Department of Philosophy, University of Chicago.

Schaffner, K. F. 1967. "Approaches to Reduction." *Philosophy of Science* 34: 137–147.

Schaffner, K. F. 1974a. "The Peripherality of Reductionism in the Development of Molecular Biology." *Journal of the History of Biology* 7: 111–139.

Schaffner, K. F. 1974b. "The Unity of Science and Theory Construction in Molecular Biology." *Boston Studies in the Philosophy of Science* 11: 497–533.

Sober, E. 1984. *The Nature of Selection*. Cambridge, Mass.: The MIT Press.

Sober, E. 1993. *Philosophy of Biology*. Boulder: Westview Press.

Sterelny, K., and Griffiths, P. E. 1999. *Sex and Death: An Introduction to Philosophy of Biology*. Chicago: University of Chicago Press.

Wimsatt, W. C. 1976. "Reductive Explanation: A Functional Account." *Boston Studies in the Philosophy of Science* 32: 671–710.

Wimsatt, W. C. 1980. "Reductionistic Research Strategies and Their Biases in the Units of Selection Controversy." In T. Nickles, ed., *Scientific Discovery: Case Studies*. Dordrecht: Reidel, pp. 213–259.

Acknowledgments

Many individuals have contributed to the ideas developed in this book. Special thanks are due to Angela Creager, James Crow, Raphael Falk, Patricia Foster, Trevon Fuller, Justin Garson, Scott Gilbert, Stephen Jay Gould, Paul Griffiths, Larry Holmes, Gregg Jaeger, Robert Josephs, Lily Kay, Evelyn Keller, Michael Lachmann, Manfred Laubichler, Joshua Lederberg, Isaac Levi, Cyrus Levinthal, Richard Lewontin, Wing Ma, Michael Martin, John Maynard Smith, Ernst Mayr, Kelly McConnell, Alexander Moffett, Vidyanand Nanjundiah, Jennifer Noonan, Jessica Pfeifer, Anya Plutynski, Hans-Jörg Rheinberger, Jason Scott Robert, Guido vH. Sandri, Ken Schaffner, Abner Shimony, John Stachel, Howard Stein, Alfred Tauber, David Thaler, Denis Thiéffry, Scott Williams, and William Wimsatt. Alex Moffett deserves special thanks for help with preparing the manuscript. Most of these individuals bear no responsibility for the errors contained in this book. Finally, work on this book was supported by the U.S. National Science Foundation, grant no. SES-0090036, 2002–2003.

Sources

Chapter 2, "Models of Reduction and Categories of Reductionism," *Synthese* 91: 167–194 (1992).

Chapter 3, "Genes versus Molecules: How To, and How Not To, Be a Reductionist," in M. van Regenmortel and D. L. Hull (eds.), *Promises and Limits of Reductionism in Biomedical Sciences* (New York: Wiley, 2002), pp. 191–206 (2002).

Chapter 4, "Reduction: A Philosophical Analysis," in S. Robertson (ed.), *Encyclopedia of Life Sciences*, vol. 16 (London: Macmillan, 2001), pp. 109–113 (see http://www.els.net/).

Chapter 5, "Reductionism and Functional Explanation in Molecular Biology," *Uroboros* 1: 67–94 (1991).

Chapter 6, "Natural Selection, Hypercycles, and the Origin of Life," in A. Fine and J. Leplin (eds.), *PSA 1988: Proceedings of the 1988 Biennial Meeting of the Philosophy of Science Association* (East Lansing: Philosophy of Science Association, 1988), vol. 1, pp. 197–206.

Chapter 7, "Form and Function in the Molecularization of Biology," in A. I. Tauber (ed.), *Aesthetics and Science: The Elusive Synthesis* (Dordrecht: Kluwer, 1996), pp. 153–168.

Chapter 8, "Decoding 'Coding': Information and DNA," *Bioscience* 46: 857–863 (1996).

Chapter 9, "Biological Information: A Skeptical Look at Some Central Dogmas of Molecular Biology," in S. Sarkar (ed.), *The Philosophy and History of Molecular Biology: New Perspectives* (Dordrecht: Kluwer, 1996), pp. 187–231.

Chapter 10, "How Genes Encode Information for Phenotypic Traits." Forthcoming in C. Hitchcock (ed.), *Current Issues in the Philosophy of Science*. (Oxford: Blackwell, 204), pp. 259–274. (A slightly revised and expanded version is printed here.)

Chapter 11, "Neo-Darwinism and the Problem of Directed Mutations," *Evolutionary Trends in Plants* 6: 73–79 (1992).

Chapter 12, "Lamarck *contre* Darwin, Reduction versus Statistics: Conceptual Issues in the Controversy over Directed Mutagenesis in Bacteria," in A. I. Tauber (ed.), *Organism and the Origins of Self* (Dordrecht: Kluwer, 1991), pp. 235–271.

Chapter 14, "From Genes as Determinants to DNA as Resource: Historical Notes on Development and Genetics." Forthcoming in E. Neumann-Held and C. Rehmann-Sutter (eds.), *Genes in Development: Re-reading the Molecular Paradigm* (Durham: Duke University Press). (A slightly revised and expanded version is printed here.)

1 Introduction

The year 1953 is one of iconic significance in the history of modern biology, perhaps even as important as 1858, when Darwin and Wallace first presented the theory of evolution by natural selection. On April 25, 1953, James D. Watson and Francis H. C. Crick published a two-page letter to *Nature* proposing a double-helix model for the structure of DNA, the icon alluded to in the last sentence.[1] The structure was constructed by piecing together an unprecedented variety of evidence: (diffraction) scattering patterns of X-rays from crystals of DNA; changes in these patterns brought about by the presence of water; knowledge that the two nucleotide base pairs, adenine (A) and thymine (T), cytosine (C) and guanine (G) (which are all components of DNA) occur in 1:1 ratios in the DNA of all species without known exception; knowledge of the lengths and strengths of chemical bonds, and so on. Equally important, the double helix was "derived" using physical models built of cardboard, wire, and wood, with hydrogen, carbon, nitrogen, oxygen, phosphorus, and other atoms simply being represented as solid balls with different radii. This modeling technique had entered the biological arena fewer than five years earlier, when Linus Pauling and Robert B. Corey had used it to propose a structural model for proteins that consisted of helical chains of the amino acid residues that composed them.[2] The success of this modeling strategy, which was widely recognized by the late 1950s and 1960s, represented the coming-of-age of the new discipline of molecular biology. Today, for good or for bad, depending on one's point of view, most of biology is molecular biology. Pauling's modeling strategy remains central to the field, though computer graphics have replaced the cardboard, wire, and wood.

The term "molecular biology" was introduced by Warren Weaver in 1938 in an internal report of the Rockefeller Foundation: "And gradually there is coming into being a new branch of science—molecular biology— ... in which delicate modern techniques are being used to investigate ever more minute details of certain life processes."[3] The next decade saw the steady increase in the use of these "delicate" techniques, in particular, X-ray crystallography, to study biological macromolecules "minutely," increasingly with an emphasis on proteins. The central problem was the elucidation of the three-dimensional structures (the relative positions of the atoms) of biological macromolecules. Proteins were singled out because they were believed to be the most important of these macromolecules. In particular, since the founding of biochemistry as a self-standing scientific discipline in the 1920s, enzymes and their interactions had been held to be the key to understanding metabolism (the catch-all term for the complex chemical reaction systems that characterize life).[4] All enzymes are proteins. Until the early 1940s it was believed that the hereditary material (of the genes) was also likely to be composed of proteins. (The nucleic acids, constructed out of only four base types, were believed to be insufficiently complex to be able to specify the immense variety of known genes.)

What is philosophically most important about Pauling's modeling strategy is that, in a sense, its success apparently resolved, for once and for all, the long standing question of whether biology (at least at the organismic level) can be reduced to physics and chemistry. The resolution was in favor of a thoroughgoing physical reductionism, which forms an underlying theme of most of the discussions of this book. However, this resolution raises foundational questions that remain pertinent today. The next section of this introduction (sec. 1.1) sets the context for three essays exploring reductionism (chapters 2–4). It discusses in some detail both the history of reductionism in biology and the developments within molecular biology that have made reductionism more plausible than ever before in the history of biology. All three chapters argue that what is both scientifically and philosophically interesting about reductionism are not such formal issues about the logical structure of explanations but, rather, substantive claims about the world that are of far-reaching significance. Can wholes be successfully decomposed into parts in models of biological systems? Do properties of the parts alone explain all properties of wholes? Such parts–wholes reductionism receives a mitigated defense in these chapters for the case of biology, while its failure at lower levels of organization (for instance, in quantum physics) is duly noted. However, a reduction to molecular interactions is not necessarily a reduction to genetics, once the dream of the hereditarian program in the biology of the twentieth century. All three chapters are skeptical of such genetic reductionism (that is, the thesis that genes alone bear the weight of explanations at the molecular level). The success of physical reductionism should not be misinterpreted as a triumph of genetic reductionism. Molecular biology provides no added ammunition to the hereditarian program in biology. Rather, molecular details show that the claims of both hereditarianism and environmentalism are misplaced. Molecular biology may be reductionist, but it is not simplistic.

As in all other areas of biology, function ascriptions are ubiquitous in molecular biology. Functional *explanations* are typically offered to answer questions of origin: why some biological feature exists (or is the way that it is). The second section of this book (chapters 5–7) turns to the analysis of function ascriptions, mainly exploring the challenge presented by functional explanations to reductionism. Chapter 5 tries to clarify exactly how functional explanations pose a problem for reductionism. It argues that functional explanations are usually invoked to answer questions of origin rather than the questions of mechanism that form the staple of molecular biology. Thus, even if there is a conflict between functional explanation and reductionism, that conflict may be confined to a narrowly circumscribed set of questions. Chapter 5 also sketches a model that provides a potential resolution of this conflict. Chapter 6 gives more details of that model. However, both chapters may be criticized for using a narrow evolutionary concept of function, which may have to be broadened to capture many functional attributions in molecular biology. Meanwhile, chapter 7 notes that concern for function—often taken to be definitive of the practice of biology—may not capture all the aims of biological research. Following analytic techniques to the limits of their power (the scientific equivalent of formalism in the visual arts) may also guide research programs. The quest for sequences in the Human Genome Project may perhaps best be understood in this way, rather than in terms of any epistemological benefit it may provide. This chapter was written as a contribution to a conference on the role of aesthetics in science, and its conclusions remain rather tentative. Section 1.2 of this introduction puts the discussions of chapters 5 through 6 in context mainly by analyzing whether they take too narrow a view of function to capture all function ascriptions in molecular biology; no final resolution of this issue is offered (as is typical of most discussions in this introduction). Section 1.2 also puts the more speculative observations of chapter 7 in context by defending the view that aesthetic considerations have always been part of scientific heuristics and the motivations that lead to the pursuit of science in the first place.

Part of the historical importance of the double-helix model was that it opened up a radically new possibility for the foundations of biology, one based on the concept of information. Section 1.3 of this introduction turns to the question of how the informational interpretation of biology interacts with the reductionist program (see chapters 8–10). Information is not a physical parameter (that is, one that occurs in physical theory),[5] and thus the stage is set for a potential conflict with reductionism. But is informational talk necessary in molecular biology? Worse, is information in molecular biology merely a metaphor masquerading as a theoretical concept? Chapter 8 summarizes the biological reasons for skepticism about information. Chapter 9 lays out the philosophical ramifications and argues that either all talk of information should be abandoned in favor of a thoroughgoing physicalism, or, a concept of information adequate for molecular biology should be properly explicated. Although both options are left open, the first is presented as being more plausible. Nevertheless, chapter 10 begins an exploration of the second option, after carefully delimiting the scope of informational talk in molecular biology. Section 1.3 provides a new analysis of the cognitive reasons for the perceived significance of the DNA double-helix model when it was first constructed. It also emphasizes problems with the highly popular but conceptually misleading metaphor of DNA as language.

"Nothing in biology makes sense except in the light of evolution," as Dobzhansky famously quipped.[6] At one level, Dobzhansky is obviously correct: without the context of the evolutionary history of any taxon, its biological features—in particular, those features that make it distinct from other taxa—are almost impossible to understand. Nevertheless, molecular biology provides a different type of unity, a different unifying framework for biology at the level of basic physical constitution and mechanisms. This perspective currently cohabits biology in uneasy tension with the received framework of evolutionary theory that was first constructed in the 1930s. The last four essays of this book (chapters 11–14) explore that tension. The first three concern one rather specialized topic: whether the molecular complexities of mutagenesis can be straightforwardly accommodated in the received framework of evolutionary theory. Chapters 11 and 12 analyze the controversy that erupted in the early 1990s about the possibility of directional mutagenesis in bacteria. In retrospect, given that the claimed phenomena were delimited to microorganisms, the vehemence of that controversy remains a topic to be explored by historians and sociologists.[7] Chapter 13 attempts to summarize the state of the controversy some fifteen years afterward—it is likely that this chapter will also be controversial. Finally, chapter 14 touches on what molecular biology, especially the developments of the last five years, may contribute to the transformation of evolutionary theory to take into account morphogenetic development: it is an essay in the emerging field of developmental evolution. These essays barely scratch the surface of the evolutionary implications of postgenomic molecular biology. Were it not likely that such an observation would be quoted out of context and otherwise misused by miscreants such as Intelligent Design creationists, one would be tempted to quip (against Dobzhansky) that much of the received framework of evolution makes no sense in light of molecular biology.

1.1 Reductionism

As chapter 2 notes, the program of reductionism in the natural sciences goes back to the mechanical philosophy of the seventeenth century, which required that all physical laws be explained by local contact interactions between impenetrable particles of matter. Mechanical explanations replaced ones involving complicated combinations of Aristotelian elements and qualities. Reduction, as mechanical explanation, is thus a type of explanation. Reductionism is the doctrine that such explanations should be pursued because they are likely to be forthcoming. Sometimes, that doctrine also encompasses the view that these explanations will exhaust all interesting phenomena; this stronger version of the doctrine will not be assumed here.

Discussions of reductionism (see sec. 1.1.1) formed an integral part of natural philosophy, especially in the nineteenth century. The linguistic turn taken by logical empiricism in the 1930s led to the demise of traditional natural philosophy—or at least its demotion to the minor leagues—and its replacement by an anorexic philosophy of science in which the only legitimate questions were those that could be reduced to questions of syntactic or semantic form.[8] This context strongly influenced the tenor of subsequent discussions of reductionism, beginning with Ernest Nagel and Joseph Henry Woodger in the late 1940s. In effect, Nagel and Woodger assumed that explanation was adequately explicated by the covering law model popularized by Hempel.[9] Reduction then becomes a form of intertheoretic explanation, in which the *explanans* and *explanandum* are both theoretical laws. This model of reduction implicitly assumes: (i) theories play a central role in reduction; and (ii) all philosophically interesting questions about reduction should be formulated as questions about various components of formal models of reduction. These assumptions dominated discussions of reduction in the 1960s and 1970s.

Nevertheless, there were several attempts by both biologists and philosophers to move discussions of reduction beyond these two limiting assumptions.[10] Chapter 2, first published in 1992, attempts a taxonomy of all the then-extant models of reduction. With hindsight, it uses a category of "explanatory reductionism" to capture those models that focus on the substantive rather than formal assumptions involved in reductionist explanation. Chapters 3 and 4, written a decade later, incorporate a sharp distinction between formal and substantive issues and focus on the latter. They defend a model of "strong" reduction in which properties of wholes are explained entirely through properties of their parts. However, these chapters gloss over some of the subtleties of the substantive assumptions made in molecular biology; section 1.1.1 tries to remove this lacuna while providing the historical context of these assumptions. Chapter 3 also argues against a facile genetic reductionism; section 1.1.2 elaborates on that point. Finally, chapter 4 uses the model of reduction developed here to analyze reductionism in other areas of biology.

1.1.1 Wholes and Parts

Mechanical explanation entered the life sciences in 1628 when Harvey described the circulatory system of animals and argued that the heart was a pump. Harvey described how arteries pumped blood from the heart, veins returned blood to the heart, and valves in veins prevented blood from flowing in the wrong direction. The new picture displaced the one inherited from Galen (ca. 130–ca. 200) in which new blood was created at the heart. Harvey's model was instrumental in providing specific support for the view, also associated with Descartes, that the body of a living organism can be viewed as a machine.[11] Mechanical explanation, and the associated program of mechanism, allowed only for what were traditionally called "efficient causes" (which temporally precede their effects) and, moreover, required that all causes be mediated entirely by local (spatially and temporally contiguous) interactions. The mechanical philosophy irreversibly altered the course of physics and chemistry.

The mechanistic view of living organisms gained support over time, but it remained controversial up to the early twentieth century. However, what was allowed to be a mechanism became broadened to include not only contact interactions, but also central forces, and eventually all chemical and physical interactions.[12] In the eighteenth century, vitalists challenged mechanism on ontological grounds by positing vital interactions in organisms beyond those that operated in inanimate matter. In the mid-nineteenth century, mechanism was challenged by the so-called teleomechanists who admitted the power of mechanistic explanations but argued that they remained incomplete for living phenomena, which required their supplementation by a teleological principle.[13] This teleological principle did not involve any ontological assumption about the existence of special substances or interactions. Rather, claiming to follow Kant's *Critique of Judgement*, teleomechanists argued that the "efficient causation" of mechanistic explanation would not suffice to explain the goal-directedness of biological organisms. Mechanistic explanation must thus be epistemologically supplemented by a teleological principle. Variants of teleomechanism, unlike their vitalist predecessors, remained influential in biology until the beginning of the twentieth century.

In the late nineteenth century, some physiologists, including Claude Bernard and Christian Bohr, also argued that the self-regulative goal-directedness of living phenomena and the "cooperativity" of the parts of organisms would not entirely succumb to mechanistic explanation. In 1904, Bohr reported the so-called the Bohr effect, which epitomizes cooperativity:[14] at very low oxygen concentration, the binding of oxygen with hemoglobin is low. However, it increases sharply as the oxygen concentration increases before leveling off, resulting in a sigmoid-shaped curve. Hemoglobin shows "cooperative" behavior: when some of it binds oxygen, it helps more hemoglobin also to bind oxygen, until saturation is reached. For antimechanists, cooperative behavior is supposed to be inexplicable using only individual properties of parts and without invoking collective properties of the whole: the sigmoid binding curve is supposed to show that "the whole is more than the sum of its parts."

Traditional mechanism is what is here being called "strong reductionism," or, more succinctly, "reductionism."[15] By the time Jan Christian Smuts coined the term "holism" in 1926 to describe the antimechanist project,[16] it had become reasonable to think that mechanism would finally emerge triumphant in this long-running dispute. As early as the first decade of the twentieth century, Jacques Loeb had laid out the mechanistic program in full detail.[17] Meanwhile, between 1900 and 1920, Frederick Gowland Hopkins, an avowed mechanist, had established biochemistry as a discipline in its own right, centered around the study of enzymes. While some physiologists such as J. S. Haldane continued to espouse the holist alternative, other biologists equally versant in physiology, such as Lancelot Hogben, were adamantly in favor of a mechanistic interpretation of all of biology.[18] By the 1930s, within biology, the philosophical dispute between mechanism and holism had become replaced by experimental research programs designed to explore the structure of biological materials at increasingly finer resolutions. Central to these was the X-ray crystallography of biological macromolecules, pioneered by J. D. Bernal, which was exploited brilliantly by Pauling and those who followed him.

For reductionists, structures were interesting because they formed the basis for biological explanations. The double helix illustrates this point beautifully. In the double helix, two phosphate chains, running antiparallel to each other, form the backbone of the helix. The DNA bases, A, C, G, and T, are stacked inside the backbone. Because of restrictions on possible hydrogen bonds, A is always coupled with T, and C with G, which explains the base pairing (or 1:1 ratios) mentioned earlier. These ratios had been reported by Chargaff in 1950;[19] for the first time they were explained. Other than base pairing, there is no restriction in the arrangement of the bases. One sequence (along one of the helices) can be entirely arbitrary; because of base pairing, it then completely determines the sequence along the other helix. Thus, a practically unlimited variety of DNA sequences and therefore of genes is possible. This explains how macromolecules built from only four nucleotide bases can specify the thousands of known genes. Finally, the structure immediately suggests how it can be reproduced: by base pairing, each helix can serve as a template for the formation of a new replica.

Introduction

In the double-helix model, wholes are explained in terms of their parts, exactly as reductionism demands. Here, some care must be taken to make sure that the reductionist claim does not become philosophically vacuous. As noted before, since at least the nineteenth century, antireductionists generally do not make any ontological claim beyond those that are admitted by reductionists. They do not claim the existence of vital forces or peculiarly living components of matter.[20] What is at stake is the epistemological question of what can legitimately be invoked in an explanation. Here, the reductionist has a more restricted repertoire available than the holist. For an explanation to be reductionist, two criteria must be satisfied:

(i) the properties invoked in explaining some feature of a whole must be properties of the parts alone, each definable without reference to some other part;[21] and

(ii) the *weight* of the explanation must be borne by these properties of the parts. The relevant explanatory factors do not include every factor that has some influence on the behavior being explained. The context determines what is explanatorily relevant, that is, what bears the weight of explanation.[22] Within epistemology, this concept of explanatory weight has proved notoriously difficult to explicate. It will be assumed here that some sort of substitutional insensitivity criterion will be adequate. If some factor in an explanation can be substituted for without significant change of behavior, that factor does not bear explanatory weight.[23]

That reductionist explanation in molecular biology can potentially fail shows that, at the very least, reductionism is not an empty doctrine.

Such "structural" reductionist explanation implicitly invokes four seemingly innocuous rules about the behavior of biological macromolecules in this context:

(i) the *weak interactions* rule: the interactions that are critical in molecular interactions are very weak;[24]

(ii) the *structure-function*[25] rule: the behavior of biological macromolecules can be explained from their structure as determined by techniques such as crystallography;

(iii) the *molecular shape* rule: these structures, in turn, can be characterized entirely by molecular size and, especially, external shape, and some general properties (such as the hydrophobicity) of the different regions of the surfaces;

(iv) the *lock-and-key fit* rule: in molecular interactions, molecules interact only when there is a lock-and-key fit between the two molecular surfaces. There is no interaction when these fits are destroyed. A lock-and-key-fit thus based on shape is one way of achieving specificity, that is, that a biologically active macromolecule interacts exactly with one (or at most a very few) other entities.

Because they explain specificity, the molecular shape and the lock-and-key fit rules are probably the most important of these rules. These four rules are all rules of macromolecular physics. They are at best only approximately derivable from physics at lower levels of organization. Thus this success of reductionism cannot be taken as a vindication of any type of fundamentalist physicalism that requires that fundamental physics at the lowest level of organization provides explanations of phenomena at all higher structural levels.[26] This situation merits much more philosophical reflection than has so far been afforded to it. Should the triumph of reductionism here be regarded as only a pyrrhic victory, at best? Or, is it that there is something very peculiar about the macromolecular and some higher levels of organization that permit the success of reductionism? Is there any connection between the ability to analyze wholes into parts successfully and the emergence of the complicated phenomena associated with life?

It is truly remarkable how powerful these apparently innocuous rules are. Only two examples will be further analyzed here, both selected because they deflate cherished examples from the holists' repertoire:

(i) Recall the discussion of the Bohr effect, in which cooperativity between the parts is supposed to show that the whole is more than the sum of the parts. In the early 1960s, in another of early molecular biology's most significant achievements, Jacques Monod and Francois Jacob developed a model of "allostery" that dispelled any doubt that such cooperative phenomena could be given standard reductionist explanations.[27] Protein molecules such as those of hemoglobin are "oligomers" consisting of several "protomers," which are single polypeptide chains. Hemoglobin, for instance, has four protomers. The allostery model starts with the spatial structure of the oligomer and makes four assumptions: (a) identical protomers occupy equivalent positions in the oligomeric protein; (b) each protomer contains exactly one receptor site for the reactant; (c) the oligomer has at least two distinct conformations available to it—the affinity of the receptor sites for the reactant may be different in these two conformations; and (d) this affinity depends on the conformational state of the oligomer and, therefore, the protomers, but not on the occupancy of the neighboring sites. From these assumptions, using standard chemical kinetics, it is trivial to derive the sigmoid binding curve of the Bohr effect. Had (d) not been satisfied, then the explanation would not have been accomplished using properties of the parts alone; there would then have been some solace for holists.

(ii) Lactose digestion in the bacterium, *Escherichia coli*, is negatively regulated by feedback. The enzyme, β-galactosidase, which digests lactose, is produced by *E. coli* only in the presence of that substrate. Feedback regulation, which used to be called "homeostasis" by physiologists, was also traditionally part of the holists' repertoire.[28] Systems exhibiting feedback regulation were supposed to have such complex interactions that it would be impossible to explain the systems' behavior by (conceptually) dissociating the systems into parts and invoking only the properties of those parts. Explanations were supposed to have to refer irreducibly to states of the whole. In the late 1950s, in yet another of molecular biology's early triumphs, Jacob and Monod constructed the "operon" model to explain feedback regulation.[29] The first critical assumption in this model is a distinction between structural and regulatory loci. Structural loci produce proteins, whereas regulatory loci are involved in the control of protein production at structural loci. In the operon model, a regulator locus is responsible for the synthesis, at a slow constant rate, of a repressor molecule (which is usually also a protein). The repressor molecule binds to an operator locus in the absence of the inducer molecule, in this case, lactose. Presumably because of steric hindrance, when the repressor molecule is bound to the operator locus, synthesis of β-galactosidase does not take place. In the presence of the inducer molecule, because of interactions between it and the repressor molecule, the latter is no longer bound to the operator locus and β-galactosidase can be produced by the usual cellular transcription and translation processes. When all the lactose has been digested, the repressor molecule binds to the operator site again, and the production of β-galactosidase stops. The operon model provides a trivial mechanistic explanation of something that holists found mysterious.

1.1.2 Genes

Proteins and nucleic acids received equal attention in the molecular biology of the 1950s and 1960s, but, since the 1970s, nucleic acids have been at the center of research in molecular biology, even as the sway of molecular genetics has been replaced by that of genomics.[30] A central concern of the philosophy of biology in the late 1960s and early 1970s was whether classical genetics was being reduced to molecular genetics (see chapter 2). Antireductionists usually argued that either classical genetics was being replaced by molecular genetics, or molecular genetics involved an extension of classical genetics that is not correctly viewed as a reduction.[31] Over the years, some measure of consensus has been reached on four points:

(i) If reduction is viewed as necessarily a relation between theories, as construed by logical empiricists, there is no question of a reduction of classical genetics to molecular genetics. The latter does not have laws and theories (as these are formalized by logical empiricists—see chapter 2); for some antireductionists even classical genetics lacks laws of the relevant sort.[32] A search for intertheoretic reduction has not played any significant role in the research strategies of molecular biology.[33]

(ii) There is no question of molecular genetics replacing classical genetics, particularly the use of Mendel's rules, not only to predict patterns of gene transmission into the future, but also retrospectively to infer patterns of evolutionary change in the past. Molecular detail—variation at the level of protein and DNA sequences—permits such retrospective inferences with greater precision than any method previously available.

(iii) There is similarly little question that molecular genetics provides explanations of classical regularities, including Mendel's rules.[34] Moreover, the molecular mechanisms that are operative also show the rather unexpected extent to which these rules may be violated. As is commonplace among reductions, the reducing rules, besides explaining, partly correct the reduced rules.

(iv) The critical problem in interpreting the development of molecular genetics as a reduction is that of providing a molecular "definition" of the classical gene. Ever since it became clear that there was no one-to-one correspondence between genes and DNA sequences (see chapters 4, 8, and 9), it also became clear that any molecular "definition" of the classical gene cannot have the logical form of a biconditional. However, biconditionals are not necessary for explanations of classical regularities from the molecular level even if explanations are supposed to have the form required by Hempel's covering law model.[35]

The upshot of these developments is that, if reduction is construed necessarily as an intertheoretic relation, and theories are construed in the traditional manner of logical empiricism, there are problems with the reduction of classical genetics to molecular genetics. However, there would also be almost no successful reduction anywhere in the history of science. If, however, attention is focused on substantive issues—as is argued for throughout this book—the case for successful reduction is compelling.

Far more philosophically interesting than these somewhat arcane questions about definability and reductionism, though surprisingly rarely discussed by philosophers, is molecular biology's contribution to the nature–nurture dispute. Chapter 4 details how the success of molecular genetics led to the hope that developmental genetics will provide a sound theory of development. Proponents of the Human Genome Project exploited that hope to initiate massive blind DNA-sequencing projects: the full sequencing of entire genomes without prior concern for the functions of the sequences.[36] The same hope led to numerous, often irresponsible, claims that complex human behaviors (including male sexual orientation, schizophrenia, alcoholism, autism, reading disability, bipolar affective disorder [or manic depression], neuroticism, adolescent vocational interests, spatial and verbal reasoning, alleged differences in intelligence, etc.) had genetic etiologies.[37] From this perspective, phenotypic traits are being explained from a genetic basis: the framework is one of genetic reductionism. Not one of these claims of genetic etiology has survived further experimentation and scrutiny (as chapter 4 notes), though it would also be irresponsible to argue that inherited biological constitution has no role in the etiology of human behavior.

As chapter 4 records, the failures of genetic reductionism are fairly transparent. Nevertheless, two points deserve more emphasis than they receive there:

(i) Molecular biology has little to do with the claims of genetic etiology mentioned in the last paragraph. They were based on classical genetic methods, sometimes using molecular markers, that is, molecular types, to distinguish phenotypes. Such a use of markers, however, does not involve explanation at the molecular level.

(ii) Molecular biology has done much to demonstrate that genetic reductionism itself is sterile by showing how complex the path is from DNA sequence to phenotype, even for ordinary morphological phenotypes, let alone complex behavioral ones. From the molecular perspective, simple genotype-phenotype determinations are exceptional; phenotypic plasticity is ubiquitous.[38] In the proteomic era, toward which postgenomic biology seems to be heading (see chapter 14), genes are but one of many interacting resources participation in the construction of phenotypes.

Finally, given what has so far been said, it may be tempting to conclude that the advent of molecular biology has shown that the traditional nature–nurture dispute itself is sterile. However, any such claim would be premature, involving an illegitimate conflation, though it may turn out to be correct in the future. Molecular developments have shown that the construction of phenotypes can receive neither a genetic nor an environmental etiology alone. Thus, there is more to biology than genetics, and the two should not be conflated. Identifying the natural with the genetic is illegitimate. However, suppose that the "nature" of the nature–nurture dispute refers to a putative biological substratum, a result of the interactions between the genes and environmental factors during development, that can be operationally distinguished from cultural factors. To the extent that such a distinction can be usefully maintained, nature–nurture questions may not be entirely misplaced.

1.2 Functions

The somewhat eccentric English naturalist and traveler, Charles Waterton, lived in what was British Guiana (now Guyana) from 1804 to 1812, returning there in 1816, 1820, and 1824.[39] Watertown was fascinated with sloths and even kept one (presumably a three-toed sloth, *Bradypus tridactylus*[40]) in his room for several months. Waterton's initial description of a sloth, in his entertaining 1825 travelogue, *Wanderings in South America*, emphasized its apparent maladaptiveness:

On comparing him to other animals ..., you could perceive deficiency, deficiency and super-abundance in his composition. He has no cutting teeth, and though four stomachs, he still wants the long intestines of ruminating animals. He has only one inferior aperture, as in birds. He has no soles to his feet, nor has he the power of moving his toes separately. His hair is flat, and puts you in mind of grass withered by the wintry blast. His legs are too short; they appear deformed by the manner in which they are joined to the body ..., and his claws are disproportionately long. Were you to mark down upon a graduated scale, the different claims to superiority amongst the four-footed animals, this poor, ill-formed creature's claim would be the last upon the lowest degree.[41]

But, Waterton wisely observes later: "This singular animal is destined by nature to be produced, to live and to die in the trees; and to do justice to him, naturalists must examine him in his upper element."[42] Once the arboreal perspective is adopted, the sloth's apparent malformations are recognized as functional for a life largely spent hanging from branches. Waterton proceeds to give one of the first reasonably accurate descriptions of the natural history of sloths and justly takes the Comte de Buffon to task for assuming that the sloth must live its life in misery because of the poverty of its design. Modern research has fully vindicated Waterton's assessment.[43]

That organismic features, the color, shape, size, and organization of parts, as well as behaviors, often serve functions was recognized at least as early as Aristotle. Ever since then, considerations of function have played a central role in analyzing biological systems. Function ascriptions are ubiquitous in molecular biology, as they are in every other area within biology. Also going back to Aristotle is the tradition of using the function of some feature to explain its origin, why it is there in the sense that it has the specific properties that it has. That tradition continues to this day, even in molecular biology, as chapter 5 attests. Arthropods (including insects) usually have a hardened cuticle with waxy waterproofing. They have them because these cuticles functioned to protect the body from desiccation when the first arthropods invaded the land after their evolutionary origin in the oceans.[44] Many mammals sweat, but humans sweat most profusely and efficiently. Sweating functions for heat tolerance.[45] Among mammals, probably only camels are more heat tolerant than humans. Questions of origin, whether of arthropod cuticles or human sweating behavior, thus receive functional answers.[46] The important distinction, here, is between such *why*-questions and *how*-questions. The latter probe how some feature is brought about, what mechanism leads to its production. For Aristotle, how-questions were to be answered by appeal to efficient causes; why-questions by appeals to final causes. This distinction is periodically rediscovered, apparently independently. For Mayr, writing in 1961, the same distinction is one between "proximate" and "ultimate causes."[47]

The distinction here is between reductionist explanations, which invoke only temporally antecedent conditions, and functional explanations, which refer to the future. The function of a feature refers to some future effect of that feature's being there. Resistance to desiccation is a result of arthropods already having hardened cuticles with waxy waterproofing. Efficient cooling is a result of the human ability to sweat profusely. However, not all effects of the possession of a feature are functions: the waxy surface of arthropod cuticles also make them light-reflective, but that is not one of its functions. Sweating temporarily reduces the weight of human bodies (though only marginally) but this reduction of weight is not one of its functions. Only those effects that somehow seem to serve a future purpose can constitute functions. Functional explanations thus seem to violate any requirement that adequate explanations must refer only to antecedent conditions. This requirement amounts to denying that there are purposes in nature, that is, denying that nature is in any fundamental sense goal-directed. Ever since the rise of the mechanical philosophy in the seventeenth century this requirement has been a metaphysical presupposition of the physical sciences. Its justification has been the spectacular success of those sciences under the new metaphysics, compared to what had been achieved in the centuries spent under the banner of teleology. This metaphysics was systematically extended to cover the biological sciences during the last few centuries with equal success, particularly with the advent of molecular biology in the mid-twentieth century. But functional attributions remained part of biology. The task of establishing consistency between functional and reductionist explanation is the problem of naturalizing function.

1.2.1 Broad-sense Functions

For most philosophers of biology, the problem of naturalizing function was effectively solved by Darwin and Wallace's theory of natural selection, though the details of various proposed solutions vary. Roughly, the solution is simply that some effect of a feature is function if it is associated with an increased fitness. Effects that do not increase fitness are not functions. Because of this increase in fitness, a feature with a function is retained during evolution once it arose by blind variation; a feature with only nonfunctional effects is not similarly retained. In fact, with continued selection, any feature with a function is likely to become more ubiquitous in that species. In this sense the function of a feature explains why it is there, thus answering the question of its origin.[48] Functions are thus intimately tied to fitness enhancement. Here, these functions will be called "functions in the narrow sense" or, simply, "narrow-sense functions." They solve the problem of restricting explanations to antecedent factors, because a feature's potential to increase fitness in a given environment is present temporally before any effect of a feature is expressed. This is the etiological view of biological functions. It is assumed in chapters 5 and 6.

A different narrow-sense view of function, but one also tied to increased fitness, is the propensity theory, according to which the functions of a feature are its adaptive effects (that is, those that increase fitness) rather than the effects for which it is an adaptation (effects favored by natural selection because of which the feature exists).[49] The propensity theory effectively denies that functional attributions arise as answers to questions of origin. Though these two views are usually presented as alternatives, there is no contradiction between the two views because they emerge from two different questions: the etiological theory asks why a trait arose irrespective of what it does, and the propensity theory asks what it does no matter how it arose. Moreover, in an important sense, the etiological view is reducible to the propensity view. Faced with a question of origin—for instance, why do birds have feathers?—the propensity view can invoke past functions that were functions at the relevant past times because they then enhanced fitness. In the early evolution of birds, feathers helped in thermal regulation and thus enhanced fitness. The etiological view thus consists of the reconstruction of past functions that may be quite heterogeneous—from helping in thermal regulation, feathers eventually came to help achieve buoyancy for flight.[50] This heterogeneity is a result of the assumption that functional attributions are necessarily made in response to answers to questions of origin.

The propensity view allows a broader category of function than the etiological view. Moreover, the propensity view has another advantage over the etiological view: it seems to have the potential to capture the common practice in biology (and elsewhere, for instance, psychology) of making functional attributions from present roles with no reference to evolutionary history.[51] Nevertheless, there is at least one reason to suspect that narrow-sense function, even with the propensity rather than the etiological view, does not fully capture such usage. Ernst Mayr, for example, distinguishes between functional biology and evolutionary biology and claims: "[t]he functional biologist is vitally concerned with the operation and interaction of structural elements, from molecules up to organs and whole individuals. His [*sic*] ever-repeated question is 'How?' How does something operate, how does it function?"[52] It is the "functional" biologist who is concerned with proximate causes. Functions, in this sense, refer to roles played by particular mechanisms in the networks of interactions by which organisms carry out their typical activities.[53] Mayr had molecular biology in mind in the passage quoted above. The function of the gene (allele) for sickle-cell hemoglobin is to encode that hemoglobin irrespective of whether it increases fitness (as, for instance, in heterozygotes for whom it reduces susceptibility to malaria) or decreases fitness (in homozygotes in whom it results in sickle-cell disease). In mammals, the function of the heart is to pump blood even in females well past their reproductive age: this function clearly does not enhance fitness.[54] Narrow-sense function is so closely tied to adaptation that it makes it impossible to attribute functions to any feature that is not adaptive. If it turns out to be true that, at the molecular level, many features of organisms are not fitness-enhancing, as the neutral theory of molecular evolution holds, there can be very few attributions of function at that level. Even at higher levels of organization, the extent to which organismic structures and behaviors are adaptations is a matter of ongoing controversy.[55] Using "function" only in its narrow sense would require a radical revision of the customary linguistic practice of contemporary biology, especially molecular biology.

Thus what seems to be required is a broader sense of "function" ("broad-sense function") so that all customary uses of the term in contemporary biology may be captured. The problem, once again, is that of naturalizing function—distinguishing functional features from nonfunctional ones and doing so with reference only to factors that are temporally antecedent to functional attributions. One possible solution—which requires much more careful elaboration than can be attempted here—is to invoke a principle of persistence: an effect of some structure, A, is a function if it contributes to the persistence of some system, B, of which A is a part; the persistence of B is to be defined contextually to allow for some state changes and disallow others.[56] Thus the function of the heart is to pump blood even in mammals past reproductive age because it contributes to the persistence of the individual with the heart. For broad-sense function, the principle of persistence plays the role that the principle of natural selection plays for narrow-sense function.[57] Note, *critically*, that the principle of persistence is also less problematic than the principle of natural selection from the perspective of having a reputable physical basis: organisms are structures that tend to persist because of their physical constitution. The mechanisms responsible for the persistence of organisms are precisely the ones that are being elaborated by molecular biology. Finally, narrow-sense function is a subcategory of broad-sense function: organisms must persist in order to reproduce and evolve by natural selection.

Recourse to broad-sense function constitutes a departure from usual practice in the philosophy of biology, but it has the advantage of keeping the discussions closer to scientific practice. Without some expansion beyond the narrow sense, it is likely that a sense of function (for instance, that used in chapters 5 and 6) cannot do justice to functional ascriptions in molecular biology.

As a final caveat, it is also not clear that, in answering questions of origin, the sharp distinction between "proximate" and "ultimate" factors will continue to be helpful. Consider a crystal: any answer to a question of its origin will refer only to the efficient mechanisms by which it was brought about. The why-question has no different answer than the how-question. In the context of biological features, the distinction between proximate and ultimate answers depends heavily on the assumption that why-questions traditionally receive a radically different answer from how-questions. Answering why-questions involves recourse to the theory of natural selection, which is conceptually disparate from the physical and chemical theories that are invoked to answer how-questions. There are two related reasons why this situation may change:

(i) If it turns out that adaptation is not quite as ubiquitous as it has traditionally been assumed, then appeal to natural selection will not provide an answer to all questions of origin. There will still be an evolutionary story to be told, but the factors invoked, such as chance production and survival or physical rules of body construction, will often be the usual proximate factors that are used to answer questions of mechanism.

(ii) Even in the presence of adaptation, recent attempts to understand development in an evolutionary context have begun to integrate these proximate factors into a theory of phenotypic evolution that looks very different from the traditional genetic theory of natural selection that goes back to the late 1920s and 1930s (see sec. 1.4 below). In contemporary evolutionary biology, the distinction between proximate and ultimate factors is becoming blurred. The loss of this distinction affects only the analysis of narrow-sense function in terms of adaptive value; the status of broad-sense function is not affected.

1.2.2 Biology beyond Functions

The original aim of the Human Genome Project (HGP) was to sequence the entire human genome with no concern for the function of any part of the sequence.[58] The project was soon expanded to include the sequences of several other species. Why bother? Going beyond the explicit epistemic aims of the HGP's proponents (about which there has always been ample room for justified skepticism), chapter 7 suggests that part of the HGP's appeal was aesthetic. However, the operative aesthetic motivation was not some ascription of beauty or taste. Rather, it was the pursuit of a technique (a "formalism"), in this case, sequencing, to its limit. Just as formalism in the visual arts of the 1920s was supposed to reveal deep features of the human spirit, the sequence was supposed to reveal the essence of humanity.

Given the dearth of work on the aesthetics of science—even while it is freely acknowledged that scientists routinely and explicitly appeal to aesthetic norms—it is hard to evaluate the soundness of the argument of chapter 7 is. Suffice it here to note that there is another interpretation of the initiation of the project that does not appeal to the explicit epistemological aims of the HGP's proponents. The various genome-sequencing projects can also be viewed as a continuation of the descriptive project of biology that goes back at least to Aristotle: describe every organism in excruciating detail simply because it is there, no matter whether the description seems capable of providing any further insight. Part of the motivation—and value—of such descriptive projects has been that such detailed descriptions have often eventually yielded other insights, as has also been claimed for the HGP.[59]

1.3 Biological Information

As noted at the beginning of this introduction, chapters 8 through 10 concern biological information. All three chapters are concerned with the

Introduction

question of whether the concept of biological, in particular, genetic, information is coherent. Chapter 8 may be regarded as an extended abstract of chapter 9 (which is rather long). Chapter 8 was commissioned as a generally accessible summary of chapter 9; there is thus much in common between the two chapters. Chapter 10 brings those discussions up to date. Consequently, this section of the introduction will turn to two other issues regarding the informational interpretation of biology, rather than elaborate on the material presented in chapters 8 through 10.

1.3.1 The Double-helix Model Revisited

This introduction started with the observation that the DNA double-helix model has iconic status in contemporary biology. Typical explanations of this status refer to the alleged beauty of the model, the popularity of Watson's entertaining and irreverent—as well as exceptionally fictionalized—account of its discovery,[60] as well as the epistemic reasons connected to reductionism discussed earlier (sec. 1.1.1). However, there are even deeper epistemic reasons why the double-helix model was so important, and these reasons shed considerable light on the use of models in science, especially how they may contribute to theoretical unification. The emergence of the informational interpretation of biology is central to this story.

What is critical about the double-helix model is that it provides a point of contact between four different research programs:

(i) Classical transmission genetics, as practiced during the first half of the twentieth century, made no commitment to the physical nature of the gene.[61] As noted earlier (at the beginning of this introduction), until the 1940s, genes were generally believed to be specified by proteins, but this was no ground for cognitive or epistemic dissonance with the rest of what was known about biology. According to the received view of evolution (see sec. 1.4.1 below), the classical gene must satisfy three constraints: (a) it must be capable of being duplicated during reproduction; (b) its duplication and transmission must obey Mendel's rules in diploids; and (c) it must be capable of occasional mutation. The received view implicitly assumed the primacy of the gene, that is, that genes produce traits. However, as with the physical nature of the gene, it was silent about how traits were so produced.

(ii) However, starting in the 1930s, and both conceptually and organizationally independent of the research program of transmission genetics, there emerged a program of studying gene action. This program began largely within biochemistry and can be interpreted as a rudimentary attempt to incorporate some developmental biology into classical genetics. From this program two new constraints were imposed on the chemical gene: (d) it was required to be connected by chemical mechanisms to other chemical constituents of cells; and (e) since the late 1930s it was becoming clear that there must be a specific relationship between genes and enzymes. What deserves emphasis is that there was potential for considerable tension—if not downright contradiction—between the classical gene (as conceptualized by the transmission geneticists) and the chemical gene (of the biochemists).

(iii) Yet another constraint was imposed on the gene by a third research program. The biophysical structural studies mentioned earlier (sec. 1.1.1) required of the physical gene that: (f) it satisfy all bond length and other stereochemical restrictions between atoms discovered from crystallographic studies.

(iv) To confound the matter further, a fourth set of constraints emerged, not from a well-defined research program, but from sporadic theoretical work from a variety of sources. These constraints were clearly articulated by Erwin Schrodinger in *What Is Life?* in 1944:[62] (g) from a physical perspective, the gene is unusually stable, remaining constant over hundreds of generations. It is more similar to an inorganic crystal than to the usual organic matter studied by biochemistry; but, (h) nevertheless, the structure of the gene must permit almost infinite variability to account for the known diversity of genes. Schrödinger presciently proposed a combinatorial solution to this problem, speculating on the possibility of a "hereditary code-script."

What is remarkable about the double-helix model is that it simultaneously satisfied all eight constraints seamlessly (that is, there was no post hoc approximation required to meet the constraints). In this sense, it is a *confluent* model (standing at the confluence of four different research programs). The unpacking of the DNA strands of the double helix immediately suggested a mechanism for the satisfaction of constraints (a) and (b). An occasional change of base sequence provided a mechanism for satisfying (c). Constraints (d) and (e) could be interpreted as that of deciphering the chemical properties of DNA sequences. All physical assumptions of the double helix model satisfy the stereochemical constraints (f). With the nucleotide bases protected inside the core of the helical cylinder, the double helix was believed to be a model of stability (constraint [g])—it was realized only much later that maintaining the structural integrity of the double helix requires a formidable array of repair mechanisms. Finally, the combinatorial explosion of possible sequences accounts for (h).

Confluent models establish a certain kind of consistency between different research programs without epistemically privileging any of them. Each program can now pursue its own agenda without worry that results from one of the others would contradict its core assumptions and render it unviable. The double helix permitted biochemical genetics to flourish without the specter of biophysics or transmission genetics haunting it, and so on, for the other research programs. Moreover, in the case of those constraints that were yet to be embedded in an identifiable research program (the fourth set above), the confluent model provided the impetus for such a program to be created—chapters 8 through 10 record the vagaries of that program. Finally, the double helix furthered Pauling's strategy of model construction, as was emphasized earlier in this introduction. It should come as no surprise that the double helix was recognized for its importance right from the beginning.

It is an open question how often confluent models occur in the history of science. Within molecular biology, both the allostery and the operon models played significant confluent roles, though not to the same extent as the double helix. The former established consistency between biophysical structural studies and a long-term research program within physiology, dating back to the 1890s. The latter established consistency between, once again, biophysical structural studies and a research program within biochemistry studying enzymatic adaptation (see chapter 9). (Beyond biology, Maxwell's mechanical models of the electromagnetic field provide one obvious example.) However, confluent models provide an epistemic account of theoretical unification very different from any that would be obtained by a focus on theories as linguistic structures. (Obviously, such an account is also fundamentally *not* a sociological one.)

1.3.2 DNA as Language

One result of the informational interpretation of molecular genetics has been a set of linguistic metaphors that pervades contemporary biology:

deletion, (RNA) editing, frameshift (mutations), insertion, messenger (RNA), missense (mutations), nonsense (mutations), open reading frames (in DNA), (DNA) proofreading, reading frames (for DNA), readthrough (of termination codons), sense and nonsense (DNA strands for transcription), translation (from DNA to protein), transcription (from DNA to RNA); and only a very few others that are not derived as directly from linguistic contexts, for instance, chaperones, housekeeping (genes), orphan (genes), and (RNA) splicing.

These metaphors congeal into an overarching metaphor of DNA as a language (or, sometimes, as the language of the genes). Even in the context of development, about which early molecular biology—like classical genetics before it—could say very little, as early as 1965, John Tyler Bonner argued that "all cells contain the directions for all cell life, written in DNA."[63] Several technical as well as popular books have been written that explicitly endorse the linguistic metaphor.[64] It also forms the basis of the discipline that calls itself "biosemiotics."[65] However, probably because molecular biology came of age simultaneously with the advent of digital computation, the metaphor of language was usually taken to refer not to natural languages—in which case compelling disanalogies would probably soon have been recognized—but to computer languages. To invoke yet another metaphor, the genome was supposed to contain a program for development. In 1961, Mayr explicitly argued that an individual was programmed by natural selection through its genome.[66] Almost simultaneously, introducing the operon model, Jacob and Monod argued that the "discovery of regulator and operator genes ... reveals that the genome contains not only a series of blue-prints but a coordinated program of protein synthesis and the means of controlling its execution."[67] And in Monod's later words: "The logic of biological regulatory systems abides not by Hegelian laws but, like the working of computers, by the propositional algebra of George Boole."[68]

The three metaphors—genetic information, DNA or genes as language, and the genetic program—reinforce each other since they are mixed and matched without producing cognitive dissonance: computer programs are written in artificial languages constructed on principles originally abstracted from natural languages, and all languages trivially facilitate the transmission of information, no matter how "information" is construed. Chapters 8 and 9 challenge the use of the metaphor of genetic information. However, following one of the strategies suggested in chapter 9, chapter 10 attempts to convert the metaphor into a model by providing a new account of what it means for DNA to encode information for traits. Chapter 10 also argues for the increasing epistemic irrelevance of the concept of information as levels of organization higher than proteins become the foci of interest. Chapter 14 argues against the relevance of the metaphor of a genetic program, noting its epistemic incompetence not only when confronted with the phenomena of development but also at the level of DNA and protein. The metaphor of DNA as language is perhaps even less compelling though it has presumably done less harm because, when it comes to designing research protocols, molecular biologists are well aware that DNA is a molecule, a physical structure, not a linguistic one. Returning to the molecular biology of the 1960s, the linguistic metaphor could perhaps be defended insofar as it suggested research programs designed to decipher the code, establish translation rules, and so on. But the unexpected complexity of eukaryotic genetics has prevented the straightforward extension of those programs beyond prokaryotes. The linguistic metaphor no longer suggests interesting research programs—consequently, in contemporary molecular biology, it is of little relevance.

Meanwhile, the main harm of the linguistic metaphor is that it obfuscates the physical complexity and developmental contingency of gene expression. The DNA sequence is not a book waiting to be read unless, perhaps, the metaphor is intended to refer to a work of fiction or to a religious text, the deep meaning of which can be understood only by detailed knowledge of context. (Defending the HGP, Walter Gilbert endorsed such an interpretation by describing the human DNA sequence as the Holy Grail of biology.)[69] The linguistic metaphor suggests far too static a picture of the genome than is warranted; it ignores the dynamic use, conversion, and transformation of DNA and its immediate products during even the production of protein from DNA (as chapter 14 will detail). At the level of traits further removed from DNA, there is no question of encoding or translation (see chapter 10). But, perhaps most unfortunately, the linguistic metaphor also suggests an unsound determinism in which all biological features are supposed to be dependent on the DNA sequence. As section 1.4 (sec. 1.4.3) will emphasize, part of the conceptual excitement of genomics has been that, especially from an evolutionary point of view, the organization of DNA can be understood only from the perspective of the evolving organism.

1.4 Evolution

The evolution of populations requires the existence of variation, as well as change in the intensity and extent of this variation. Conceptually, for evolution to occur, there is no requirement of statistical independence between the production of variation, its intensity (in different "individuals" of a population), and its extent (that is, its frequency in the population). As chapter 12 notes, Lamarck's evolutionary model posited a positive correlation between all three factors. In contrast, Darwin's original (1859) model of evolution by natural selection (which is equally due to Wallace) explicitly assumed independence between the first and third factors.

Darwin's model of evolution by natural selection makes three assumptions about the process of the change of the extent of variation:[70]

(i) intrapopulation variation: that such variation exists at the level of entities or "individuals" within a population. (these "individuals" could be cells, organelles, biological individuals, groups, etc. Although Darwin focused mainly on biological individuals, the model is not committed to any single level of selection);

(ii) differential fitness: that different variants contribute different numbers of offspring to the next generation; and

(iii) inheritance of fitness-determining properties: that there is a correlation between parent and offspring in the properties that, by varying, produce differential fitness.

It is instructive to note what this model does not assume: (a) any particular mechanism of hereditary transmission.[71] Darwinian evolution by natural selection can even occur through the cultural inheritance of traits. Moreover, it does not assume (b) a genotype-phenotype distinction, which Mendel implicitly introduced in 1866 by distinguishing between characters and factors (which were transmitted during reproduction and were responsible for producing the corresponding characters). The modern form of the last distinction was introduced only around 1909 by W. Johannsen, who conceptualized the genotype as an abstract entity defining the total genetic profile of an individual organism.[72]

1.4.1 Mathematical Mendelism

The received view of evolution, sometimes called neo-Darwinism (or the "Modern Synthesis"), emerged in the late 1920s and early 1930s primarily through the work of three individuals, R. A. Fisher, J. B. S. Haldane, and Sewall Wright.[73] At the core of the received view are mathematical models of heredity; for diploids these are based on Mendel's rules as modified by linkage. For expository simplicity, attention will be restricted here to the diploid case.[74] These models predict evolutionary stasis in the null model when reproduction takes place in isolated populations large enough for chance effects to be irrelevant, no variation through mutation, and no selection: this stasis is captured by the Hardy–Weinberg rules, which state that allelic frequencies do not change from generation to generation. For a single locus, if mating is random with respect to that locus, genotypic frequencies also do not change. These other factors that can lead to change are modeled as deviations from Hardy–Weinberg predictions. It is more accurate to call the received framework of evolution mathematical Mendelism rather than neo-Darwinism (in contrast to the proposal of chapter 12), because the Hardy–Weinberg rules are a result of Mendelian inheritance, not Darwinism in general.[75] Philosophers have paid little attention to how this framework fares in the light of modern molecular biology when the abstract genotype is confronted with the dynamics of the material genome.

Going beyond Darwin's original model, neo-Mendelism assumes:

(i) a sharp genotype–phenotype distinction, with inheritance occurring only through the genotype, thus usually preventing the inheritance of acquired characters;

(ii) that transmission of hereditary material between generations is governed by Mendel's rules for diploids, and by their analogues in the case of haploids or polyploids;

(iii) that variation arises through mutations in the genotype;

(iv) that these mutations are "random" or "blind" in the sense that the mechanisms responsible for mutagenesis are equally likely to produce a mutation in contexts where that mutation enhances or diminishes fitness;[76] and

(v) evolutionary dynamics is adequately modeled in genotypic space, which assumes that phenotypes may be predictively tracked from genotypic space; thus phenotypic evolution is reducible to genotypic evolution.

Developments within molecular biology have challenged all these assumptions. Historically, the identification of DNA as the genetic material, and the informational interpretation of the gene (see sec. 1.3), led to the interpretation of genotypic inheritance as inheritance of DNA sequences alone. All forms of epigenetic inheritance (for instance, the inheritance of methylation patterns of DNA) then violate assumption (i). Several biologists have pointed out that some types of epigenetic inheritance can be interpreted as the inheritance of acquired characteristics. In particular, Eva Jablonka and Marion Lamb have urged the evolutionary significance of epigenetic inheritance. However, they have also generated confusion by claiming to defend Lamarckianism.[77] As chapter 12 notes, Lamarck was not unique in assuming that acquired characteristics are inherited—in fact it is a mistake (unfortunately, one prevalent among biologists) to associate Lamarck with that position. Rather, Lamarck's name should be associated with the position that the generation of hereditary variation itself is preferentially adaptive. Moreover, as noted at the beginning of this section, the inheritance of acquired characteristics is logically consistent with Darwin's own model of evolution by natural selection. Such inheritance falls afoul of the received view, not of Darwinism.[78]

Most types of epigenetic inheritance also violate assumptions (ii) and (iii). Those who would hold that the received view of evolution requires no major modification in light of the molecular data can perhaps argue that epigenetic inheritance may be of only limited importance to evolution. But assumptions (ii) and (iii) are also routinely violated at the level of DNA. Three examples will be noted here, each consisting of phenomena recognized during the era of classical genetics but regarded as being anomalous and probably of no great evolutionary significance:

(a) As early as 1951, Barbara McClintock reported the presence of transposable DNA elements in maize that inserted themselves into different regions of the genome,[79] violating (ii), and generated new genotypes, violating (iii). McClintock's original report was greeted with skepticism bordering on derision.[80] In the molecular era, mobile genetic elements (such as transposons) have been found in virtually every species examined, prokaryotic or eukaryotic.[81] They result in chromosome rearrangements and enable horizontal gene transfer. In bacteria, they enable the rapid spread of antibiotic resistance.

(b) Meiotic drive, or the preferential transmission of one of the two homologous alleles, in violation of Mendel's rule of independent segregation, has been known since 1928.[82] Thus, assumption (ii) of the received view is violated. Meiotic drive systems can be chromosomal or genic.[83] In the former case, some property of the entire chromosome must give one homologue a replication or orientation advantage in the spindle during meiosis. This chromosome is preferentially represented in the gametes. In the latter case, gametes containing the relevant allele are at an advantage over other gametic types. Most cases of genic meiotic drive are believed to be explained by the segregation distortion model. Most proponents of the received view of evolution see meiotic drive as an instance of intragenomic conflict, also showing that selection acts at different levels of organization. Even if this interpretation is accepted, it nevertheless requires an expansion of the received view beyond assumption (ii). More than Mendel's rules are necessary.

(c) Gene conversion, defined as the nonreciprocal transfer of "information" between homologous DNA sequences, has been known since Carl Lindegren's work on the yeast *Saccharomyces cerevisiae* in the 1950s.[84] Unlike crossovers, in which the exchange of DNA between homologous chromosomes is reciprocal, in gene conversion, one chromosome may donate its sequence while the other loses its sequence, during meiosis (and, more rarely, mitosis). Assumption (ii) of the received view is violated; arguably, so is assumption (iii), at least on those occasions when the resulting genotype is new. Mechanistic models of meiotic gene conversion go back to a model by Robin Holliday in 1964.[85] One major effect of gene conversion is that it enables concerted evolution: that duplicated (paralogous) sequences within a species become more closely related to each other than to orthologous sequences from other species. Gene conversion is one of two mechanisms that lead to the situation in which repeated homologous sequences within a genome tend to get homogenized.[86] There are at least three ways in which this process is evolutionarily significant and will require moving beyond the received view of evolution: conversion can facilitate the spread of a mutation in a population; it erases that trace of evolutionary history from the genome that consists in the use of the extent of divergence to estimate the time of the duplication; and it decreases intraspecific divergence while potentially increasing interspecific divergence.

Molecular biology also calls assumption (iv) into question—this issue is dealt with in chapters 11 through 13.

Finally, one topic that is not broached in this book is the neutral theory of evolution, the first challenge presented by molecular biology to the received view of evolution. In the early 1960s, the development of gel electrophoresis led to the realization that variation at the level of proteins was ubiquitous in natural populations.[87] Given the degeneracy of the genetic code, this observation implied that variation at the DNA level was even more pronounced. In 1968 Motoo Kimura argued, on the basis of J. B. S. Haldane's thesis of there being a cost to selection (in terms of the required loss of the less fit individuals), that natural selection could not maintain this degree of variation.[88] Much of this variation must, therefore, be neutral. A year later, drawing on Kimura's calculation, J. L. King and T. H. Jukes announced the advent of a "non-Darwinian" model of evolution in which drift of neutral alleles replaced selection as the driving force of evolutionary change.[89] Partly because of King and Jukes's rhetoric, there immediately began an acrimonious debate about the etiology of molecular variation in natural populations, which has not ended yet.

However, the received view of evolution does allow for the existence of other factors of evolutionary change, besides selection, including random change (called "drift"). Consequently, in one straightforward sense, Kimura's position is one that can be formulated within the received view: it states that drift rather than selection best explains the patterns of diversity and variation seen at the molecular level. The conflict is between neutrality and selectionism, the latter being the position that selection is the dominant mechanism of evolutionary change at all levels of organization. Both positions are correctly viewed as alternatives within the received view.

As such, Kimura's theory contrasts earlier alternative global theories of evolution proposed by R. A. Fisher and Sewall Wright to account for the patterns of change seen during evolutionary history.[90] Fisher argued for the importance of selection acting on individual mutations with small fitness effects in large panmictic populations, whereas Wright emphasized the importance of population structure and, to some extent, drift in small populations. Kimura and most other neutralists have also maintained that neutrality at the molecular level is compatible with ubiquitous selection at higher levels of organization; however, since entities at these higher levels of organization are composed of their constituent molecules, there is a paradox here. For future reference, this will be called the "neutrality paradox." Any resolution of this paradox probably requires that development screens off the molecular level from the effects of selection acting, for instance, on entire organisms, a position that is not inconsistent with the received view of evolution, though only because the latter is entirely silent about development—a point emphasized in chapter 14.

1.4.2 Developmental Evolution

A theory of development was the central goal of several research programs of nineteenth-century biology. As chapter 14 notes, in 1926, T. H. Morgan initiated an operational divorce between studies of heredity (that is, the new science of genetics) and development on the basis of the assumption (justified in its own context) that Mendelism could be pursued using traits that had a simple genotype–phenotype relationship. Three consequences of this strategy are worth emphasis: (i) the received view of evolution, which emerged from the genetics of the 1920s, largely ignored development (as noted at the end of sec. 1.4.1); (ii) attention came to be restricted to those traits that did not display complex developmental origins;[91] and (iii), partly as a result of (ii), genetic reductionism (which has already been criticized in sec. 1.1.2) gained plausibility within biology.

In contrast, by elucidating the complexity of the molecular mechanisms required for gene expression, let alone the control of phenotypic expression at even higher levels of organization, molecular biology has shown the potential complexity of the relationship between phenotypes and genotypes. Developmental genetics, starting in the 1960s, promised to bring some simplicity and theoretical order to the field by pursuing explanations from a genetic basis, but it promised much more than it ever came close to delivering. Until the 1990s there seemed to be little plausible prospect for a theoretical understanding of development.

Although it would be rash, even now, to suggest that a *theory of development* is forthcoming in the foreseeable future, a suite of related recent results makes it plausible to expect that there will at least be piecemeal theoretical insights into development. First, there has been a shift of focus from the DNA to the cellular level, with proteins dominating attention at the latter (see sec. 1.4.4). Second, partly as a result of this shift, it has been realized that there is a convergence—or universality—of molecular structures and mechanisms at the cellular level that is not discernible at the DNA level. Third, this universality holds across wide classes of taxa. For instance, there are hundreds of different genes and their primary products that are known to be involved in transducing signals across the plasma membranes of cells; yet the mechanisms of signal transduction can all be classified into only sixteen signaling systems in metazoans.[92] At least arguably, taxa are more similar at the level of cells and protein-mediated biochemical mechanisms than at the level of their DNA sequences. Thus, and this point has not been widely appreciated, the new molecular biology resolves the neutrality paradox mentioned earlier (sec. 1.4.1): neutral variation at the molecular level (in particular, the DNA level) may be consistent with selection for the small number of possible functional complexes at the cellular and higher levels of organization.

The emergence of these molecular insights into development has been accompanied by the resurgent hope that the time has come for reconciliation, an end to the divorce between development and evolution promulgated by Morgan by the contruction of an integrated framework for both disciplines. Several variant research programs have emerged that all embody this hope in different ways:[93]

(i) studies of the evolution of development,[94] perhaps intellectually the most traditional of these programs, which treats developmental features (such as life-history traits) as standard phenotypes to be studied using the usual techniques of the received view of evolution;

(ii) evolutionary developmental biology,[95] which uses phylogenetic relationships to elucidate developmental mechanisms in individual species; and

(iii) developmental evolution, intellectually the most radical of these programs,[96] which calls for the modeling of which calls for the modeling of evolution in phenotypic space (besides genotypic space) to incorporate constraints and opportunities for phenotypes; it thus denies assumption (v) of the Mendelian received view of evolution (described in sec. 1.4.1).

Only one of the many innovations of developmental evolution will be noted here: in that research program, the genotype–phenotype relationship itself becomes a target of evolutionary change. For instance, mutational bias because of physical constraints on chromosomal dynamics may lead to directional evolution in infinite populations in the absence of selection, a phenomenon not permitted by the received view.[97] The program of developmental evolution offers perhaps the most hope for constructing an evolutionary theory that gives a detailed account of phenotypic evolution (including both morphology and behavior). Although the hope of such a theory is hardly new—Darwin clearly stated it and attempted to construct such a story for many special cases, including the functional morphology of orchids[98]—it is only since the early 1990s that there has been some plausible prospect of success.

1.4.3 From Genetics to Genomics

As chapter 14 records, the massive human and other genome sequencing projects of the 1990s have led to an unexpected transformation of the conceptual terrain of molecular biology. The sequencing projects were supposed to inaugurate a triumphant new era of genetic reductionism in both biology and medicine. Instead, the sequences that emerged exposed the impotence of sequence-gazing.[99] The study of DNA migrated from genetics to genomics, with an emphasis in the latter on computational tools and informatics. One fact has become clear: the peculiar properties of eukaryotic genomes—in particular, the G-value paradox (that there is no clear correlation between the number of genes and organismic complexity)[100]—seem to require the formulation of a framework for the study of heredity very different from classical genetics, in which genes are seen as indivisibile units generally posited to have one-to-one maps to proteins (if not more complex phenotypes).

Enthusiasts of traditional genetic reductionism have not entirely disappeared. Even in 2003, in an introductory contribution to a book entitled *Behavioral Genetics in the Postgenomic Era*,[101] Watson endorses sociobiology and genetic reductionism, and, for good measure, indulges in an ad hominem attack on the political left: "Though human twin studies ... had provided incontrovertible evidence for genetic involvement in personality and intelligence differences [by 1975, when E. O. Wilson's *Sociobiology* was first published], those on the radical left continued to shout 'not in your genes.' Sadly, many of their students enthusiastically accepted their antigenetics diatribes, wanting futures determined by free will as opposed to genetics."[102] The bloated rhetoric is not matched by the substantive contributions in the book, which, by and large, while maintaining an official party line of the primacy of the gene, emphasize the complexity of genotype–environment interactions and unpredictability of individual behavior from DNA sequences.

By now, views such as Watson's border on scientific irrelevance. A recent encyclopedia article notes that, in fully sequenced microbial genomes, 30–60 percent of protein-coding regions (identified as open reading frames, or

ORFs) are "orphan" genes to which no function can be assigned: "sequence information alone provides no clue as to the molecular or cellular functions of these hypothetical proteins."[103] A new framework for the study of heredity must rescue such orphans and establish functional roles for them within cells. Inspection of genes alone, or even predominantly, holds little promise of success. The future will show what the new framework for modeling heredity looks like, how it embraces development, and whether it may even be regarded as a partial connotation of the older tradition of classical genetics. Chapter 14 presents one speculative model. It is likely that a period of intellectual ferment and excitement awaits the study of heredity and development during the first few decades of the twenty-first century. The likely center of that study is the new discipline of proteomics.

1.4.4 Toward Proteomics?

The term "proteome" was introduced only in 1994 to describe the total protein content of a cell produced from its genome.[104] Unlike the latter, it is not even approximately a fixed feature of a cell (let alone an organism), changing as it does during development. Deciphering the proteome, and following its temporal development during the life cycle of each tissue of an organism, has emerged as the major challenge for molecular biology in the postgenomic era.[105] The discovery of universality of developmental processes at the level of cells and proteins mentioned earlier (in sec. 1.4.2) has contributed to the perceived promise of proteomics. The emergence of proteomics in the wake of the various sequencing projects signals an acceptance of the position that studying processes largely at the DNA level will not suffice to explain phenomena at the cellular and higher levels of organization. Even genomics did not go far enough; a sharper break with the past will be necessary.

In one important sense, the emergence of proteomics recaptures the spirit of early molecular biology, when all molecular types, but especially proteins, were the foci of interest, and the deification of DNA had not replaced a pluralist vision of the molecular basis for life.[106] In the late 1960s, Sidney Brenner and Crick proposed "Project K," "the complete solution of *E. coli*." *E. coli* (strain K-12) was selected as a model organism because of its simplicity (as a unicellular prokaryote) and ease of laboratory manipulation. Project K included: (i) a "detailed test-tube study of the structure and chemical action of biological molecules (especially proteins)";[107] (ii) completion of the models of protein synthesis; (iii) work on the structure and function of cell membranes; (iv) the study of control mechanisms at every level of organization; and (iv) the study of the behavior of natural populations, including population genetics. Once *E. coli* was solved, biology could move on to more complex organisms.

Notice that: (i) DNA receives no preferential attention at the expense of other molecular components in Project K; and (ii) the centrality of proteins as the most important active molecules in a cell is fully recognized. Project K accepts that there is much more to the cell than DNA; it accepts that no simple solution of the cell's behavior can be read from the genomic sequence. After a generation of infatuation with DNA—which chapter 3 interprets as an infatuation with genetic, rather than physical, reductionism—the aims of proteomics return in part to the vision of biology incorporated in Project K. However, in at least one important way, that project went beyond even proteomics as currently understood: it emphasized all levels of organization, whereas the explicit aims of proteomics are limited to the protein level. The future will probably require further expansion, consistent with Project K, but proteomics is a good beginning and serves as a healthy antidote to the deification of DNA. Physical reductionism is not abandoned in this vision; rather, it presumes that, unlike the myopic genetic reductionism of the last two decades, the physical reductionism of molecular biology will produce a complete theoretical biology, at least at the level of individual organisms. This is the biology of the future. The purpose of this book is to encourage philosophical reflection on it.

Acknowledgments

Thanks are due to Justin Garson and Alex Moffett for comments on earlier drafts of this chapter. This work was supported by the United States National Science Foundation, Grant No. SES-0090036, 2002–2003.

Notes

1. See Watson and Crick (1953). Judson (1979) provides a scintillating history.

2. Pauling and Corey (1950); this is so-called α-helix model of protein structure. Unlike the DNA double helix it is not nearly universal and is only one of several structural motifs found in proteins.

3. As quoted by Olby (1974, p. 442).

4. Kohler (1973) has emphasized the point that the study of enzymes was central to the establishment of biochemistry as a discipline separate from the older organic chemistry.

5. This is true, at least in the traditional intepretations of physics, notwithstanding the many attempts to apply information theory to physics—see, for instance, Brillouin (1956).

6. See Dobzhansky (1973).

7. See, in this context, Keller (1992).

8. It may seem odd to cast semantics as a study of form rather than content. However, formal semantics involves a reduction of questions of content to those of form—as the logician, Church (1956, p. 65) insightfully noted. See also Sarkar (1992).

9. See Nagel (1949, 1951, 1961), Woodger (1952), and Hempel and Oppenheim (1948). For a detailed historical analysis, see Sarkar (1989).

10. See, for instance, Kauffman (1972) and Wimsatt (1976).

11. See Harvey (1628). Descartes, of course, muddied this view by insisting that humans were special, insofar as they had a soul besides being machines.

12. This broadening reflected the ultimate failure of mechanism within classical physics itself: gravitation, the empirically most successful theory within classical physics, required action-at-a-distance and could not, even after many repeated attempts, be given an interpretation in terms of contact interactions of material particles.

13. For a history, see, for instance, Lenoir (1982).

14. See Bohr, Hasselbalch, and Krogh (1904); for a modern review, see Riggs (1988).

15. For other forms of reductionism, see chapter 4 and Sarkar (1998).

16. See Smuts (1926). No general history of the emergence of holism in biology exists.

17. See Loeb (1912); for historical details, see Pauly (1987).

18. See J. S. Haldane (1931) and Hogben (1930).

19. Chargaff (1950); Judson (1979) particulary emphasizes the significance of Chargaff's ratios as a constraint on model-building.

20. There are exceptions—see, for instance, Bergson (1911).

21. This does not mean that these properties do not depend on some other part: the weight of a piece of matter depends on the gravitational pull of other pieces of matter, but, nevertheless, it is definable by a single parameter referring to nothing else; this type of dependence does not force a relational defintion. However, consider some entity that is a member of a set of cardinality n, and consider some property of that entity that depends on n. A (not entirely unproblematic) biological example comes from population ecology when some property of an individual—say, its fertility—may depend on the density of the population in its habitat and, therefore, on population size. A reductionist explanation need not invoke entities only at the lowest possible level of organization. Reductionist explanation of cellular behavior need not invoke only individually defined properties of molecules; it can (and routinely does) refer to those of organelles within cells. But, to be a *reductionist* explanation of a cell, it may not refer to properties defined using the entire cell or entities at higher hierarchical levels.

22. The atomic weight of the carbon in it has some bearing on exactly how a strand of DNA behaves but, in most contexts, it does not bear explanatory weight. A different isotope of carbon would not have made any relevant difference. However, in the Meselson–Stahl experiment (Meselson and Stahl 1958) to demonstrate semiconservative replication, it is exactly the atomic weight that mattered. DNA composed of one radioactive and one nonradioactive strand migrated differently in the centrifuge from how DNA composed of two radioactive or two nonradioactive strands would have. The context determines what bears explanatory weight.

23. Suppose that a biological system consists of a network of reactions and some property it has is explained by the network topology. Then one may substitute individual molecular types without altering the behavior so long as the relevant topological features are maintained. In this case, the explanation of the whole is not in terms of its parts because individual properties of the parts do not bear the explanatory weight. One of the standard features of diploid organisms is that some traits show dominance. An unsolved problem of molecular biology is to provide an empirically adequate molecular account of dominance. None is immediately forthcoming. One intriguing proposal has been that what matters are not specific properties of alleles, or the proteins produced from them, but the toplogy of the reaction networks in which they participate (see Kacser and Burns 1981 and Kacser 1987). If this model is correct, it will provide a very interesting explanation of dominance but, nevertheless, not one that is reductionist in the sense being explicated here—and it will be the first such example in molecular biology.

24. For instance, covalent bonds of ordinary chemistry have a bond strength of about 90 kcal/mole. (It takes 90 kilocalories of energy to break 1 mole [6.022×10^{23}] of these bonds.) Covalent bonds help maintain the gross structure of biological macromolecules (for instance, the backbone of the DNA double helix), but these bonds are not of much explanatory relevance for the biological behaviors that are of

interest. Ordinary ionic bonds (in nonaqeuous environments) have a bond strength of about 80 kcal/mole. These bonds are also not usually of explanatory relevance. Rather, the bonds that are relevant are hydrogen bonds, ionic bonds in aqueous environments, and what are called hydrophic bonds but which are not actual bonds. They represent molecular regions brought into contiguity by the hydrophobic effect. In the aqueous cellular environment, hydrogen bonds have a strength of about 1 kcal/mole, and ionic bonds about 4 kcal/mole. Hydrophobic "bonds" have a strength of about 1 kcal/mole.

25. The term "function" is being used here because it is ordinarily so used in molecular biology—the philosophical significance of such ascriptions of function will be discussed in sec. 1.2.

26. Perhaps the clearest statement of such a position is Oppenheim and Putnam (1958).

27. Monod, Changeux, and Jacob (1963); the discussion in the text is based on Monod, Wyman, and Changeux (1965).

28. See, for instance, von Bertalanffy (1975).

29. See Jacob et al. (1960); the account here follows Monod (1971).

30. Perhaps, if, during the next few decades, proteomics becomes central to molecular biology, as many have predicted, proteins will once again return to center stage.

31. See, for instance, Hull (1972, 1974) and Kitcher (1984).

32. See, for instance, Kitcher (1984).

33. Schaffner (1974) was the first to emphasize this important point, while defending the view that what was (nevertheless) achieved constituted intertheoretic reduction.

34. Sarkar (1998, chapter 6) elaborates this point in detail.

35. Nagel (1961) explicitly recognized this elementary logical point though it was lost by almost all subsequent advocates of models of intertheoretic reduction. Sarkar (1998, chapter 2) emphasizes this point.

36. See Tauber and Sarkar (1992, 1993), which are highly critical of blind sequencing.

37. For a survey, including references to the literature, see Sarkar (1998, chapter 1).

38. See Pigliucci (2001) for an overview.

39. Among other things, Waterton, during his South American travels, transported a fourteen-foot "Coulacanara" snake in wreaths around his body and rode a cayman

along the bank of a river. For more on Waterton's eccentricities, see Barber (1980, chapter 7).

40. Waterton does not give scientific names of the species he describes. This identification is based on the geographical distribution of the various species of sloth (Eisenberg 1989, p. 56).

41. Waterton (1973, pp. 5–6).

42. Ibid., p. 93.

43. See Goffart (1971).

44. See Strickberger (2000, p. 382).

45. Human sweat is produced by eccrine glands and has low salt and almost no fat content compared to other mammals in which sweat is produced by apocrine glands. The low mineral content of human sweat and the general hairlessness of the human body ensure rapid evaporation of the sweat. The result is a very efficient way of cooling the body. See Baker (1992) for more detail.

46. Note, however, that this leaves open the logical possibility that functional attributions can also be made in other contexts, that is, not necessarily in response to a question of origin.

47. Mayr (1976, pp. 362–363).

48. See Wimsatt (1971), Wright (1973), Millikan (1989), and Neander (1991) for discussions of variants.

49. See Bigelow and Pargetter (1987).

50. Thus, the etiological view may fall back equally on adaptations and exaptations. See Gould and Vrba (1982).

51. However, this feature does leave the propensity account open to the charge of vagueness: at present there is no clear delimitation of what may be considered to be functional (see Millikan 1989).

52. Mayr (1976, p. 360).

53. To take such usage into account, some philosophers have argued for a "causal role" theory of function—see Cummins (1973), Godfrey-Smith (1993), and Amundson and Lauder (1994). As is often the case, the term "causal" here does no work: no particular explication of causality, or even a presumption that causal talk is necessary or justified, is required in the "causal role" theory.

54. Technically, the problem here is more with the propensity theory of narrow-sense functions than with the etiological theory since such effects are trivially no longer adaptive. But even the etiological theory has to explain why these effects

should be regarded as having arisen through selection since a female that no longer had a heart for pumping blood after the postreproductive period would be no more selectively favored than the one that did. (At best, the etiological theory can say that some selected features persist because of accident—but this brings the etiological theory close to the "broad-sense" account of function explicated later in the text.)

55. See Gould and Lewontin (1979) and Maynard Smith (1978).

56. For a similar account, see Nagel (1961, pp. 410–414). Note the contrast to Cummins (1973) who puts no restriction on what features (including potentially pathological behaviors) may be invoked when making functional ascriptions. The account given here is much more restrictive.

57. However, Wimsatt (1972) argues that such a position is too broad since it would allow function ascriptions to parts of inanimate entities such as autocatalytic reactions. From the perspective of the position being advocated here, this is a virtue, because there is no principled distinction between living and nonliving matter.

58. This was blind sequencing—see Tauber and Sarkar (1992, 1993). There were many problems with the original project, including the fact that the concept of *the* human genome is incoherent given the ubiquitous variation of intraspecific DNA sequences. See Sarkar and Tauber (1991).

59. See, for instance, Gilbert (1992).

60. See Watson (1968).

61. Sarkar (1998) emphasizes the point that classical genetics was a formal science based on statistical laws. This was explicitly recognized by the geneticists—see chapter 3 (sec. 3.2) of this volume. The best account of the conceptual structure of classical genetics in the 1940s, including the biochemical aspects referred to in the next paragraph, is Haldane (1942).

62. See Schrödinger (1944). There have been many historical assessments of this book; for a conceptual assessment of what it achieved, see Sarkar (1991).

63. See Bonner (1965, p. v).

64. At least four books use "language" and "DNA," "genes," or "genetics" in their titles—Beadle's (1966) *The Language of Life: An Introduction to the Science of Genetics*; Berg and Singer's (1992) *Dealing with Genes: The Language of Heredity*; Jones's (1994) *The Language of the Genes*; and Pollack's (1994) *Signs of Life: The Language and Meanings of DNA*. Restricting attention to just the last five years (since 1998), at least thirteen books invoke the "language" of "genes"—Dikotter (1998, p. 73); Hademenos and Fried (1998, p. 98); Mawer (1998, p. 115); Coen (1999, pp. 37, 66, 54); Gross (1999, p. 62); Mahoney (1999, p. 166); McGinn (1999, p. 225); Enriquez (2001, p. 5); Fortey (2000, p. 87); Lock, Young, and Cambrosio (2000, p. 24); Martineau (2001,

p. 29); Bellavite, Signorini, and Steele (2002, p. 136); and Little (2002, p. 18). At least fifty books from the same period refer to DNA as providing a language.

65. See Hoffmeyer (1996) and Merrell (1996). Emmeche (1999) tries to address the criticisms made here and in chapter 9.

66. Mayr (1961); see chapter 14 for further discussion.

67. Jacob and Monod (1961, p. 354). For an interesting historical commentary, see Keller (2000, chapter 3).

68. Monod (1971, p. 77).

69. See Gilbert (1992).

70. This reconstruction of evolution by natural selection is a modification of the one originally proposed by Lewontin (1970).

71. Given that Darwin did not have any initial model of heredity, it was perhaps inevitable that the model of evolution in the 1859 (first) edition of the *Origin* was neutral about the nature of heredity.

72. See, for instance, Johannsen (1909).

73. This date (the late 1920s and early 1930s) is not universally accepted by historians, some of whom place it a decade later—see Sarkar (2004) for a critical perspective on the historiography of the received view. For the work that established the received view of evolution, see, in particular, Fisher (1930), Wright (1931), and Haldane (1932). Provine (1971) provides a succinct account of the developments until 1932. There is no reliable comprehensive history of later developments in evolutionary biology.

74. Other models of heredity, for instance, haploidy, sex-linked inheritance, and polyploidy, raise technical complications but no new conceptual problems.

75. From a historical perspective, it is also more accurate to describe the received view as Mendelism rather than Darwinism. Bowler (1989) emphasizes this point.

76. Note that this formulation is supposed to avoid the following problem: in chapter 12, a mutation is defined as random "if and only if the probability of its occurrence in an environment has no correlation with the fitness of the phenotype associated with it in that environment." However, this definition falls afoul of the fact that most mutations are harmful, and that, consequently, there will be a negative correlation between mutations and the fitness changes they induce. The new formulation given in the text is supposed to be immune to this objection.

77. See, for instance, Jablonka and Lamb (1995). Despite the unfortunate use of "Lamarckian" in its title, this book remains an important theoretical contribution to evolutionary biology.

78. One of Darwin's immediate followers, George Romanes, fully appreciated this point, coining the term "neo-Darwinism" to distinguish between the position that denies the inheritance of acquired characteristics from Darwinism proper. See Romanes (1896).

79. See McClintock (1951).

80. See the discussion in Comfort (2001).

81. See Braam and Reznikoff (1998).

82. See Gershenson (1928) and Sandler and Novitski (1957).

83. See Lyttle (1993) for more details of the mechanisms being discussed.

84. See Lindegren (1953) and Hurles (2002).

85. This is the Holliday junction model, also used to explain recombination through crossovers without conversion. See Holliday (1964).

86. The other is unequal crossover, which can act only on tandemly duplicated repeats.

87. This was originally established by Lewontin and Hubby (1966) for *Drosophila pseudoobscura* and Harris (1966) for humans.

88. See Kimura (1968) and Haldane (1957).

89. See King and Jukes (1969), entitled "Non-Darwinian Evolution."

90. Kimura (1983) presents an early synthesis of the neutral theory.

91. As a result, for several generations, biology, especially in the West, ignored phenotypic plasticity (as, for instance, represented by variable norms of reaction)—see Sarkar (1999) for further discussion of this issue.

92. These are: (i) transmembrane tyrosine kineasas; (ii) receptors linked to cytoplasmic tyrosine kinases; (iii) transmembrane serine/threonine kinases; (iv) transmembrane protein phosphatases; (v) Wnt receptors; (vi) IL-1, toll receptors; (vii) G-protein linked receptors; (viii) hedgehog receptors; (ix) Notch/Delta; (x) nuclear hormone receptors; (xi) integrins; (xii) ligand-gated cation channels; (xiii) gap junctions; (xiv) nitric oxide receptors; (xv) cadherins; and (xvi) receptor guanylate cyclases. See Gerhart and Kirschner (1997, pp. 102–103).

93. For a historical discussion, see Sarkar and Robert (2003).

94. See, for instance, Purugganan (1998).

95. See Hall (1998, 2000) and Arthur (2002) for a slightly varying statement of this program.

96. See Wagner (2000, 2001) for a statement of the program.

97. See Stadler et al. (2001) and Garson, Wang, and Sarkar (2003). Of course, what still remains subject to debate is the relative significance of such a departure from the received view.

98. See Darwin (1862); Hall (1998) and Sarkar and Robert (2003) provide histories.

99. See, for instance, Stephens (1998).

100. See Hahn and Wray (2002). The G-value paradox is the genomic analog of the earlier C-value paradox, that there is no correlation betwen organismic complexity and the size of the geonome (measured by the length of DNA sequences)—see Cavalier-Smith (1985). That paradox was resolved by the discovery of ubiquitous noncoding DNA in euklaryotic genomes.

101. See Plomin et al. (2003); the book's contributions come from a 2001 conference and are already outdated.

102. Watson (2003, p. xxi), referring to Wilson (1975). It is peculiar that the reference is to *Not in Our Genes* (Lewontin, Rose, and Kamin 1984), an author of which was one of the most prominent geneticists of his generation.

103. See Hung and Kim (2000), writing in the *Nature Encyclopedia of Life Sciences*.

104. See Williams and Hochstrasser (1997).

105. For an accessible overview, see Hung and Kim (2000).

106. For more on the deification of DNA, see Tauber and Sarkar (1992, 1993) as well as Lewontin (1992).

107. See Crick (1973, p. 67); this is the only published account of the project initially proposed to the European Molecular Biology Organisation (EMBO) by Brenner and Crick.

References

Amundson, R., and Lauder, G. V. 1994. "Function without Purpose: The Uses of Causal Role Function in Evolutionary Biology." *Biology and Philosophy* 9: 443–469.

Arthur, W. 2002. "The Emerging Conceptual Framework of Evolutionary Developmental Biology." *Nature* 415: 757–763.

Baker, P. T. 1992. "Sweating—The Human Response to Heat." In S. Jones, R. Martin, and D. Pilbeam, eds., *The Cambridge Encyclopedia of Evolution*. Cambridge: Cambridge University Press, p. 48.

Barber, L. 1980. *The Heyday of Natural History*, 1820–1870. New York: Doubleday.

Beadle, G. 1966. *The Language of Life: An Introduction to the Science of Genetics*. Garden City, N.Y.: Doubleday.

Bellavite, P., Signorini, A., and Steele, A. 2002. *The Emerging Science of Homeopathy: Complexity, Biodynamics, and Nanopharmacology*. New York: North Atlantic Books.

Berg, P., and Singer, M. 1992. *Dealing with Genes: The Language of Heredity*. Mill Valley, Calif.: University Science Books.

Bergson, H. 1911. *Creative Evolution*. New York: Holt.

Bigelow, J., and Pargetter, R. 1987. "Functions." *Journal of Philosophy* 54: 181–196.

Bohr, C., Hasselbalch, K., and Krogh, A. 1904. "Ueber einen in biologischer Beziehung wichtigen Einfluss, den die Kohlensäurespannung des Blutes auf dessen Sauerstoffbindung übt." *Skandinavisches Archiv für Physiologie* 16: 402–412.

Bonner, J. T. 1965. *The Molecular Biology of Development*. Oxford: Oxford University Press.

Bowler, P. J. 1989. *The Mendellan Revolution: The Emergence of Hereditarian Concepts in Science and Society*. Baltimore: Johns Hopkins University Press.

Braam, L. A. M., and Reznikoff, W. S. 1998. "DNA Transposition: Classes and Mechanisms." In *Nature Encyclopedia of Life Sciences*. London: Nature Publishing Group. Available at ⟨http://www.els.net/⟩.

Brillouin, L. 1956. *Science and Information Theory*. New York: Academic Press.

Cavalier-Smith, T. 1985. "Introduction: The Evolutionary Significance of Genome Size." In T. Cavalier-Smith, ed., *The Evolution of Genome Size*. Chichester: John Wiley, pp. 1–36.

Chargaff, E. 1950. "Chemical Specificity of Nucleic Acids and the Mechanisms of Enzymatic Degradation." *Experentia* 6: 201–209.

Church, A. 1956. *Introduction to Mathematical Logic*. Princeton: Princeton University Press.

Coen, E. 1999. *The Art of Genes. How Organisms Make Themselves*. Oxford: Oxford University Press.

Comfort, N. C. 2001. *The Tangled Field: Barbara McClintock's Search for the Patterns of Genetic Control*. Cambridge, Mass.: Harvard University Press.

Crick, F. H. C. 1973. "Project K: 'The Complete Solution of *E. coli*.'" *Perspectives in Biology and Medicine* 17: 67–70.

Cummins, R. 1973. "Functional Analysis." *Journal of Philosophy* 72: 741–764.

Darwin, C. 1862. *On the Various Contrivances by which British and Foreign Orchids are Fertilised by Insects, and on the Good Effects of Intercrossing*. London: J. Murray.

Dikotter, F. 1998. *Imperfect Conceptions: Medical Knowledge, Birth Defects, and Eugenics in China*. New York: Columbia University Press.

Dobzhansky, T. 1973. "Nothing in Biology Makes Sense Except in the Light of Evolution." *American Biology Teacher* 35: 125–129.

Eisenberg, J. F. 1989. *Mammals of the Neotropics: The Northern Neotropics,* volume 1: *Panama, Colombia, Venezuela, Guyana, Suriname, French Guiana.* Chicago: University of Chicago Press.

Emmeche, C. 1999. "The Sarkar Challenge to Biosemiotics: Is There Any Information in a Cell?" *Semiotica* 127: 273–293.

Enrico, C. 1999. *The Art of the Genes: How Organisms Make Themselves.* Oxford: Oxford University Press.

Enriquez, J. 2001. *As the Future Catches You: How Genomics and Other Forces and Changing Your Work, Health, and Wealth.* New York: Crown Business.

Fisher, R. A. 1930. *The Genetical Theory of Natural Selection.* Oxford: Clarendon Press.

Fortey, R. 2000. *Trilobite! An Eyewitness to Evolution.* New York: Alfred Knopf.

Garson, J., Wang, L., and Sarkar, S. 2003. "How Development May Direct Evolution." *Biology and Philosophy* 18: 353–370.

Gerhart, J., and Kirschner, M. 1997. *Cells, Embryos, and Evolution.* Oxford: Blackwell Science.

Gershenson, S. 1928. "A New Sex Ratio Abrnormality in *Drosophila obscura.*" *Genetics* 13: 488–507.

Gilbert, W. 1992. "A Vision of the Grail." In D. J. Kevles and L. Hood, eds., *The Code of Codes.* Cambridge, Mass.: Harvard University Press, pp. 83–97.

Godfrey-Smith, P. 1993. "Functions: Consensus without Unity." *Pacific Philosophical Quarterly* 74: 196–208.

Goffart, M. 1971. *Function and Form in the Sloth.* Oxford: Pergamon Press.

Gould, S. J., and Lewontin, R. C. 1979. "The Spandrels of San Marco and the Panglossian Paradigm." *Proceedings of the Royal Society of London B* 205: 581–598.

Gould, S. J., and Vrba, E. S. 1982. "Exaptation—A Missing Term in the Science of Form." *Paleobiology* 8: 4–15.

Gross, M. 1999. *Travels to the Nanoworld: Miniature Machinery in Nature and Technology.* New York: Plenum Trade.

Hademenos, G., and Fried, G. 1998. *Schaum's Outline of Biology.* New York: McGraw Hill.

Hahn, M. W., and Wray, G. A. 2002. "The G-value Paradox." *Evolution and Development* 4: 73–75.

Haldane, J. B. S. 1932. *The Causes of Evolution*. London: Harper.

Haldane, J. B. S. 1942. *New Paths in Genetics*. New York: Random House.

Haldane, J. B. S. 1957. "The Cost of Natural Selection." *Journal of Genetics* 55: 511–524.

Haldane, J. S. 1931. *The Philosophical Basis of Biology*. London: Hodder and Stoughton.

Hall, B. K. 1998. *Evolutionary Developmental Biology*, second ed. London: Chapman and Hall.

Hall, B. K. 2000. "Evo-devo or Devo-evo—Does It Matter?" *Evolution and Development* 2: 177–178.

Harris, H. 1966. "Enzyme Polymorphism in Man." *Proceedings of the Royal Society Series B* 164: 298–310.

Harvey, W. 1628. *Exercitatio Anatomica de Motu Cordis et Sanguinis in Animalibus*. Frankfort: Guilielmi Fitzeri.

Hempel, C. G., and Oppenheim, P. 1948. "Studies in the Logic of Explanation." *Philosophy of Science* 15: 135–175.

Hoffmeyer, J. 1996. *Signs of Meaning in the Universe*. Bloomington: Indiana University Press.

Hogben, L. 1930. *The Nature of Living Matter*. London: Kegan Paul, Trench, Trubner.

Holliday, R. 1964. "A Mechanism for Gene Conversion in Fungi." *Genetical Research* 5: 282–303.

Hull, D. 1972. "Reduction in Genetics—Biology or Philosophy?" *Philosophy of Science* 39: 491–499.

Hull, D. 1974. *Philosophy of Biological Science*. Englewood Cliffs: Prentice-Hall.

Hung, L.-W., and Kim, S. 2000. "Genome, Proteome, and the Quest for a Full Structure–function Description of an Organism." In *Nature Encyclopedia of Life Sciences*. London: Nature Publishing Group. Available at ⟨http://www.els.net/⟩.

Hurles, M. E. 2002. "Gene Conversion." In *Nature Encyclopedia of Life Sciences*. London: Nature Publishing Group. Available at ⟨http://www.els.net/⟩.

Jablonka, E., and Lamb, M. J. 1995. *Epigenetic Inheritance and Evolution: The Lamarckian Dimension*. Oxford: Oxford University Press.

Jacob, F., and Monod, J. 1961. "Genetic Regulatory Mechanisms in the Synthesis of Proteins." *Journal of Molecular Biology* 3: 318–356.

Jacob, F., Perrin, D., Sanchez, C., and Monod, J. 1960. "L'opéron: Groupe de Gènes à Expression Coodinée par un Opérateur." *Comptes Rendus des Séances de l'Academie des Sciences* 250: 1727–1729.

Johannsen, W. 1909. *Elemente der Exacten Erblichkeitslehre*. Jena: Gustav Fischer.

Jones, S. 1994. *The Language of the Genes: Biology, History, and the Evolutionary Future*. London: Flamingo.

Judson, H. F. 1979. *The Eighth Day of Creation*. New York: Simon and Schuster.

Kacser, H. 1987. "Dominance Not Inevitable but Very Likely." *Journal of Theoretical Biology* 126: 505–506.

Kacser, H., and Burns, J. A. 1981. "The Molecular Basis of Dominance." *Genetics* 97: 639–666.

Kauffman, S. A. 1972. "Articulation of Parts Explanation in Biology and the Rational Search for Them." *Boston Studies in the Philosophy of Science* 8: 257–272.

Keller, E. F. 1992. "Between Language and Science: The Question of Directed Mutation in Molecular Genetics." *Perspectives in Biology and Medicine* 35: 292–306.

Keller, E. F. 2000. *The Century of the Gene*. Cambridge, Mass.: Harvard University Press.

Kimura, M. 1968. "Evolutionary Rate at the Molecular Level." *Nature* 217: 624–626.

Kimura, M. 1983. *The Neutral Theory of Molecular Evolution*. Cambridge: Cambridge University Press.

King, J. L., and Jukes, T. H. 1969. "Non-Darwinian Evolution: Most Evolutionary Change in Proteins May Be due to Neutral Mutations and Genetic Drift." *Science* 164: 788–798.

Kitcher, P. 1984. "1953 and All That: A Tale of Two Sciences." *Philosophical Review* 93: 335–373.

Kohler, R. E. 1973. "The Enzyme Theory and Biochemistry." *Isis* 64: 181–196.

Lenoir, T. 1982. *The Strategy of Life: Teleology and Mechanics in Nineteenth Century German Biology*. Dordrecht: Reidel.

Lewontin, R. C. 1970. "The Units of Selection." *Annual Review of Ecology and Systematics* 1: 1–18.

Lewontin, R. C. 1992. *Biology as Ideology: The Doctrine of DNA*. New York: HarperPerennial.

Lewontin, R. C., and Hubby, J. L. 1966. "A Molecular Approach to Genic Heterozygosity in Natural Populations. II. Amount of Variation and Degree of Heterozygosity in Natural Populations of *Drosophila Pseudoobscura*." *Genetics* 54: 595–609.

Lewontin, R. C., Rose, S., and Kamin, L. J. 1984. *Not in Our Genes: Biology, Ideology, and Human Nature*. Cambridge, Mass.: Harvard University Press.

Lindegren, C. C. 1953. "Gene Conversion in Saccharomyces." *Journal of Genetics* 51: 625–637.

Little, P. 2002. *Genetic Destinies*. Oxford: Oxford University Press.

Lock, M., Young, A., and Cambrosio, A., eds. 2000. *Living and Working with the New Medical Technologies: Intersections of Inquiry*. Cambridge: Cambridge University Press.

Loeb, J. 1912. *The Mechanistic Conception of Life*. Chicago: University of Chicago Press.

Lyttle, T. W. 1993. "Cheaters Sometimes Prosper: Distortion of Mendelian Segregation by Meiotic Drive." *Trends in Genetics* 9: 205–210.

Mahoney, P., ed. 1999. *Nature, Risk, and Responsibility: Discourses on Biotechnology*. Hampshire: Macmillan.

Martineau, B. 2001. *First Fruit: The Creation of the Flavr Savr Tomato and the Birth of Genetically Engineered Food*. New York: McGraw-Hill.

Mawer, S. 1998. *Mendel's Dwarf*. New York: Harmony Books.

Maynard Smith, J. 1978. "Optimization Theory in Evolution." *Annual Review of Ecology and Systematics* 9: 31–56.

Mayr, E. 1961. "Cause and Effect in Biology." *Science* 134: 1501–1506.

Mayr, E. 1976. *Evolution and the Diversity of Life: Selected Essays*. Cambridge, Mass.: Harvard University Press.

McClintock, B. 1951. "Chromosome Organization and Gene Expression." *Cold Spring Harbor Symposia in Quantitative Biology* 16: 13–47.

McGinn, C. 1999. *The Mysterious Flame: Conscious Minds in a Material World*. N.Y.: Basic Books.

Merrell, F. 1996. *Signs Grow: Semiosis and Life Processes*. Toronto: University of Toronto Press.

Meselson, M., and Stahl, F. W. 1958. "The Replication of DNA in *Escherichia coli*." *Proceedings of the National Academy of Sciences (USA)* 44: 671–682.

Millikan, R. G. 1989. "In Defense of Proper Functions." *Philosophy of Science* 56: 288–302.

Monod, J. 1971. *Chance and Necessity: An Essay on the Natural Philosophy of Modern Biology*. New York: Knopf.

Monod, J., Changeux, J.-P., and Jacob, F. 1963. "Allosteric Proteins and Cellular Control Systems." *Journal of Molecular Biology* 6: 306–329.

Monod, J., Wyman, J., and Changeux, J.-P. 1965. "On the Nature of Allosteric Transitions: A Plausible Model." *Journal of Molecular Biology* 12: 88–118.

Nagel, E. 1949. "The Meaning of Reduction in the Natural Sciences." In R. C. Stauffer, ed., *Science and Civilization*. Madison: University of Wisconsin Press, pp. 99–135.

Nagel, E. 1951. "Mechanistic Explanation and Organismic Biology." *Philosophy and Phenomenological Research* 11: 327–338.

Nagel, E. 1961. *The Structure of Science*. New York: Harcourt, Brace, and World.

Neander, K. 1991. "Functions as Selected Effects: The Conceptual Analyst's Defense." *Philosophy of Science* 58: 168–184.

Olby, R. C. 1974. *The Path to the Double Helix*. Seattle: University of Washington Press.

Oppenheim, P., and Putnam, H. 1958. "The Unity of Science as a Working Hypothesis." In H. Feigl, M. Scriven, and G. Maxwell, eds., *Concepts, Theories, and the Mind–body Problem*. Minneapolis: University of Minnesota Press, pp. 3–36.

Pauling, L., and Corey, R. B. 1950. "Two Hydrogen-bonded Spiral Configurations of the Polypeptide Chains." *Journal of the American Chemical Society* 71: 5349.

Pauly, P. J. 1987. *Controlling Life: Jacques Loeb and the Engineering Ideal in Biology*. Oxford: Oxford University Press.

Pigliucci, M. 2001. *Phenotypic Plasticity: Beyond Nature and Nurture*. Baltimore: Johns Hopkins University Press.

Plomin, R., DeFries, J. C., Craig, I. W., and McGuffin, P., eds. 2003. *Behavioral Genetics in the Postgenomic Era*. Washington, D.C.: American Psychological Association.

Pollak, R. 1994. *Signs of Life: The Language and Meaning of DNA*. Boston: Houghton Mifflin.

Provine, W. B. 1971. *The Origins of Theoretical Population Genetics*. Chicago: University of Chicago Press.

Purugganan, M. D. 1998. "The Molecular Evolution of Development." *BioEssays* 20: 700–711.

Riggs, A. F. 1988. "The Bohr Effect." *Annual Review of Physiology* 50: 181–204.

Romanes, G. J. 1896. *Life and Letters*. London: Longmans, Green.

Ryan, F. 1997. *Virus-X: Tracking the New Killer Plagues: Out of the Present and Into the Future*. Boston: Little, Brown.

Sandler, L., and Novitski, E. 1957. "Meiotic Drive as an Evolutionary Force." *American Naturalist* 91: 105–110.

Sarkar, S. 1989. "Reductionism and Molecular Biology: A Reappraisal." Ph.D. Dissertation, Department of Philosophy, University of Chicago.

Sarkar, S. 1991. "*What Is Life?* Revisited." *BioScience* 41: 631–634.

Sarkar, S. 1992. "'The Boundless Ocean of Infinite Possibilities': Logic in Carnap's *Logical Syntax of Language*." *Synthese* 93: 191–237.

Sarkar, S. 1998. *Genetics and Reductionism*. New York: Cambridge University Press.

Sarkar, S. 1999. "From the *Reaktionsnorm* to the Adaptive Norm: The Norm of Reaction, 1909–1960." *Biology and Philosophy* 14: 235–252.

Sarkar, S. 2004. "Evolutionary Theory in the 1920s: The Nature of the 'Synthesis.'" *Philosophy of Science*. In press.

Sarkar, S., and Robert, J. S. 2003. "Introduction." *Biology and Philosophy* 18: 209–217.

Sarkar, S., and Tauber, A. I. 1991. "Fallacious Claims for HGP." *Nature* 353: 691.

Schaffner, K. F. 1974. "The Peripherality of Reductionism in the Development of Molecular Biology." *Journal of the History of Biology* 7: 111–139.

Schrödinger, E. 1944. *What Is Life? The Physical Aspect of the Living Cell*. Cambridge: Cambridge University Press.

Smuts, J. C. 1926. *Holism and Evolution*. New York: Macmillan.

Stadler, B. M. R., Stadler, P. F., Wagner, G. P., and Fontana, W. 2001. "The Topology of the Possible: Formal Spaces Underlying Patterns of Evolutionary Change." *Journal of Theoretical Biology* 213: 241–274.

Stephens, C. 1998. "Bacterial Sporulation: A Question of Commitment?" *Current Biology* 8: R45–R48.

Strickberger, M. W. 2000. *Evolution*, third ed. Sudbury, Mass.: Jones and Bartlett.

Tauber, A. I., and Sarkar, S. 1992. "The Human Genome Project: Has Blind Reductionism Gone Too Far?" *Perspectives on Biology and Medicine* 35(2): 220–235.

Tauber, A. I., and Sarkar, S. 1993. "The Ideology of the Human Genome Project." *Journal of the Royal Society of Medicine* 86: 537–540.

von Bertalanffy, L. 1975. *Perspectives on General Systems Theory*. New York: George Braziller.

Wagner, G. P. 2000. "What Is the Promise of Developmental Evolution? Part I: Why Is Developmental Biology Necessary to Explain Evolutionary Innovations?" *Journal of Experimental Zoology (Molecular Development and Evolution)* 288: 95–98.

Wagner, G. P. 2001. "What Is the Promise of Developmental Evolution? Part II: A Causal Explanation of Evolutionary Innovations May Be Impossible." *Journal of Experimental Zoology (Molecular Development and Evolution)* 291: 305–309.

Waterton, C. 1973. *Wanderings in South America, the North-west of the United States, and the Antilles, in the years 1812, 1816, 1820, and 1824*. Oxford: Oxford University Press.

Watson, J. D. 1968. *The Double Helix: A Personal Account of the Discovery of the Structure of DNA*. New York: Atheneum.

Watson, J. D. 2003. "A Molecular Genetics Perspective." In R. Plomin, J. C. DeFries, I. W. Craig, and P. McGuffin, eds., *Behavioral Genetics in the Postgenomic Era*. Washington, D.C.: American Psychological Association, pp. xxi–xxii.

Watson, J. D., and Crick, F. H. C. 1953. "Molecular Structure of Nucleic Acids—A Structure for Deoxyribose Nucleic Acid." *Nature* 171: 737–738.

Williams, K. L., and Hochstrasser, D. F. 1997. "Introduction to the Proteome." In M. R. Wilkins, K. L. Williams, R. D. Appel, and D. F. Hochstrasser, eds., *Proteome Research: New Frontiers in Functional Genomics*. Berlin: Springer, pp. 1–12.

Wilson, E. O. 1975. *Sociobiology: The New Synthesis*. Cambridge, Mass.: Harvard University Press.

Wimsatt, W. C. 1971. "Some Problems with the Concept of 'Feedback.'" *Boston Studies in the Philosophy of Science* 8: 241–256.

Wimsatt, W. C. 1972. "Teleology and the Logical Structure of Function Statements." *Studies in the History and Philosophy of Science* 3: 1–80.

Wimsatt, W. C. 1976. "Reductive Explanation: A Functional Account." *Boston Studies in the Philosophy of Science* 32: 671–710.

Woodger, J. H. 1952. *Biology and Language*. Cambridge: Cambridge University Press.

Wright, L. 1973. "Functions." *Philosophical Review* 82: 139–168.

Wright, S. 1931. "Evolution in Mendelian Populations." *Genetics* 16: 97–159.

Part I Reduction

2 Models of Reduction and Categories of Reductionism

2.1 Introduction

The idea of reduction in the empirical sciences is at least as old as the mechanical philosophy of the seventeenth century, which, in a sense, simply required that all physical laws and facts be explained by, or "reduced" to, local contact interactions between impenetrable particles of matter. The mechanical philosophy ultimately failed to provide an adequate basis for physical theory.[1] However, the program of accounting for the theories, laws, and empirical facts of one scientific discipline by those of another, that is, "reducing" the former to the latter, continued to play a significant role in scientific research. At least partial success was obtained in some cases, such as the explanation of the laws of physical optics by the laws of classical electromagnetism and the laws of thermodynamics by the principles of statistical mechanics, just to cite two nineteenth-century examples. The logical empiricist program in the philosophy of science in the early part of the century acknowledged implicitly the importance of reduction in the development of science. Significant extended analysis of the process, however, only began with Nagel (1949), who regarded reduction as a species of intertheoretic explanation.[2] The reduced theory is explained by a reducing theory which is presumed to be more fundamental.

The Nagel model was almost immediately challenged on three fronts. *First*, Kemeny and Oppenheim (1956) presented a model in which reduction is a form of theory replacement where the reduced theory is replaced by a reducing theory which is superior to it because it explains the same (if not more) phenomena, and is more "systematized." *Second*, Suppes (1957) argued that all that is necessary in reduction is the establishment of an isomorphism between the entity and predicate terms of two theories axiomatized using set theory. *Third*, and finally, Feyerabend (1962) and others argued that reduction was basically impossible because the alleged reducing theory showed how the ostensibly reduced one was false insofar as it was less accurate.[3] These early analyses were extended and afforded a unified treatment in a seminal paper by Schaffner (1967). Schaffner presents a model of reduction which modifies Nagel's approach to allow the reducing theory to correct the reduced one thus meeting the objections raised by Feyerabend. What can be derived from the reducing theory is not the reduced theory itself but a corrected version which remains "strongly analogous" to it. Schaffner also argues, very persuasively, that the Kemeny–Oppenheim model captures theory replacement rather than reduction, and that the Suppes approach is too weak to capture what occurs in a reduction.[4]

All these approaches assume that reduction is necessarily a relation between theories and that it is a form of explanation.[5] Both of these assumptions have since been systematically challenged. *First*, some critics have been motivated by a desire to capture, through a model of reduction, actual scientific activity and the structure of explanations *within a scientific context* rather than what obtains after their rational reconstruction by philosophers.[6] In general, they have denied that most cases of interlevel scientific reductions (i.e., where the properties of wholes are explained in terms of those of their parts) involve theories at all. *Second*, a modified version of the Suppes proposal has been developed which escapes the strictures of Schaffner (Sneed 1971). Yet others have argued for a notion of reduction which is basically little more than an assertion of physicalism (Rosenberg 1985). Both of the latter two approaches seem to deny that explanation is necessarily involved in reduction. Furthermore, within each of these approaches, variants differing in many significant details have emerged. Since no comprehensive critical account of these developments has yet been attempted, the relations between the various analyses of reduction remain obscure.[7] It is no longer even clear that there is a single concept of reduction that is being explicated in all these approaches.

This general confusion has also routinely been aggravated by two additional factors. *First*, the two questions, (i) whether a particular model of reduction is intended to capture the structure of an explanation (if reduction is being construed as explanation), and (ii) whether it is intended to describe research strategy, are quite routinely conflated, though this distinction is explicitly recognized by Schaffner (1974) and Wimsatt (1976b). *Second*, there has often been a failure to keep epistemological and ontological questions separate. For instance, the questions, whether reduction is being attempted in order to explain some theories, laws or facts by others, or whether reduction is intended to show what entities are composed of other, perhaps more "fundamental," entities, have often been routinely confused.[8] The former is an epistemological question, the latter is an ontological one—they are obviously not the same.[9] This is not, of course, to deny that there might well be deep connections between ontological and epistemological assumptions that may be made during the course of a scientific investigation. What entities (if any) that one is committed to might well influence what type of explanation one seeks. Conversely, the explanatory success of a theory or a research program might well increase one's confidence in "its ontology," that is, the entities it used to state its propositions. But these are not necessary connections. One might well prefer an ontology that only accepts the most fundamental objects of physics as "real" and yet pursue upper-level explanations (as, for instance, afforded by chemistry) for pragmatic reasons such as ease of computation or representation. Explanation, after all, has a pragmatic component; otherwise one would not know when to stop explaining. Conversely, one could pursue different modes of explanation and have no concern for ontology whatsoever. Thus, since the connections between ontological and epistemological commitments are not necessary in any usual sense, it is always possible to distinguish between them. In the context of discussion of reduction, moreover, scrupulously maintaining this distinction leads to a considerable clarification of the issues that divide various explications of reduction, as the rest of this paper purports to show.

The primary purpose of this paper is to offer a classification of models of reduction using as the basis distinctions between three broad categories of reductionism: theory reductionism, explanatory reductionism and constitutive reductionism.[10] After these categories are introduced and discussed (section 2.2), models of reduction are organized into them (section 2.3). Then, the paper very briefly considers the potential applicability of these models to the ostensible reduction of classical genetics to molecular biology, a case which has received considerable philosophical attention (section 2.4). It is argued there that the classification of models of reduction offered here casts significant light on this case. The concluding section

(section 2.5) draws the morals that the issue of reduction is much more complicated than is usually recognized and that there is no reason to suppose that any single model of reduction captures all the cases of reduction in the sciences. Throughout the discussions a conscious attempt has been made to keep epistemological and ontological issues separate and to scrutinize carefully, for each model of reduction, whether it is an appropriate candidate to capture research strategy or the structure of explanation or, possibly, neither.

2.2 Categories of Reductionism

For the purpose of the classification of models of reduction in a manner that helps elucidate their interrelations, it is helpful, first, to distinguish between three broad categories: *theory* reductionism, *explanatory* reductionism and *constitutive* reductionism. Models of reduction can then be organized into these categories. These categories are to be construed in quite general terms. Each category includes several different models that share some very general features. These are the features that define the category in question. The individual models differ in the more specific features.

The category of theory reductionism consists of those models of reduction that view it necessarily as a relation between theories. Explanation might also be involved, in the sense that the reduced theory is explained by the reducing one, but this is not necessary. However, most of the models of reduction that fall into this category satisfy this additional criterion. The category of explanatory reductionism consists of those models of reduction that construe it as a relation of explanation in the sense that the reduced entity is explained by the reducing entity no matter whether these entities are theories, laws, empirical generalizations or even individual observation reports. To the category of constitutive reductionism belong those models of reduction that assert, at least, that upper-level (intuitively larger) systems are composed of lower-level (intuitively smaller) systems and conform to the laws governing the latter. Unlike the cases of the previous two categories, models from this category of reductionism necessarily involve the separation of their domains into levels of organization.[11] For brevity, models of reduction which fall into the categories of theory, explanatory and constitutive reductionism will be called models of "theory reduction," "explanatory reduction," and "constitutive reduction," respectively.

At first sight, it might appear that these three categories are connected in the sense that each is weaker than the one that preceded it. Thus, models of theory reduction would be making stronger assumptions than models of explanatory reduction which would, in turn, be making stronger assumptions than models of constitutive reduction. However, it is worth emphasis that, as the categories have been construed here, no such relation of increasing strength necessarily holds. In fact, as will now be argued, the assumption that such a relation holds is often yet another manifestation of the common conflation of epistemological and ontological issues. It is only when it is assumed that reduction necessarily involves explanation (an epistemological assumption) that models of theory reduction *usually* make stronger assumptions than models of explanatory reduction.[12] The former are committed to the notion that explanation is a relation between theories; the latter are not. If, further, some assumption is made about the constitution of entities where entities at a higher level are somehow composed of entities from a lower level (an ontological assumption), models of theory and explanatory reduction necessarily make stronger assumptions than models of constitutive reduction. Only if both assumptions are made can an approximate hierarchy of these categories be constructed. Neither assumption need be made as the next paragraph and, especially, the next section demonstrate.

To drive the point home, constitutive reductionism is ontological reductionism without any necessary epistemological pretension, no more, no less.[13] Explanatory reductionism is epistemological reductionism. However, it need not entail ontological commitments, as the discussion of "bridge laws" (connecting different levels) in the next section illustrates. Theory reductionism might construe reduction as involving explanation; it might not. Irrespective of whether reduction is construed as involving explanation, it, too, might or might not carry ontological assumptions. The interpretation of the relation between the reduced and reducing theories (for example, the "bridge laws" just mentioned) determines whether such assumptions are being made. Finally, as should be apparent, the three categories of reductionism are not mutually exclusive. Models of theory reduction that construe reduction as involving explanation also fall within the category of explanatory reductionism. Models, either of theory or explanatory reduction, that make ontological assumptions also fall in the category of constitutive reductionism.

2.3 Models of Reduction

2.3.1 Theory Reduction

All models of reduction that construe reduction as a relation between theories fall in the category of theory reductionism. Currently, there are at least three types of such models. The most common, and in many ways the most plausible, models of theory reduction derive from the work of Nagel and Schaffner.[14] In Nagel's approach, the reduction of one theory to another is the explanation of the former (reduced) theory by the latter (reducing) one, where explanation is construed as deductive-nomological (Hempel and Oppenheim 1948). The reduced and reducing theories are assumed to be formalized in first-order logic. Ultimately, the relation of reduction holds between these theories when the reduced theory is derived from the reducing one—Nagel calls this requirement the *condition of derivability* (Nagel 1961, p. 354). If the reduced theory contains no terms not occurring in the reducing theory, such a derivation can immediately be attempted. According to Nagel, such reductions are *homogeneous* (ibid., p. 339). If, however, the reduced theory does contain such terms, before a derivation is attempted, terms in the reduced theory must somehow be connected to the terms of the reducing theory. Such reductions are *heterogeneous*, according to Nagel (ibid., p. 342), and the requirement of connecting the relevant terms comprises the *condition of connectability* (ibid., p. 345). The nature of the connections between the terms of the reduced and reducing theories, often called "bridge laws," is treated in detail by Nagel (ibid., p. 354). Their syntactic form could be that of conditionals or biconditionals and they were to be interpreted as conventions or factual statements depending on the context. Thus, reduction is for Nagel purely an epistemological issue with no necessary ontological commitment.[15]

Schaffner (1967) modifies Nagel's model to allow that what can be exactly derivable from the reducing theory is not the reduced theory itself but a corrected version that is "strongly analogous" to it.[16] The notion of "strong analogy" is left intuitive. Later (Schaffner 1974), this model was further modified to allow the reducing theory to be corrected by the reduced one since, after all, the claims made by the reduced theory form part of the evidence that the reducing theory must take into account. Thus, the condition of derivability is now satisfied only between the corrected reduced theory and the corrected reducing one. More important, Schaffner interprets the relations between terms of the reduced and reducing theories to be *synthetic identities*.[17] These are established between individuals or groups of individuals in the two theories which are represented by entity terms. Predicates are then construed extensionally, thus allowing the identities to be immediately extended to them. Since these identities are naturally interpreted as requiring ontological commitment, Schaffner's model differs here from Nagel's.[18] However, since Schaffner, like Nagel, also assumes that reduction, ultimately, is explanation, the same epistemological issues are also at stake. Many variants of Schaffner's general model of theory reduction are available.[19] Most put different kinds of conditions on the synthetic identities involved.[20]

The second type of model of theory reduction is due to Nickles (1973), who made an important distinction which was later more fully explicated by Wimsatt as one between "interlevel" and "intralevel" reductions. The former kind of reduction occurs when a theory at one level of organization is reduced to one at another, presumably lower, level. Such reductions are usually "domain-combining" in the sense that they might combine the domains of two theories. For such reductions Nickles admits that the Nagel (or Schaffner) model might hold, even if approximately, presumably with the same epistemological and ontological implications.[21] The latter kind of reductions, which are often "domain-preserving," occur between successive reductions at the same level.[22] For Nickles, the succeeding theories get reduced to the preceding ones, such as in the case of special relativity being reduced to classical mechanics when an appropriate limit is taken, for instance, the speed of light going to infinity.[23] In such cases, there is clearly no question of the derivation of the more general reduced theory (special relativity in the example) from the less general reducing theory (classical mechanics in the example). Thus, a Nagel–Schaffner type of theory reduction simply cannot be obtained. Nickles argues, instead, that what is necessary for a reduction of this sort to hold is the existence of a well-defined transformation between the theories. The nature of this transformation is left unexplicated though Nickles gives the examples of taking limits and making approximations. Routinely, such approximations and limits are counterfactual, such as in the case of the speed of light taken to be infinite.

Nickles does not explicitly consider the ontological and epistemological issues raised by intralevel reductions, but some elementary consequences easily follow from his analysis. When the relevant limits and approximations in intralevel reductions are counterfactual, ontological connections obviously cannot be made. Further, the reduced and reducing theories are competitors since their domains intersect. There is, therefore, a possibility that the successor theory simply replaces the predecessor. Presumably, reduction only obtains when such a replacement either permits the construction of the transformation that Nickles requires or the predecessor theory is sufficiently accurate in its domain, generically smaller than that of the successor theory, so that no replacement occurs for many applications. As far as epistemological issues are concerned, since the more general theory reduces to the less general one, the reducing theory cannot, in any ordinary sense, explain the reduced one. At most, all that can be maintained is that some sort of explanation, in a very mitigated sense, is operative in the converse direction. Ultimately, however, because of the lack of a specification of permissible transformations, Nickles's model of theory reduction, though interesting because it captures a kind of theory reduction that the Nagel–Schaffner type of model cannot capture, is still left imprecise. Wimsatt (1976b) has, however, pointed out that the sort of transformation invoked by Nickles provides an illuminating partial explication of Schaffner's unexplicated notion of "strong analogy."

An entirely different type of model of theory reduction has recently been advocated by Balzer and Sneed (1977) and Balzer and Dawe (1985, 1986). They formulate theories using set-theoretic predicates (rather than in first-order logic), as advocated by Sneed (1971) and Stegmueller (1976) following the initial suggestion of Suppes (1957). What is gained, thereby, is a considerable simplicity of the formalization. Essentially, their intuition is that one theory reduces to another if, given models of the two theories, there exists a relation between such models—the reduction relation—which ensures that every model of the reduced theory turns out to be a model of the reducing theory. The existence of such a formal relation, however, does not ensure that the two theories are not about entirely disparate subjects.[24] Balzer and Dawe (1986) therefore construe it as a necessary but not sufficient condition of reduction. In order to ensure sufficiency, Balzer and Dawe impose two further conditions. The *first* is an informal one: they simply stipulate that the two theories must be about the same thing (Balzer and Dawe 1986, p. 86). The *second* is more formal. Though so far reduction has only been conceived as a relation between models of theories rather than theories themselves, they now require that the reduced theory be derivable from the reducing one. However, this is done not at the syntactic level but by putting two conditions on the models which they show to hold. The first of these is exactly the existence of the reduction relation noted above.[25] The second is that there exists, for all models of the reducing theory, at least one model of the reduced theory that has the reduction relation to it.[26] However, these two conditions are sufficient to ensure derivability only if the theories in question can be formalized in first-order logic or something slightly more general.[27] But, if this has to be proved, and there seems to be no straightforward way to do it without actually providing the formalizations in question, the attractive simplicity of this set-theoretic approach is simply lost in the bargain (Sarkar 1989, p. 77n.).

2.3.2 Explanatory Reduction

The Nagel and Schaffner models of reduction, since they construe reduction as a form of explanation, also fall into the category of explanatory reductionism. However, more interesting are those models of explanatory reduction that are not also models of theory reduction because, as will be argued in the next section, these are the models that often seem best suited to capture the flavor of actual scientific reductions. There are two intuitions that motivate this interest. *First*, scientific explanation is often "messy" and involves the use of semiempirical rules and the invocation of mechanisms that do not form part of any fully explicated theory. At best, in such cases, explanation involves the use of a fragment of a theory. Reduction, in such circumstances, can only be construed as a relation between such rules, mechanisms or fragments, not as a relation between theories. However, even in such cases, certain kinds of explanation are regarded as being reductive, and a general account of types of reduction needs to be able to incorporate these. *Second*, the Nagel and Schaffner models, applied to interlevel reductions and explanations, do not explicitly incorporate the basic idea that a reductive explanation is the explanation *of a whole in terms of its parts*. There is, of course, no reason to suggest that, in principle, models of theory reduction cannot incorporate such explanations. However, the models of explanatory reduction considered here do take cognizance of this factor and this adds to their interest.

There are at least three different attempts to construct models of explanatory reductionism. The best-known of these is that due to Wimsatt (1976b). Wimsatt entirely rejects the point of view that interlevel reduction must be construed as a relation between theories. Instead, he partially adopts an analysis of scientific explanation, due to Salmon (1971), which confines itself to explanation of individual events. Scientific explanations, in Salmon's analysis, are a search for statistically relevant factors, and an explanation is best seen as a list of such factors, with associated conditional probabilities for the occurrence of the event to be explained. Deductive-nomological explanations, on this account, are the degenerate cases when one of these probabilities is equal to 1 and all the others are equal to 0. The list of factors occurring in the explanation are obtained by a partition of the reference class of events by some property. This property can be "screened off" by some other property which provides a better partition of the reference class in the sense that, if the second partition is effected, a further partition using the first property does not affect the associated probabilities while the converse is not true. The "screened off" property and the partition it produced now become irrelevant because the new property, in a sense, provides a deeper explanation of the phenomena (Salmon 1971, p. 55).

Wimsatt adopts this model of explanation with two critical changes. *First*, for him, explanation is a search for causally relevant, rather than just statistically relevant, factors.[28] Deductive-nomological explanation is, once again, simply a degenerate case of statistical explanation. *Second*, in order to avoid the conclusion that causal factors at one level are eliminated by causal factors at a lower level when a reductive explanation is effected, Wimsatt introduces a notion of "effective screening off" in which the cost of explanation by the second property is much higher than that by the first (Wimsatt 1976b, p. 702). This criterion provides pragmatic grounds for the retention of the causal factors at the higher level. That the new factors come from a lower level is ensured by Wimsatt's requirement that the new property involves a "compositional redescription" of the old (ibid., p. 702). Thus, Wimsatt's model of explanatory reduction is wedded both to this notion of "compositional redescription," which is not explicated any further, and to his modified version of the Salmon account of explanation.

In 1970 Kauffman (1971) offered an analysis of a certain type of explanation in biology, which he called "articulation of parts" explanation, that turns out to constitute a new model of explanatory reduction. This model is presented as a way of conducting a search for such an explanation, that is, as a research strategy. There are three stages. *First*, given "an adequate description of an organism as doing a thing, we will use that description to help us decompose the organism into particular parts and processes which articulate together to cause it to behave as described" (Kaufmann 1971, p. 258). *Second*, the

> use of an adequate description of an organism seen as doing a particular thing to guide our decomposition of it into interrelated parts and processes, and indeed part of the logic of search, is intimately connected with the sufficient conditions for the adequate description. In particular we can use the sufficient conditions to generate a cybernetic model showing how symbolic parts might articulate together to cause a symbolic version of the described behavior. (Ibid., p. 258)

Third, and finally, "[w]e can use such a cybernetic model to help find an isomorphic causal model showing how presumptive parts and processes of the real system might articulate to cause the described behavior" (ibid., p. 258). What results from this is a causal explanation using the properties of the parts and the relations between them. Note, however, for the appropriate causal relations to hold, the isomorphism that Kauffman requires is not really necessary. An entailment from the causal model to the cybernetic one would suffice.

Liberated from its biological context, and viewing the model as being as descriptive of the structure of a reductive explanation as for their search, Kauffman's model of explanatory reduction consists of the satisfaction of three criteria. *First*, the explanation is an explanation of a whole using its parts. While this seems intuitively obvious to the notion of a reductive explanation in cases of interlevel reductions, it is not explicitly captured by any of the models of theory reduction, as has been noted above. At least Wimsatt's requirement of compositional redescription is close to the spirit of this requirement.[29] *Second*, Kauffman requires that these parts and processes somehow articulate together to provide the explanation. Unfortunately, this notion of articulation is made no more precise. *Third*, an explanation is complete when Kauffman's criterion of sufficiency is met. These depend on the question asked, that is, on what is to be explained.

A third model of explanatory reduction is given by Sarkar (1989) in the context of a defence of explanatory reductionism in molecular biology. Sarkar argues that though reduction is a form of explanation, it is important, if possible, to keep models of reduction independent of any particular account of explanation. This position is motivated by two connected factors. *First*, there is no universally accepted explication of explanation.

Therefore, a particular model of explanatory reduction might turn out to be unacceptable because of a discovered difficulty with the notion of explanation that forms part of it. In such a circumstance, it might yet be possible to salvage those parts of the model that directly concern the issue of reduction independent of the more general issue of explanation. For example, it has been pointed out that the Salmon account of explanation faces considerable difficulties in many cases which can be reasonably accommodated in some deductive account of explanation (Glymour 1984, p. 180). Moreover, this account renders all deductive explanations intuitively unnatural by regarding them as degenerate cases of statistical explanations. Finally, it simply assumes that what has to be explained is a single event, and it would require considerable modification to deal with laws or lawlike generalizations that have to be explained. These difficulties carry over, at least in part, to Wimsatt's model of reduction though they have nothing to do with reduction.[30] Similarly, any model of explanatory reduction that necessarily requires deduction from general laws falls afoul of the possibility that reductive explanations might sometimes only permit statistical explanations. Further, in the case of explanations using quantum mechanics, the most fundamental theory of nature to date, certain explanations might well be irreducibly statistical. Once again, these difficulties would trouble, for instance, the Schaffner model (when construed as a model of explanatory reduction), though they have nothing necessarily to do with the issue of reduction. *Second*, much of what is explained routinely involves the systematic use of approximations.[31] However, there does not seem to be any completely context-independent notion of a "good approximation." If explanation is supposed to conform, broadly, to the deductive-nomological model, the derivations in question are usually approximate. Further, the "goodness" of an approximate derivation may depend on the question posed, the standard of rigor that is reasonable in a given scientific field, and other such factors. It thus has a pragmatic dimension to it.[32] This further underscores the importance of keeping the issue of explanation separate from that of reduction even while construing reduction as a form of explanation. In order to avoid these difficulties, all Sarkar requires is that a case of purported explanatory reduction already satisfies the strictures imposed on it by the requirement that it be an acceptable case of scientific explanation without specifying the criteria to be used to judge this acceptability. However, this attempt to keep the issue of explanation distinct from that of reduction requires that the other criteria he imposes remain somewhat general.

Sarkar proposes three other criteria, two formal, and one informal. The two formal criteria are (i) "the explanation of the behavior of some entity ... must be accomplished using only the interactions of parts of that entity [and (ii)] the interactions involved in the explanation have physical warrant" (Sarkar 1989, pp. 306–307). The informal criterion is that the completeness of a reductive explanation depends on the question that was posed. The first criterion involves a deliberate extension of Kauffman's requirement that reductive explanations involve parts. Sarkar insists that it be possible to specify the parts of the entity independent of each other and then specify how they interact: this requirement will henceforth be called that of "strict separability." Though apparently innocuous, this requirement will turn out to have nontrivial consequences (as explained in the next paragraph). The third criterion is virtually identical to Kauffman's. In the second, however, there is a crucial difference. Rather than relying on Kauffman's unexplicated "articulation" of the parts, Sarkar specifies that the interactions of the parts that can be invoked be only those that have a physical warrant in the sense that they are derivable from physical law or have been empirically ascertained through purely physical experimentation (ibid., pp. 307–308). This restriction to the "physical" is motivated by the fact that this model is offered only as a model of explanatory reduction in molecular biology. If the model is liberated from this context, only the second formal criterion needs to be modified. Instead of physical warrants, "lower-level warrants," determined from theories at lower levels or by experimentation at those levels would be required.

The general requirement of separability of parts that both Kauffman and Sarkar impose can actually make their models of explanatory reduction stricter than the Nagel–Schaffner type models of theory reduction. For example, if the chemical properties of some body are explained by the physical properties of that body itself (and not its parts), and these properties are explicated in terms of theories, such an explanation would satisfy Nagel's and Schaffner's requirements but not those of Kauffman or Sarkar. An even more interesting case that makes the same point arises from quantum mechanics.[33] There routinely occur, in quantum mechanics, certain states of composite systems, often called "entangled states," which cannot, in principle, be represented as the sum or any algebraic function of individual states representing each component independently. If some higher-level properties of a composite whole has to be explained at the quantum-mechanical level by invoking such states, Sarkar's requirement of strict separability cannot be satisfied. However, the Nagel–Schaffner type of model of theory reduction (after modification from its usual deductive-nomological format to include statistical explanations) would still be able to incorporate such cases. Thus, when Kauffman's or Sarkar's model of explanatory reduction is used, some quite strong empirical claims may be being made. This is to their advantage. It makes the issue of reduction scientifically interesting.[34]

All three models of explanatory reduction considered here not only make the epistemological claim that reductive explanation is possible in the cases where reductions take place, but also the ontological assumption that the whole be construed as consisting only of its parts. Wimsatt and Kauffman do this implicitly while Sarkar does it explicitly by invoking synthetic identities as explicating the relations between an entity and its parts (ibid., p. 307). However, in the case of Sarkar's model, the ontological assumptions being made can be partly weakened by requiring that the explanation of the behavior of the entity be in terms of parts of it that are specified somehow, but not requiring that the resultant connections need be in the form of identities. In particular, if this model of explanatory reduction is coupled with the deductive-nomological account of explanation, the connections between the specified parts and upper-level entities could even have the syntactic form of conditionals, a possibility that Nagel (1961) left open. However, Sarkar's model, like Kauffman's makes a strong assumption about strict separability and this involves additional ontological assumptions that cannot be dropped without modifying the model beyond recognition.

2.3.3 Constitutive Reduction

Proponents of reduction who are interested in making an ontological claim, but only in as weak a sense as that of constitutive reductionism, hardly present analyses that are specific enough to qualify as "models." However, treating these analyses with some rigor does permit interpreting them, though perhaps only in a very mitigated sense, as models. Two such models have so far been presented, the second being slightly stronger than the first. The *first* is "token physicalism." As explained by Fodor

(1974), all it requires is that higher-level phenomena—mental phenomena, in Fodor's case—do not violate physical law. If this model is generalized somewhat, all it requires is that higher-level phenomena remain consistent with lower-level ones. No epistemological claim is being broached.

A *second* model of constitutive reduction is one based on the concept of supervenience as espoused by Davidson (1970). Davidson presents the notion in the context of the relation between mental and physical events. It has since been advocated in a biological context, especially by Rosenberg (1984, 1985). Presented in a context-independent fashion, the idea is that if two events are identical with respect to their specification at a lower level, they cannot differ at a higher level. Further, an entity cannot alter at a higher level without also altering at the lower levels. Dependence of this kind is called supervenience: the higher-level events (or entities) are supervenient on lower-level ones. This is obviously an ontological claim. However, no epistemological claim is necessarily being made about the possibility of accounting for the change at the higher level from the changes at the lower-level. According to Davidson (1970) and Rosenberg (1984, 1985), in general, such explanation is not possible, the source of the difficulty usually being complexity in the sense that apparently innumerably many lower-level states (or specifications) at the lower level correspond to a single one at the higher. Note, however, that this model is stronger than the last one in the sense that it seems to require a causal dependence of higher-level phenomena on lower-level ones whereas token physicalism only required consistency.

Obviously, all models of explanatory and theory reduction that make ontological claims usually satisfy the requirements of both of these models. Proponents of models of constitutive reduction, however, usually advocate these models in the belief that no model of explanatory reduction would also obtain for the cases considered (Fodor 1974; Rosenberg 1984, 1985). This position, however, is problematic. Any assumption of constitutive reductionism immediately generates a research program, namely, that of attempting to explain the properties of higher-level entities in terms of the properties of the lower-level entities that constitute them.[35] To say that, in principle, not model of explanatory reduction is applicable in such cases is to say that such explanation is impossible, that is, the research program mentioned fails. This could happen for two reasons. *First*, it could be the case that the research program was not pursued actively enough or was premature. However, this situation does not permit the "in principle" claim that no model of explanatory reduction would ever be applicable.[36] *Second*, it could be the case that the research program is itself fundamentally mistaken. However, this would probably bring into question whether even any model of constitutive reduction is itself applicable. The only plausible example where so fundamentally mistaken a research program might be being pursued is the case of the attempted reduction of the psychological realm to the physical and, here, it is easy to see that, should the program fail, the ontological assumption involved, that is, constitutive reductionism itself, is exactly what would have to be questioned.

2.4 An Application: Reduction in Molecular Biology

Given these models of reduction, it is natural to ask which of these models, if any, apply to actual cases of scientific reductions. In discussions of this sort, the questions whether a model of reduction is supposed to capture the strategy of research that was or is being pursued, or whether it is intended to capture the structure of the explanations that might result, must be kept distinct. For the sake of brevity, only one case will be examined here: the potential reduction of classical genetics to molecular biology.[37] There are at least four reasons for making this choice. *First*, the question whether there has been a reduction of classical genetics to molecular biology continues to be a controversial topic with opinions about equally divided on both sides.[38] *Second*, the disputes over the cogency of reductionist claims in molecular biology have often been made because of differing concepts of reduction, corresponding to differing models of reduction held by the disputants, thus illustrating why it is important to study systematically these models as is attempted here. *Third*, the history of molecular biology has been such that the questions of the strategy of research and the structure of explanations receives different answers, thus emphasizing the necessity of keeping these questions separate as is also been underscored here. *Fourth*, it turns out that the particular classification offered here does very clearly elucidate not only the nature of reduction that might be defended in molecular biology, but also the philosophical debate that has persisted since the late 1960s. Thus, it argues strongly not only for analyses of the sort attempted here, but also for this specific one.

The issue of reduction in biology is an old one. It is already present in the mechanical philosophy of the seventeenth century with the living organism being imagined as a complicated machine.[39] It is really what is at stake in the mechanist-vitalist disputes of the eighteenth century. In the conflict between the mechanists and the "teleomechanists" of the nineteenth century, the issues already become complex.[40] The dispute becomes one where the mechanists assert that all phenomena involving living organisms would find complete explanations in mechanical terms, while the teleomechanists, under the influence of Kant, argued for the necessity of a second "teleological" mode of explanation of biological phenomena. In the twentieth century, though the issue of vitalism, that is, the existence of forces peculiar to living matter, has largely been irrelevant (with a few exceptions such as Bergson 1911), the nineteenth-century dispute persisted because of the once-popular notion of complementarity in molecular biology. In 1932, the physicist, Bohr (1933) argued that there were certain living phenomena that could not be entirely explained by physical considerations alone. This was so not because there were operative in living matter any non-physical force of interaction but because experiments on living organisms had certain limitations imposed on them by the necessity of keeping the organism alive (Bohr 1933, p. 458).[41] These limitations would require a "complementary" mode of explanation beyond the physical one. For Bohr, it was another application of the principle of complementarity that he had introduced as part of his interpretation of quantum mechanics and which he continued to pursue as an independent epistemological principle.[42] Bohr's speculative arguments provided the impetus for another theoretical physicist, Delbrück, to begin research in biology with the hope of demonstrating complementarity by discovering biological phenomena incapable of physical explanation.[43] Delbrück's work was critical to the establishment of an organized discipline of molecular biology. Complementarity, however, was never discovered—no biological phenomena have yet been excavated that are incapable of, at least approximate, physical explanation. Judged by its own terms, then, the Bohr–Delbrück program was a failure. Molecular biology thus illustrates the point that the research strategy followed and the structure of explanations obtained might well be different.

Philosophical attention to the impressive developments of molecular biology since the early 1950s began with the work of Schaffner (1969). Schaffner initially argued that what was being achieved in molecular biology were reductions, for example, of classical genetics to molecular genetics.[44] Further, according to Schaffner, his model of theory reduction captured both the research program of molecular genetics and the structure of explanations afforded by it. Later, Schaffner (1974) withdrew the first of these two claims and argued that carrying out reductions was only peripheral to the program of molecular biology, but what was achieved by it, nevertheless, was reduction, for instance, of classical genetics to molecular genetics. This shift in position was motivated by his observation that molecular biologists did not seem to design their research strategy around explicit satisfaction of the principles of connectability and deducibility. In effect, Schaffner's position amounts to a claim that the structure of explanations in molecular biology, but not the research program, is captured by his model of reduction. Schaffner's position was rejected immediately by Hull (1972, 1974, 1976) who argued, among other things, that explanations in molecular biology were routinely given by the citation of mechanisms responsible for particular observed phenomena rather than by anything remotely like derivations. Further, Hull (1972, 1976) pointed out that many different mechanisms at the molecular level all routinely account for the same phenomenon at the "classical" level. Moreover, according to Hull (1976), not only do scientists not seem to pursue reductions in the sense of Schaffner (1969, 1976), but philosophers who claimed that reductions were occurring in molecular biology had failed to exhibit the purported laws of classical and molecular genetics in suitably precise form and the required deductive relations between them. On these grounds, Hull (1972, 1976) argued that classical genetics was being replaced by molecular genetics rather than being reduced to it.

 Wimsatt (1976b), however, argued that reduction was occurring in molecular biology, both in the sense that molecular biology provided reductive explanations of biological phenomena and in the sense that scientists were actively pursuing reduction in their research. However, Wimsatt (1976b) explicates reduction using the model of explanatory reduction described in section 2.3.2, which is radically different from Schaffner's. The causal factors required by that model, in its account of explanation, are nothing other than the mechanisms that molecular biologists invoke in order to explain observed phenomena. Kitcher (1982, 1984), however, revives Nagel's model of reduction and discards it on the grounds that classical genetics did not have laws that could be reduced to molecular genetics. Meanwhile, Rosenberg (1985) abandons hope of explanatory reduction in molecular biology on the grounds of the complexity of molecular explanations and resorts to supervenience instead. Sarkar (1989), however, argues that at least the structure of explanations in molecular biology is reductionist and he provides the model discussed in section 2.3.2 in order to capture the sense in which it is. He also argues, using the examples of Bohr–Delbrück program and the theoretical attempts to decipher the genetic code in the 1950s as evidence, that the research strategy of molecular biology was not uniformly reductionist.[45]

The classification of models of reduction that has been attempted here sheds considerable light on these disputes regarding reduction in molecular biology. The differences between the disputants turn out to be less radical than they supposed, though they remain important after their nature is clarified. The differences between Schaffner and Hull depend critically on what constitutes a scientific theory. The last question is controversial, to say the least, but what seems uncontroversial is that a scientific theory should *at least* be a reasonably small body of laws that explain a much greater number of empirical generalizations.[46] If this minimal constraint on scientific theories is accepted, it is apparent that classical genetics does have a theory: the laws of classical genetics are simply Mendel's laws corrected by a host of discoveries, mainly in the 1920s, long before the advent of molecular biology.[47] Molecular genetics, however, does not yet have any such unifying set of general laws, though there are generalizations of quite wide scope, such as those expressing the rules of DNA replication. At most, at present, all that can be found in molecular biology are fragments of theories. Explanations using these, such as the explanations of chromosome duplication using the rules of DNA replication, are notoriously complicated, invoking a large number of mechanisms, and caveats, such as those about the minimal requirements of environmental consistency necessary for accurate DNA replication. These hardly have anything like the pattern of derivations. Thus, Schaffner (1974) cannot be completely correct in claiming that the structure of explanations in molecular biology can be captured by his model of reduction. In fact, the absence of a fully explicated theory at the molecular level entails that no model of theory reduction can, at present, capture the relation between classical and molecular genetics. Consequently, attempts to reconstruct theories in molecular biology (for example, Balzer and Dawe 1985, 1986) are artificial and largely irrelevant. The criticisms leveled by Hull (1972, 1974, 1976) against Schaffner are justified insofar as they note the absence of a genetical theory at the molecular level and point to the explanatory mechanisms instead. The criticisms of Kitcher (1982, 1984), however, are unjustified since they are based on the claim that classical genetics does not have a theory.

What is most important, however, is that these criticisms leave the categories of explanatory and constitutive reductionism untouched. Thus, the models of explanatory reductionism advocated by Wimsatt (1976b) and Sarkar (1989) escape these criticisms. Had the Bohr–Delbrück expectation of complementarity in molecular biology been realized, then even models of explanatory reduction would have been doomed. All that could have been asserted was a model of constitutive reduction since, in accepting that no non-physical laws are operating in living matter, Bohr (1933) and Delbrück (1949) embrace at least this minimal ontological commitment. One might then have had to fall back on mere supervenience, but the success of molecular biology in accounting for all biological phenomena investigated by it in purely physical terms makes this move premature, contrary to Rosenberg (1985). Supervenience, at this stage, is the counsel of unnecessary despair: it involves, at best, a self-denial of scientific curiosity; at worst, an abrogation of scientific responsibility. However, the existence of the Bohr–Delbrück program and the critical influence of Delbrück on the development of molecular biology do show that models of explanatory reduction, such as those of Wimsatt (1976b) and Sarkar (1989), do not capture all of the research strategy of molecular biology. Rather, their applicability is limited to the structure of explanations in molecular biology and to a part of its research program.

The foregoing considerations at least suggest that some model of explanatory reduction is far more likely to capture the sense of reduction appropriate in the context of molecular biology than any model of theory reduction. The same conclusion might well hold in many other cases. Shimony (1987) has given several cases in low energy physics—the properties of atoms and molecules, Coulomb systems in their thermodynamic limit, normal fluids and spin systems (including ferromagnets)—where the properties of the composite system can be explained not using only "first principles," that is, the theory of the lower level system, but by also using "secondary principles," including semiempirical ones. Approximations are also routine.[48] All these cases seem to be ones in which explanatory reduction is far more plausible than theory reduction.[49] This is not to suggest, however, that models of theory reduction are totally irrelevant. In certain contexts, such as the reduction of physical optics to electromagnetism, theory reduction as modelled by Schaffner (1967) seems reasonably adequate.[50] Moreover, for intralevel reduction between theories, as Nickles (1973) has pointed out, his model of intralevel theory reduction seems to capture quite well the relations between some theories, such as classical mechanics and special relativity. What the considerations adduced in this section do suggest, however, is that confining attention to theory reduction misses the potential for capturing many cases of scientific reductions that are much more conceptually interesting than those cases that seem to be captured by models of theory reduction. This point will be taken up in the next section.

2.5 Conclusion

In his stimulating and important series of papers, Hooker (1981a, 1981b, 1981c) argues that of all potential relations that might be construed as reductions, "the richest and most precise relationship [is] that of the reduction of one theory to another" (Hooker 1981a, p. 42). Further, "all other cases [are] embeddable, and requiring embedding, in these" (ibid., p. 44). If the discussions of the last section are remotely correct, Hooker is fundamentally mistaken. Models of theory reduction that involve detailed specification of certain relations, such as those of Nagel (1961), Schaffner (1967, 1974, 1988) or Balzer and Sneed (1977), are indeed precise, but so is the model of explanatory reduction due to Wimsatt (1976b). But formal precision does not translate into strength. In section 2.3 it was argued that the formally less specific models of Kauffman (1971) and Sarkar (1989) are more difficult to satisfy than the models of Nagel (1961) and Schaffner (1967, 1974) *and this is so because of additional empirical claims they require*. Moreover, formal precision might well be achieved at the cost of describing few, if any, actual cases of scientific reductions.[51] Consequently, explications of reduction that rely on such precision are exactly the opposite of "rich." Formal precision, for its own sake, is hardly a desideratum to be taken seriously.[52] More important, however, the considerations of the last section show that it is often quite inappropriate even to conceive of reductions as occurring between theories. *First*, it can be the case, as in the example of molecular biology, that reductions occur without well-defined theories being operative, if even some minimal criteria are imposed on what constitutes "theories." *Second*, even in cases where theories might be involved, secondary principles are often needed for reductions to obtain, as emphasized by Shimony (1987).

The moral to be drawn from these observations, however, is not that reduction should never be construed as a relation between theories. Indeed, as was observed in the last section, there are circumstances in which certain models of theory reduction seem to be applicable. The moral to be drawn, instead, is that discussions of reduction pay sufficient attention to its complexity. There is no a priori reason to assume that all cases of reduction are so similar that they can all be captured by any single model of reduction. Where reduction occurs between theories, models of theory reduction might look for their adaptive niche. Where reductions occur without theories, and yet explain, some models of explanatory reduction might yet save the phenomena. Where reductive explanation fails altogether, though ontological connections between the entities involved remain plausible, some model of constitutive reduction might be the only one that is applicable. And, finally, there can be situations where no reductions obtain. A priori considerations cannot eliminate any of these possibilities. What is required is a detailed investigation of the context.

Acknowledgments

Thanks are due to David Hull, Michael Martin, Ken Schaffner, Abner Shimony and William Wimsatt for many valuable discussions of these issues and for comments and criticism of an earlier version of this paper. This paper was partly written during the tenure of a grant from the Boston University Graduate School.

Notes

1. For a masterly account of the development and eventual demise of the mechanical philosophy, see, e.g., Stein (1958).

2. A very similar approach was independently (and almost simultaneously) advocated by Woodger (1952).

3. In other words, the allegedly reduced theory cannot be derived from the reducing one. Feyerabend (1962) also argued that the two theories cannot be connected by any such reduction relation because of an alleged "incommensurability" between

the terms of the two theories. The metaphor of incommensurability has long run its course in the philosophy of science, arguably without contributing any insight whatsoever. Rather than explicitly argue against it, this paper will simply ignore issues connected with incommensurability on grounds of irrelevance.

4. The trouble with the Suppes (1957) approach seems to be that what it does is to take the usual mathematical notion of reduction (which involves nothing more than the establishment of the type of isomorphisms required by Suppes) and apply it in the context of an empirical science. There is no attempt to take cognizance of the different epistemological and ontological concerns that arise with this change of context.

5. It is not clear, however, that Suppes (1957) intended his model of reduction necessarily to be a form of explanation. However, that is how he is usually regarded (e.g., by Schaffner 1967).

6. See Kauffman (1971), Sarkar (1989) and, especially, Wimsatt (1976b).

7. When reduction is construed as an explanatory relation between theories, Hooker (1981a, 1981b, 1981c) has, at least, attempted a unified treatment. However, as the discussion in the text shows, confining attention to theories leaves out most of the more interesting attempts at explicating reduction since Schaffner (1967).

8. Shimony (1987) emphasizes this point.

9. These are, of course, not the only epistemological and ontological questions that can be asked. For instance, a much more interesting ontological question is simply whether the interactions of the more fundamental entities determine the behavior of the composite ones (determination, here, not being taken necessarily to permit prediction). For a list of epistemological and ontological questions, scrupulously kept separate, see Shimony (1987, pp. 399-400).

10. These categories are based partly on some distinctions introduced (though not very clearly) by Mayr (1982).

11. See, however, Wimsatt (1976a) for possible counterinstances.

12. Exceptions include models of explanatory reductionism due to Kauffman (1971) and Sarkar (1989) which have a strong requirement of separability that models of theory reduction might violate. These are discussed in the next section.

13. Of course, different models of constitutive reduction might make different ontological assumptions as will be discussed in the text.

14. Nagel's initial treatment (Nagel 1949) was systematically elaborated in Nagel (1961) without any significant change. The latter account will be followed here. Attention will, however, be confined only to Nagel's formal conditions for reduction assuming that the more interesting nonformal conditions—that the reducing theory

is not ad hoc, that it is fecund, and that both theories be at a stage of development that makes reduction worthwhile—are all satisfied (Nagel 1961, pp. 358–360). See Sarkar (1989) for a critical discussion of these issues.

15. This is pointed out, and criticized, by Hempel (1969) who, however, sees in Nagel's avoidance of ontological commitment only an unfortunate linguistic emphasis in his treatment of reduction.

16. The motivation for this modification was to meet objections, such as those of Feyerabend (1962), as was noted in section 2.1.

17. That this connection be one of synthetic identity has been underscored by virtually all proponents of this type of theory reduction, especially Sklar (1967), Causey (1972a, 1972b) and Hooker (1981a, 1981b, 1981c).

18. Martin (1972) has emphasized the role the requirement of identity plays in the construal of reductionist claims as ontological ones.

19. See Hooker (1981a, 1981b, 1981c) and Schaffner (1993) for further details.

20. The strongest of these would be to require that these identities connect "natural kinds" (Fodor 1974). However, in the absence of any reasonably noncontroversial explication of the notion of a "natural kind," these further subtleties and issues will be ignored here.

21. Perhaps ironically, Wimsatt (1976a, 1976b), who clarified the same distinction that Nickles is making, comes to the exact opposite conclusion with respect to the potential applicability of the Nagel–Schaffner model. Wimsatt denies that it holds for interlevel reductions while admitting that something like it might be appropriate for intralevel reductions.

22. Nagel's "homogeneous" reductions are, thus, a special case of domain-preserving intralevel reductions. Of course, intralevel reductions need not necessarily preserve domains since they might, for instance, at least partially extend the domain of applicability of prior theories, as Hull (personal communication) has rightly pointed out.

23. Shimony (personal communication), however, has argued that Nickles's use of the notion of reduction in such cases is not standard. The issue remains controversial. Moreover, to the extent that Nickles is explicating a notion of reduction that is used at all in a scientific concept, it seems to be one that (as with Suppes 1957) is merely parasitic upon the standard mathematical use of "reduction" where all that is meant is the demonstration of an isomorphism between two mathematical structures, as has been noted above. When the speed of light is taken to be infinite, the equations of special relativity do become formally identical (and, therefore, trivially isomorphic) to the equations of Newtonian mechanics.

24. This is a problem that originates with the initial attempt of Suppes (1957) to interpret reduction in terms of isomorphisms (mentioned in section 2.1). It was first noted by Schaffner (1967).

25. The necessity of this condition was already noted by Adams (1959).

26. This condition is a simpler version of one introduced by Bourbaki (1968, p. 267).

27. Balzer and Dawe (1986, p. 185n) attribute this result to Feferman.

28. In a later version of his model, Salmon (1984) also makes this move.

29. See, e.g., Wimsatt (1976a, 1976b). However, Wimsatt seems to assume, implicitly, that such description of entities in terms of their parts is trivially possible and, therefore, pays no further attention to this issue. However, the situation can be far more complex, as the discussion of Sarkar's requirement of strict separability (see below) will show.

30. Wimsatt (personal communication) believes, however, that these difficulties are not insurmountable. The first and third, for instance, could be met by requiring reductive explanations of each case covered by the law. However, such a move would violate the initial requirement that it is the law itself, not only its consequences, that has to be explained. This objection, in turn, could be met by assuming that the set of instances of a general law would have some internal structure that would be maintained through the reduction. While this is plausible, this idea needs more elaboration before its merit can be judged.

31. Wimsatt (1976b) and Shimony (1987) have noted this point.

32. This point is connected to Kauffman's requirement that the completeness of an explanation depends on the question posed, a criterion that Sarkar explicitly adopts (see next paragraph).

33. For an illuminating discussion of these issues, including their relationship to the Einstein–Podolsky–Rosen "paradox" in quantum mechanics, see Shimony (1989).

34. In the case of molecular biology, where Sarkar (1989) claims that his model of explanatory reductionism holds (except for functional explanations), the empirical claim being asserted includes, for example, that quantum mechanical entanglement need not be invoked to account for any of the phenomena being investigated.

35. Shimony (personal communication) has made this point.

36. If Rosenberg (1985) is correct for the case of molecular biology, and it is argued in the next section that he is not, then this possibility would be the one that would have to be invoked.

37. For a comprehensive bibliography of discussions of scientific reductions, then up to date, see Wimsatt (1979). No better bibliography still seems to exist.

38. Kitcher (1982, 1984) and Hull (1972, 1974, 1976) are prominent critics of reductionist claims; Schaffner (1969, 1974, 1988), Wimsatt (1976b) and Sarkar (1989) continue to defend them, though using very different models of reduction. Details will be discussed in the text.

39. For an examination of the connections between the mechanical philosophy and the program of research of contemporary molecular biology, see Sarkar (1989).

40. For details, see Lenoir (1982).

41. Bohr gives no argument for the necessity of maintaining life in the organisms during all experiments. For a critical evaluation of Bohr's arguments, see Sarkar (1989, pp. 96–105).

42. Even within the standard Copenhagen interpretation of quantum mechanics, which was chiefly due to Bohr, complementarity is not required as Stein (1972) points out.

43. For an account of Delbrück's career and the role played by complementarity in it, see Fischer and Lipson (1988) and Sarkar (1990).

44. The entire debate about reduction in molecular biology has been centered around genetics. There are several reasons for this, the main one being that it is in molecular genetics that the most impressive early triumphs of molecular biology have so far been achieved.

45. The second example is one of nonreductive research because of its reliance on a naïve nonphysical concept of "information," as for example, in the notion of a "comma-free" code introduced by Crick, Griffith and Orgel (1957). For details, see Sarkar (1989, pp. 168–225).

46. This minimal characterization was advocated by Rosenberg (1985, p. 121).

47. See Sarkar (1989) for details.

48. Shimony (1987, pp. 419–421) also gives a penetrating discussion of how approximations might *systematically* be used in approximate derivations.

49. Note, however, that Sarkar's model would need modification to relax strict separability in the context of physics because of the routine occurrence of entangled states.

50. See Schaffner (1967) for details.

51. The dispute over the relation between classical genetics and molecular biology described in the last section amply illustrates this point.

52. Wimsatt (1976b) has also made this point, though, perhaps somewhat less vehemently. Formal precision, for him, is undesirable whenever the cost (in terms of

effort and complexity) is not worth the benefit, as is usually the case in logical reconstructions of complicated explanatory patterns.

References

Adams, E. W. 1959. "The Foundations of Rigid Body Mechanics and the Derivation of its Laws from those of Particle Mechanics," in L. Henkin et al. (eds.), *The Axiomatic Method*. North-Holland, Amsterdam, pp. 250–265.

Balzer, W., and Dawe, C. M. 1986. "Structure and Comparison of Genetic Theories: (1) Character-Factor Genetics," *British Journal for the Philosophy of Science* 37: 55–69.

Balzer, W., and Dawe, C. M. 1986. "Structure and Comparison of Genetic Theories: (2) The Reduction of Character-Factor Genetics to Molecular Genetics," *British Journal for the Philosophy of Science* 37: 177–191.

Balzer, W., and Sneed, J. 1977. "Generalized Net Structures of Empirical Theories I," *Studia Logica* 36: 195–211.

Bergson, H. 1911. *Creative Evolution*. Macmillan, London.

Bohr, N. 1933. "Light and Life," *Nature* 131: 421–423, 457–459.

Bourbaki, N. 1968. *Theory of Sets*. Hermann, Paris.

Causey, R. L. 1972a. "Attribute-Identities in Microreduction," *Journal of Philosophy* 69: 407–422.

Causey, R. L. 1972b. "Uniform Microreductions," *Synthese* 25: 176–218.

Crick, F. H. C., Griffith, J. S., and Orgel, L. 1957. "Codes without Commas," *Proceedings of the National Academy of Sciences* 43: 416–421.

Davidson, D. 1970. "Mental Events," in S. Foster et al. (eds.), *Experience and Theory*. University of Massachusetts, Amherst, pp. 79–101.

Delbrück, M. 1949. "A Physicist Looks at Biology," *Transactions of the Connecticut Academy of Science* 38: 173–190.

Feyerabend, P. K. 1962. "Explanation, Reduction, and Empiricism," *Minnesota Studies in the Philosophy of Science* 3: 28–97.

Fischer, E. P., and Lipson, C. 1988. *Thinking about Science: Max Delbrück and the Origins of Moelcular Biology*. Knopf, New York.

Fodor, J. 1974. "Special Sciences (or: the Disunity of Science as a Working Hypothesis)," *Synthese* 28: 97–115.

Glymour, C. 1984. "Explanation and Realism," in J. Leplin (ed.), *Scientific Realism*. University of California Press, Berkeley, pp. 173–192.

Hempel, C. G. 1969. "Reduction: Linguistic and Ontological Issues," in S. Morgenbesser et al. (eds.), *Philosophy, Science, and Method*. St. Martin's Press, New York, pp. 179–199.

Hempel, C. G., and Oppenheim, P. 1948. "Studies in the Logic of Explanation," *Philosophy of Science* 15: 491–499.

Hooker, C. A. 1981a. "Towards a General Theory of Reduction. Part I: Historical and Scientific Setting," *Dialogue* 20: 38–59.

Hooker, C. A. 1981b. "Towards a General Theory of Reduction. Part II: Identity in Reduction," *Dialogue* 20: 201–236.

Hooker, C. A. 1981c. "Towards a General Theory of Reduction. Part III: Cross-Categorical Reduction," *Dialogue* 20: 496–529.

Hull, D. 1972. "Reduction in Genetics—Biology or Philosophy?" *Philosophy of Science* 39: 491–499.

Hull, D. 1974. *Philosophy of Biological Science*. Prentice-Hall, Englewood Cliffs.

Hull, D. 1976. "Informal Aspects of Theory Reduction," *Boston Studies in the Philosophy of Science* 32: 653–670.

Kauffman, S. A. 1971. "Articulation of Parts Explanation in Biology and the Rational Search for Them," *Boston Studies in the Philosophy of Science* 8: 257–272.

Kemeny, J., and Oppenheim, P. 1956. "On Reduction," *Philosophical Studies* 7: 6–19.

Kitcher, P. 1982. "Genes," *British Journal for the Philosophy of Science* 33: 337–359.

Kitcher, P. 1984. "1953 and All That: A Tale of Two Sciences," *Philosophical Review* 93: 335–373.

Lenoir, T. 1982. *The Strategy of Life*. D. Reidel, Dordrecht.

Martin, M. 1972. "The Body–mind Problem and Neurophysiological Reduction," *Theoria* 37: 1–14.

Mayr, E. 1982. *The Growth of Biological Thought*. Harvard University Press, Cambridge.

Nagel, E. 1949. "The Meaning of Reduction in the Natural Sciences," in R. C. Stauffer (ed.), *Science and Civilization*. University of Wisconsin Press, Madison, pp. 99–135.

Nagel, E. 1961. *The Structure of Science*. Harcourt, Brace and World, New York.

Nickles, T. 1973. "Two Concepts of Inter-Theoretic Reduction," *Journal of Philosophy* 70: 181–201.

Rosenberg, A. 1984. "The Supervenience of Biological Concepts," in E. Sober (ed.), *Conceptual Issues in Evolutionary Biology*. The MIT Press, Cambridge, pp. 99–115.

Rosenberg, R. 1985. *The Structure of Biological Science*. Cambridge University Press, Cambridge.

Salmon, W. 1971. *Statistical Explanation and Statistical Relevance*. University of Pittsburgh Press, Pittsburgh.

Salmon, W. 1984. *Scientific Explanation and the Causal Structure of the World*. Princeton University Press, Princeton.

Sarkar, S. 1989. "Reductionism and Molecular Biology: A Reappraisal," Ph.D. Dissertation, Department of Philosophy, University of Chicago.

Sarkar, S. 1990. "Review of E. P. Fischer and C. Lipson's *Thinking about Science: Max Delbrück and the Origins of Molecular Biology*," *Perspectives on Biology and Medicine* 33: 612–616.

Schaffner, K. 1967. "Approaches to Reduction," *Philosophy of Science* 34: 137–147.

Schaffner, K. 1969. "The Watson–Crick Model and Reductionism," *British Journal for the Philosophy of Science* 20: 325–348.

Schaffner, K. 1974. "The Peripherality of Reductionism in the Development of Molecular Biology," *Journal for the History of Biology* 7: 111–129.

Schaffner, K. 1993. *Discovery and Explanation in Biology and Medicine*. Chicago: University of Chicago Press.

Shimony, A. 1987. "*The Methodology of Synthesis: Part and Wholes in Low-energy Physics*," in R. Kargon and P. Achinstein (eds.), *Kelvin's Baltimore Lectures and Modern Theoretical Physics*. The MIT Press, Cambridge, pp. 399–423.

Shimony, A. 1989. "Conceptual Foundations of Quantum Mechanics," in P. Davies (ed.), *The New Physics*. Cambridge University Press, Cambridge, pp. 373–395.

Sklar, L. 1967. "Types of Inter-theoretic Reductions," *British Journal for the Philosophy of Science* 18: 119–120.

Sneed, J. D. 1971. *The Logical Structure of Mathematical Physics*. D. Reidel, Dordrecht.

Stegmueller, W. 1976. *The Structure and Dynamics of Theories*. North-Holland, Amsterdam.

Stein, H. 1958. "Some Philosophical Aspects of Natural Science," Ph.D. Dissertation, Department of Philosophy, University of Chicago.

Stein, H. 1972. "On the Conceptual Structure of Quantum Mechanics," in R. G. Colodny (ed.), *Paradigms and Paradoxes: Philosophical Challenges of the Quantum Domain*. University of Pittsburgh Press, Pittsburgh, pp. 367–438.

Suppes, P. 1957. *Introduction to Logic*. Van Norstrand, New York.

Wimsatt, W. C. 1976a. "Reductionism, Levels of Organization, and the Mind–body Problem," in G. Globus, I. Savodnik and G. Maxwell (eds.), *Consciousness and the Brain*. Plenum, New York, pp. 199–267.

Wimsatt, W. C. 1976b. "Reductive Explanation: A Functional Account," *Boston Studies in the Philosophy of Science* 32: 671–710.

Wimsatt, W. C. 1979. "Reduction and Reductionism," in P. D. Asquith and H. Kyburg (eds.), *Current Research in the Philosophy of Science*. Philosophy of Science Association, East Lansing, pp. 352–377.

Woodger, J. H. 1952. *Biology and Language*. Cambridge University Press, Cambridge.

3 Genes versus Molecules: How To, and How Not To, Be a Reductionist

3.1 Introduction

The molecular revolution in biology, beginning in the late 1940s and early 1950s, must surely be regarded as one of the last century's most significant scientific developments. It transformed not only the practice, but the very conceptual framework of much of biology at the organismic and lower levels of organization to such an extent that it is sometimes difficult even to find continuity within the same research schools. Central to this transformation was the molecular characterization of the gene. Before the molecular era, what is now called "classical" genetics consisted of models of transmission as well as models of expression. The models of transmission were generally highly successful: simple duplication for haploid genomes, Mendel's rules as modified by linkage requirements for diploids, generalizations for polyploids, and special constructions for haplodiploid and other odd genetic structures. The models of expression were largely vacuous: genes somehow produced traits, with luck singly, otherwise acting in concert (physiological epistasis), and sometimes not very reliably (variable degrees of expressivity and penetrance). The many failures of the classical genetic account of expression and, concomitantly, of organismic development are well known and need no repetition here; their significance will be assessed in the next section.

The critical point, here, is that, within the theories in which it was embedded, the classical gene was an abstract entity. Its transmission properties were captured by rules such as Mendel's rules, essentially probabilistic rules asserting the statistical independence of the transmission of alleles during reproduction. Its expression properties were supposed to be captured in the rules connecting genotype to phenotype. The chemical nature of the gene was irrelevant to both these sets of rules. Indeed, until well into the 1940s it was widely believed that genes were in the protein parts of chromosomes. Even those who were committed to finding and exploring the physical basis of heredity—most notably, the Morgan school—admitted that classical genetics required no particular commitment to the physical nature of the gene.[1] As late as the 1950s, Lederberg and his collaborators interpreted data from bacterial conjugation crosses to produce a branched linkage map which, as they explicitly noted, should not be interpreted as a branched chromosome (Lederberg et al. 1951, p. 417). Classical genetics was a formal science about an abstract entity, the classical gene. Each gene came in different versions or alleles; allelic specificity was inferred from phenotypic differences. Genes, in this sense, were "diagnostic" entities inferred indirectly from phenotypic differences.

All that changed with the double helix. Genes were now concrete entities, identified with segments of DNA. What mattered was not the double-helical shape of the molecule but the sequence of bases, and the complementarity between them on the two DNA strands.[2] Instead of a conformational account of behavior and specificity, which had become the dominant mode of molecular explanation of biological behavior thanks to Pauling's pioneering work, gene specificity was determined by sequence identity. By 1958, this view had been incorporated into an informational account of biological behavior, at least at the molecular level.[3] Of the two central theoretical innovations of early molecular biology, one was about genes—the operon model of gene regulation; the other, however, was about proteins—the allostery model of hemoglobin function. Nevertheless, in the late 1960s and 1970s, molecular genetics became the ascendant sub-field within molecular biology and, as eukaryotic genetics began to yield puzzles and surprises, a continued source of intellectual excitement.[4]

It should come as no surprise that, once the gene was physically characterized, the "molecular gene" came to be routinely conflated with the "classical gene." For molecular biologists, "gene" usually refers to bits of DNA no matter whether the context is classical or molecular. Molecular biology is an immensely successful enterprise. Beyond genetics, it has revolutionized biological sub-disciplines from cytology to immunology, it has significantly permeated evolutionary studies, and it is now beginning to give a successful even, if as yet only rudimentary, mechanistic account of development that has remained elusive ever since Roux first propounded his *Entwickslungsmechanik* program in the 1890s. The trouble is that, thanks to the conflation mentioned earlier, the success of molecular genetics has been misinterpreted to have shown the success of classical genetics. For transmission, this is undeniably true although many details remain to be worked out; for expression, this is equally false. Perhaps the most significant conceptual confusion that has resulted from this conflation is that the gene, when conceived of as a bit of DNA, need not even be something that makes any phenotypic difference at the organismic level. In the program of deciphering what, if anything, a bit of DNA does, sometimes called "reverse genetics," the gene is a "constructive" rather than a diagnostic entity. It is at least arguable that it is only because of the potential for reverse genetics that talk of genes remains heuristically useful in molecular biology and cannot be entirely replaced by talk of DNA. Beyond conceptual issues, this conflation has had significant negative consequences for biology; it has been part of the rhetoric that was used to initiate the Human Genome Project (HGP).[5] More importantly it has led to the popularity of facile claims of genetic etiology for complex human behavioral traits; this obviously has significant sociopolitical implications, which, however, remain beyond the scope of this present contribution.

The purpose of this chapter is to argue against a conflation of molecular biology (even molecular genetics) with classical genetics. In philosophical jargon—which will be discarded after this paragraph—the epistemological program of classical genetics is *genetic reductionism*, the explanation of phenogenesis (the production of phenotypes) on the *basis of (inferred or diagnosed) classical genes*.[6] In contrast, the epistemological program of molecular biology (including molecular genetics) is physical reductionism, the explanation of all biological phenomena on the basis of the *physical properties of their constituent parts* at the level of molecules and macromolecules. By "reduction" is meant an explanation of phenomena in one domain from the principles of a presumably more fundamental domain. Both genetic and physical reductionism assume the existence of a hierarchical model of the systems of interest in which lower levels of the hierarchy are presumed to be progressively more fundamental than upper ones. For genetic reductionism, this is an abstract hierarchy: at the bottom lie alleles obeying laws of transmission, then come loci, linkage groups, and genotypes and, finally, the phenotypes which have to be explained. For physical reductionism, the hierarchy is given by the physical structure of the organism. The take-home messages of this chapter are (i) the facile genetic reductionism inherited from the heyday of classical genetics is vacuous if the aim is to understand phenogenesis (rather than only hereditary transmission of traits), and (ii) the physical reductionism of molecular biology continues to be a fecund research program of tremendous epistemological power and interest. However, it, too, must be treated with some caution since some recalcitrant problems remain.[7]

3.2 Hegemonic Geneticism

Mendel chose to study traits with modes of inheritance simple enough for algebraic characterization.[8] Only one locus was implicated in the etiology of a trait, there was complete dominance at each locus, only two alleles, and no linkage. After the recovery of Mendel's work around 1900, each of these assumptions was demonstrated to be violable within the first three years.[9] Nevertheless, departures from Mendelism were interpreted as deviations from the presumed resilient basic model, particularly in Britain. Why this strategy was adopted with very little explicit methodological discussion will require sociohistorical investigation of a sort that is yet to be seriously attempted.[10] The routine complexity of the relationship between genotype and phenotype (a distinction explicitly articulated by Johannsen in 1909) remained unappreciated.

The diagnostic analysis of conventional Mendelism found its most fertile home in the United States in the work of the Morgan school. While segregation analysis (that is, the analysis of pedigrees to determine whether the pattern of inheritance of a trait could be subsumed under Mendelian expectations) has been a part of Mendelian genetics from the very beginning—in the human case, Bateson (1906) used it to infer a genetic etiology for congenital cataract and brachydactyly in 1906—the first main innovation of the Morgan school was linkage analysis. Starting shortly after 1910, the Morgan school used failures of Mendel's second rule (independent assortment of alleles at different loci) to place loci in linear order; by 1925, 400 loci of *Drosophila melanogaster* had been placed in four disjoint linear structures. Critical to the invention of linkage analysis was Morgan's 1910 interpretation of linkage as representing physical contiguity of loci on chromosomes (Morgan 1910). The chromosome model, as the physical basis for Mendelism, was always the central heuristic of the Morgan school.

Moreover, at least for Muller, the physical nature of the gene was the biological problem of central interest.[11] Nevertheless, the Morgan school was well aware that the results of linkage analysis could be interpreted purely formally. Even after associating almost 400 loci with chromosomes, they observed:

> Were there no information as to the relation between the visible chromosomes and the linkage group it would still be possible to deal with the situation exactly in the same way by treating the linkage groups as a series of points held together in definite relations to each other. We might then speak of such groups as genetic [as opposed to physical] chromosomes. (Morgan et al. 1925, p. 88)

Lederberg's construction of a branched chromosome map on the basis of linkage analysis was mentioned in the last section. As late as the 1950s, Delbrück continued to hope that the physical order of loci on chromosomes would contradict their order as determined by linkage analysis: this was supposed to show the limitation of reductionist explanation in biology.[12]

While the use of linkage analysis to map loci on to chromosomes may be the most scientifically important—and uncontroversial—use of such analysis, the other traditional application of linkage analysis has been its use to posit a genetic etiology for traits. This is genetic reductionism in its purest form: if there is a statistical correlation between the inheritance pattern of a trait and the inheritance of a known locus, the trait is supposed to have, at the very least, a partial genetic etiology. If there is a well-defined simple relationship between alleles at a single locus and the trait, such an inference is unproblematic.

The customary use of linkage analysis is far less straightforward: it is the preferred strategy of the genetic reductionist to posit a genetic etiology for "complex" traits, sometimes called "polygenic" traits, such as human behavioral traits. The adjectives used to describe these traits—"complex" suggesting that "simple" means "genetic," and "polygenic" suggesting that the only etiological alternatives available are single genes or several—already indicate a prior commitment to genes as the source of traits. That phenogenesis will not so easily succumb to genetic reduction is clearly indicated by the history of such attempts. Restricting attention to human behavioral traits, linkage analysis has been used to claim a genetic etiology for, among other traits, alcoholism, autism, bipolar affective disorder, and schizophrenia (see Sarkar 1998, p. 2). Initial positive reports have been greeted with much fanfare and publicity. In *every* single case, later reports indicated a decrease and, usually, a loss of linkage. Negative reports have so far never received the same publicity as the positive ones. To the extent that incorrect public perceptions harmfully influence the medical choices made by individuals, genetic reductionism has become a public health menace.

One example will suffice to indicate the general nature of the problem with linkage analysis. In some families of the Amish community in Pennsylvania (USA), bipolar affective disorder (manic depression) appeared to be inherited. In 1987, a linkage was reported between this trait and the H-*ras* locus on the short arm of chromosome 11 (Egeland et al. 1987). Bipolar affective disorder was thus supposed to be explained on the basis of genes (alleles) at this locus. However, studies of other pedigrees failed to confirm this result (Baron et al. 1990, 1993). Finally, additional information on the original pedigree, when two previously unaffected individuals succumbed to the disease, undermined the statistical basis for the original assertion of linkage (Kelsoe et al. 1989). The negative result generated little publicity compared to the fanfare surrounding the original positive report.[13]

Unless one is a committed genetic reductionist, these negative results are not unexpected. An organism with its complement of traits is a product of its complex history-dependent developmental processes in which it uses internal resources (genes and other inherited units), along with external resources, to produce its phenotypic traits. Trivially, no organism would exist or have the form it has, without its genes. Equally trivially, no organism would exist, or have the form it has, without its particular environmental history. What genetic reductionism assumes is that genes *alone* bear the epistemological weight in explanations of phenogenesis. That this is unlikely except in a tiny minority of cases has been obvious since almost the earliest days of genetics.

As early as 1909, Woltereck, while studying pure lines of morphologically distinct strains of *Daphnia* and *Hyalodaphnia*, showed that continuous traits (such as helmet size) varied between pure lines and were affected by a spectrum of environmental parameters (Woltereck 1909).[14] Woltereck introduced the term *"Reaktionsnorm"* to capture these relationships; later the "norm of reaction" came to be used to describe the curve showing a single genotype's phenotypic response to an environmental parameter. A constant norm of reaction showed uniform genetic etiology across varying environments. The separation of distinct genetic and environmental components of the etiology of a trait required parallel norms of reaction for all of the genotypes in a population. The sensitivity of the norm of reaction to gene–environment interactions was noted as early as 1933 by Hogben who also observed that, even in experimental populations of Drosophila, selected and maintained for generic uniformity (becoming what are now called "model organisms"), norms of reaction were not parallel (see Hogben 1933). In the West, the norm of reaction was generally ignored as classical genetics focused on phenotypes with constant norms; these were phenotypes which were easiest to fit into genetic expectations. However, the norm of reaction came to be viewed as the unit of inheritance in the Soviet Union as geneticists there—no doubt, at least partially because of the Soviet state's ideological bias towards environmental rather than genetic determinism—struggled for an interpretation of genetics less rigid than the conventional Mendelism of the West. (The concept of the norm of reaction was eventually repatriated to the West by Dobzhansky in the 1930s; see, e.g., Dobzhansky 1937.)

Meanwhile, instead of addressing the variability embedded in nonconstant norms of reaction, genetics in the West came to endorse a reticulation of its basic conceptual structure to maintain the epistemological primacy of the gene (see Sarkar 1999, and Laubichler and Sarkar 2002). The experimental work initially came from the Soviet Union but its interpretation within the constraints of conventional classical genetics was due to the German neuroanatomist, Vogt (see Vogt 1926). In the Soviet Union, Romashoff and the Timoféeff-Ressovskys studied different mutations of *Drosophila funebris* (Romashoff 1925; Timoféeff-Ressovsky and Timoféeff-Ressovsky 1926). Romashoff found that homozygous mutants sometimes exhibited the mutant phenotype to different degrees. The Timoféeff-Ressovskys even found that a pure-line that was homozygous for a mutation sometimes bred untrue, and that the fraction of deviants appeared to be fixed *for each pure line*. Vogt ignored the relevance of the background pure line and interpreted these results by positing two new properties of genes (alleles) beyond the traditional (and sometimes problematic) properties such as dominance—(variable) degrees of *expressivity* and *penetrance*. The former is supposed to be the degree of manifestation of a gene (note how the gene is the sole repository of agency), while the latter is the probability of any manifestation at all of a gene.

By now, expressivity and penetrance have become part of the standard repertoire of genetics, particularly medical genetics. Those who would maintain the primacy of genetic etiology in the face of phenotypic complexity have recourse to variable expressivity and incomplete penetrance to maintain the posited primacy of the gene. Yet few scientific concepts are less well founded. The degree of expressivity is often impossible to distinguish from the degree of dominance. Without a background pure line— ethically, if not technologically, difficult to create for *Homo sapiens*—there is no empirical reason to expect that two different instances of the same gene have the same probability of being manifested as a trait (because of having the same penetrance). Experimentally measured penetrances are no more than empirical frequencies masquerading as theoretical propensities. It is hard to escape the conclusion that the role played by "penetrance" and "expressivity" in contemporary genetics, particularly human behavioral genetics, is ideological: to maintain a genetic etiology in the face of recalcitrant detail. Penetrance and expressivity rescue genetic agency in the face of equivocal, even absent, data. If there is a putative gene for a trait, but the presence of the trait is nevertheless capricious in the presence of the gene, there is still a gene *for* that trait, but the gene has *variable expressivity*. If the trait does not deign to manifest itself at all, there is still a gene *for* the trait: it is just that the gene is *incompletely penetrant*.

In recent years, the repertoire of diagnostic genetic techniques has been extended from traditional segregation and linkage analyses to include new techniques such as quantitative trait locus (QTL) mapping, allelic association studies and the allele-sharing method (see, e.g., Lander and Schork 1994). These have undoubtedly increased the degree of diagnostic resolution possible but the interpretative problems noted in this section remain unresolved.[15]

3.3 The Molecular Vision of Life

When the molecular revolution in biology began, segregation and linkage analysis were perhaps the most quantitative parts of biology other than population genetics and ecology. Consequently, it should hardly come as a surprise that the genes (alleles at specific loci) that became targets for molecular elucidation were the ones with the simple phenotypic effects that were routinely studied by linkage and segregation analysis. Nevertheless, molecularization initially paid as much, if not more, attention to proteins as to nucleic acids. This is not in the least surprising; even if genes were the focus of interest, until the mid-1940s it was generally believed that genes consisted of proteins.[16] Nucleic acids, consisting of only four base types, were presumed not to have sufficient variability to be able to specify the huge variety of genes that were known. Pauling's α-helix model for secondary protein structure set the stage for successful physical reductionism in molecular biology (see Pauling and Corey 1950, 1951). Most important, from Pauling's work, there emerged a new model of biological specificity determined by molecular structure, more precisely, by the shape of active sites. This is how enzymes catalyzed their substrates, how antibodies recognized their antigens, and how the chains of hemoglobin interacted with each other in the allostery model for cooperative protein behavior. There was no chemically necessary relation; in 1970, Monod introduced the concept of *gratuity* to capture the type of stereospecificity that lay at the theoretical core of molecular biology.[17]

While the DNA double helix, showing how gene replication must occur during cell reproduction, and the operon model of bacterial gene regulation were undoubtedly intellectually interesting developments, it is far from clear that they should have sufficed to force a myopic focus of molecular biology on genetics, especially in the 1970s. Once again, no proper sociological history of these developments has been reconstructed and any claim of historical explanation must remain speculative. However, a reasonable conjecture is that the apparent simplicity of the gene–protein relationship—captured by the genetic code—and an expectation that the protein folding problem would soon be solved, led to a heady confidence that molecular genetics, having largely successfully provided a mechanistic account of heredity, would now similarly provide one for phenogenesis. The aim was a predictively robust model of both heredity and development in which epistemic primacy resided in the genome.

That expectation was not borne out; within genetics, it fell foul of what has aptly been called the "unexpected complexity of eukaryotic genetics" (Watson et al. 1992). Starting in the 1970s, five sets of developments destroyed the simple informational model of the genome in molecular biology:[18]

(i) Although the genetic code is nearly universal, it is not entirely so. Consequently, before any prediction using the code can be made in a truly novel context, it must be determined which variant is in use.

(ii) There is no natural synchronization of transcription, and frameshift mutations are known to exist, thus allowing the same DNA to be used in a variety of ways. Once again, for prediction, the exact point of initiation of DNA transcription must be known.

(iii) Even within coding regions of DNA there are regions that are transcribed but not translated (introns). Consequently, all intron–exon boundaries must be known for prediction.

(iv) Similarly, much of eukaryotic DNA occurs between genes and has neither a structural nor a regulatory role. In fact it has no known functional role (hence, it is sometimes, perhaps unfairly, referred to as "junk DNA"). The boundaries of genes must also be determined for each case for prediction (although there are patterns here and the initiation regions are not random).

(v) Finally, mRNA is sometimes copiously edited before translation, sometimes to such an extreme extent that it becomes misleading even to say that the corresponding DNA "codes for" the eventual protein product.

The upshot of these developments is that the molecular "gene" or segment of DNA is at best a constructive gene (in the terminology of the opening section above) to be studied by using reverse genetics. Whether it should even be called a "gene," thus contributing to the conflation of classical and molecular genetics that is being criticized in this paper, is largely a matter of taste. What is important is that the molecular gene *qua* piece of DNA lacks agency; it is a molecular tool used by organisms for a variety of purposes.[19] Perhaps the most radical version of this position is to view the genome as a sequestered molecular template used by cells to transfer specificities to subsequent (cellular) generations.[20] There are subtle questions here, for instance, should the cell or the individual organism be regarded as the primary repository of agency? Intuitions from "higher" animals suggest the latter, but "higher" animals form a tiny subclass of the biological world. Questions such as these now form part of the research program of the emerging field of developmental evolution.[21]

The new focus—and some success—of molecular accounts of development has also led away from the gene. Although molecular developmental biology has its historical origins, at the level of experimental technique, in molecular developmental genetics, recent progress has revealed surprising elucidatory patterns at the protein level. Now, if developmental interactions required protein specificity in the way it seemed central to molecular biology in the 1950s and 1960s, then, because of the coding relation between DNA and protein, however enfeebled that relation has become (recall the discussion of eukaryotic genetics above), one could still maintain some epistemic primacy for DNA over proteins in accounting for development. However, confronting development, in particular, the study of development comparatively across phyla has led to the replacement of specificity by the ubiquitous *tolerance, redundancy* and *genericity* of molecular interactions.[22]

These developments began with the realization of the rather remarkable conservation of many cellular structures and processes at the protein level across almost all phyla. For instance, Cdc2 is a protein that is central to the regulation of the cell cycle of yeast (*Saccharomyces cerevisiae*). However, the human homolog of Cdc2 can take over all of the latter's functions in the yeast cell cycle, including yeast-specific functions such as sensitivity to nutritional signals and response to hormonal signals for mating, meiosis, and mitosis (Gerhart and Kirschner 1997, p. 30). The globins from all five kingdoms show functional conservatism in spite of sequence divergence. Among eukaryotes, of the 226 globin sequences that are known, there are only two invariant amino acid residues and between any two members of a pair, agreement of residues at corresponding positions can be as low as 16 percent (Gerhart and Kirschner 1997, p. 26). Yet, functionally, these are all globins. Because of the degeneracy of the genetic code, divergence at the genetic level is even greater.[23] Tolerance of some structural differences lies at the core of such functional conservatism at the protein level.

Redundancy is seen when cells have available more than one protein for the same function. In such a situation, during evolution, one of the proteins—and the corresponding DNA—can be co-opted for another function. For instance, the prokaryotic protein FtsA and the eukaryotic protein actin are believed to be derived from a common ancestor. Both bind ATP but, in spite of structural similarity, are only 20 percent identical in sequence.[24] They are not known to have similar functions at present: actin forms filaments and is a major cytoskeletal protein of eukaryotic cells; FtsA is present in all prokaryotic cells where it is involved in septum formation.

In eukaryotes, actin is not required for this function for which it has become redundant. Genericity is seen when a small subset of mechanisms are used for a variety of purposes, especially when relatively simple and robust physical interactions are used instead of highly specific—and, presumably, highly evolved—mechanisms. A standard way of linking intracellular biochemical reactions to extracellular signals is through the use of ion channels that depend on the membrane potential, that is, a difference in the electrostatic potential across a membrane because of the presence of different polyelectrolytes on the two sides. There are only six kinds of channels which can even be structurally distinguished (Gerhart and Kirschner 1997, p. 100). Only four types of ions are routinely used for signaling, i.e. Na^+, K^+, Ca^{2+} and Cl^-.

These are exciting times for molecular biology, now that the beginnings of a molecular account of development seem within reach. From this point of view, the role of research on DNA was similar to the role of the collection of facts of natural history in the formulation of the theory of evolution, an important stage but, ultimately, of little theoretical significance. The interesting structures and the interactions that make them possible all occur at the protein level. The cell co-opts for its use whatever resources it has available in its inherited DNA (and other units of inheritance).

3.4 Discussion

It is a truism that there is more to biology than molecular biology. The levels of biological organization that were of concern in the last section are far from the communities and ecosystems studied by ecologists and conservation biologists. If "molecularization" of these disciplines is supposed to mean their representation by molecular models, not only is this impossible in practice, it is hard to see why the results would even be interesting. The interesting questions that are asked, and the answers that are expected, are at levels far from the molecular. Moving to the molecular level will hardly help resolve the stability–complexity or other such debates in community ecology. There is an interesting reductionist question here, that of methodological individualism, whether the individual-based models, typical of population ecology, can explain all the phenomena of community ecology (see Huston et al. 1988, and Sarkar 1996a). However, that is a far cry from molecular/physical reductionism. Even if it is accepted that reductions successively go to lower levels or organization, it is still hard to imagine how the molecular level would add to ecological insight.

Nevertheless, even ecology and conservation biology cannot altogether avoid questions of development at the molecular level. For instance, only research at the molecular level can reveal the mechanisms and the extent to which UV radiation can be implicated in the apparently global decline of amphibian populations.[25] The response of the mammalian endocrine system to molecular cues affecting reproduction underscores the importance of *some* analysis at the molecular level in population studies that are relevant to ecology and conservation biology. Molecular knowledge may thus well be critical to determining the environmental parameters that are critical for the continued survival of populations. At least to this limited extent, the molecular level is relevant even for the study of ecological systems.

However, leaving ecology aside, it is equally hard to imagine how the future of biology can be anything but molecular.[26] This is a reductionist vision but this type of reduction has nothing to do with the primacy of genes, the obsessive deification of DNA that increasingly marked the biology of the late twentieth century. Rather, in a sense, it is a return to the mechanistic vision of the seventeenth century, attempting to understand complex phenomena from simple structural and interactive principles. Reductionism, in this version, is a piecemeal vision: choosing problems that seem tractable in a given context and explaining wholes in terms of parts and parts of parts, recursively, but often simultaneously employing a multiplicity of levels, that is, molecular moieties, molecules, macromolecules and macromolecular assemblages. There is a rich tapestry of molecules, with epistemic efficacy always percolating upwards, although often starting at different levels depending on the experimental context. It is surprising how successful this vision has so far been for biology up to the organismic level.

However, one should nevertheless resist the temptation to dub this as a triumph of physical reductionism for at least two reasons:

(i) Even in the context of development, at present, systemic properties—for instance, properties of a developing organ as a whole—are often just as necessary for the explanation of morphogenesis as local molecular interactions. It is possible that these systemic properties will themselves eventually succumb to molecular reduction. However, it is also possible that irreducibly systemic properties—for instance, the number or frequency of a particular molecular assemblage in a developing tissue—may continue to be required for explanation. The important methodological point that should be made is that complexity should be embraced, not avoided merely to save the reductionist cause.[27] This may well be the most important lesson to be learnt from the failure of genetic reductionism.

(ii) Some robust organismic phenomena have stubbornly resisted all attempts at physical reduction. Dominance is a common property of traits. There is, as yet, no satisfactory molecular account of dominance. To the extent there is at all a molecular account, it relies on the topological properties of biochemical reaction networks.[28] Topological properties are not physical properties; consequently, the epistemological weight in such explanations is not borne by the physical interactions involved. Topological accounts provide systemic explanations. The future will show the extent to which they are necessary, even at the molecular level.

There is thus no reason for a reductionist triumphalism. Reductionism is an empirical issue and the evidence for or against it is not all in: only the future will show whether all biological phenomena at any higher level of organization will succumb to the lure of physical reduction. Genetic reductionism is demonstrably vacuous as a research program at present, no matter what its past history of success has been. The jury is still out on physical reductionism. One way to investigate its limitations is to push it to its limits; this is the tenor of much of biological research today. However, the exploration of nonreductionist research strategies is another way to test the limits of physical reductionism. It deserves more support than what the biological community affords it at present.

Notes

1. See Morgan et al. (1925). This point has been extensively discussed by Sarkar (1998, chapter 5).

2. This point has been emphasized by Lederberg (1993).

3. See Crick (1958). That information was a new theory of specificity was recognized by Lederberg as early as 1956 (Lederberg 1956).

4. Unfortunately, a systematic history of these developments still remains to be written.

5. See Sarkar (2001) for a discussion of the role of reductionism in the HGP.

6. For details of the positions sketched in this paragraph, see Sarkar (1998).

7. See the last paragraph of this chapter.

8. This is not a historical paper; modern terminology is intentionally being used.

9. See Carlson (1966) for a history; unless explicitly stated otherwise, all historical material in this section is from this source.

10. It is not as if the Mendelians were unaware of this move that they made. Their opponents, the biometricians (particularly Pearson) objected vehemently but were ignored—see Provine (1971) for a history. In the German context Sapp (1987) provides some relevant historical detail, although his main focus is on non-genetic inheritance rather than alternative modes of non-Mendelian genetic inheritance.

11. See Carlson (1971) on this point.

12. See Fischer and Lipscomb (1988); Sarkar (1989) reconstructs the history of Delbrück's idiosyncratic antireductionism and the influence on him of Bohr's hope for the discovery of "complementarity" in biology.

13. See Sarkar (1998) for more detail and other examples.

14. The history of the norm or reaction has been reconstructed by Sarkar (1999)—historical details in the rest of this paragraph are from this source.

15. See Sarkar (1998, chapter 5) for a detailed appraisal of these methods from the point of view of establishing genetic etiologies.

16. The shift toward viewing nucleic acids as the genetic material begins with the critical paper by Avery et al. (1944).

17. See Monod (1971); this work also provides accessible and interesting accounts of the allostery and operon models.

18. For details, see Sarkar (1996a, 1996b).

19. An interesting possible consequence of this position is that the classical gene lacks agency (as noted in the last section), at least partly because its molecular substrate does no better. The development of this argument will be left for another occasion.

20. This position is developed in chapter 14.

21. See Wagner (2000) for "developmental evolution" and chapter 14 for a treatment of this question.

22. On tolerance and redundancy, see, especially, Gerhart and Kirschner (1997); on genericity, see Newman and Comper (1990).

23. This is known to be true of globin genes; however, because of the possibility of RNA editing, divergence at the amino acid sequence level need not necessarily imply diversity at the DNA sequence level.

24. Details are from Gerhart and Kirschner (1997, p. 25).

25. See Sarkar (1996a) for more discussions of this point.

26. Note that even evolutionary studies have moved to the molecular level over the last generation (mainly because this is the level at which genetics is now practiced). Using these studies to understand phenotypic evolution will require a molecular or, at least, a physical account of development.

27. For an elaboration of this argument, see Gilbert and Sarkar (2000).

28. See Kacser and Burns (1981); this example is discussed in detail by Sarkar (1998, pp. 168–173).

References

Avery, O. T., MacLeod, C. M., and McCarty, M. (1944). "Studies on the chemical nature of the substance inducing transformation of Pneumococcal types: induction of transformation by a deoxyribonucleic acid fraction isolated from Pneumococcus III," *Journal of Experimental Medicine*, 79: 137–157.

Baron, M., Endicott, J., and Ott, J. (1990). "Genetic linkage in mental illness," *British Journal of Psychiatry*, 157: 645–655.

Baron, M., Freimer, N. F., Risch, N., Lerer, B., Alexander, J. R., Straub, R. E., Asokan, A., Das, K., Peterson, A., Amos, A., Endicott, J., Ott, J., and Gilliam, T. C. (1993). "Diminished support for linkage between manic depressive illness and X-chromosome markers in three Israeli pedigrees," *Nature (Genetics) (London)*, 3: 49–55.

Bateson, W. (1906). "An address on Mendelian heredity and its application to man," *Brain*, 29: 157–179.

Carlson, E. A. (1966). *The Gene: A Critical History*. W. B. Saunders, Philadelphia.

Carlson, E. A. (1971). "An unacknowledged founding of molecular biology: H. J. Muller's contributions to gene theory, 1910–1936," *Journal of the History of Biology*, 4: 149–170.

Crick, F. H. C. (1958). "On protein synthesis," *Symposia of the Society for Experimental Biology*, 12: 138–163.

Dobzhansky, T. (1937). *Genetics and the Origin of Species*. Columbia University Press, New York.

Egeland, J. E., Gerhard, D. S., Pauls, D. L., Sussex, J. N., Kidd, K. K., Allen, C. R., Hostetter, A. M., and Housman, D. E. (1987). "Bipolar affective disorders linked to DNA markers linked on chromosome 11," *Nature (London)*, 325: 783–787.

Fischer, E. P., and Lipson, C. (1988). *Thinking about Science: Max Delbrück and the Origins of Molecular Biology*. Norton, New York.

Gerhart, J., and Kirschner, M. (1997). *Cells, Embryos, and Evolution*. Blackwell Science, Oxford.

Gilbert, S. F., and Sarkar, S. (2000). "Embracing complexity: Organicism for the 21st century," *Developmental Biology*, 219: 1–9.

Hogben, L. (1933). *Nature and Nurture*. W. W. Norton, New York.

Huston, M., DeAngelis, D., and Post, W. (1988). "New computer models unify ecological theory," *BioScience*, 38: 682–691.

Johannsen, W. (1909). *Elemente der exacten Erblichkeitslehre*. Gustav Fischer, Jena.

Kacser, H., and Burns, J. A. (1981). "The molecular basis of dominance," *Genetics*, 97: 639–666.

Kelsoe, J. R., Ginns, E. I., Egeland, J. A., Gerhard, D. S., Goldstein, A. M., Bale, S. J., Pauls, D. L., Long, R. T., Kidd, K. K., Conte, G., Housman, D. E., and Paul, S. M. (1989). "Re-evaluation of the linkage relationship between chromosome 11p loci and the gene for bipolar affective disorder in the Old Order Amish," *Nature (London)*, 342: 238–243.

Lander, E. S., and Schork, N. J. (1994). "Genetic dissection of complex traits," *Science*, 265: 2037–2048.

Laubichler, M., and Sarkar, S. (2002). "Flies, genes, and brains: Oskar Vogt, Nikolai Timoféeff-Ressovsky, and the origin of the concepts of penetrance and expressivity," in L. S. Parker and R. Ankeny (eds.), *Medical Genetics, Conceptual Foundations and Classic Questions*, Kluwer, Dordrecht, pp. 63–85.

Lederberg, J. (1956). "Comments on the gene–enzyme relationship," in Gaebler, O. H. (ed.), *Enzymes: Units of Biological Structure and Function*. Academic Press, New York, pp. 161–169.

Lederberg, J. (1993). "What the double helix has meant for basic biomedical science: A personal commentary," *Journal of the American Medical Association*, 269: 1981–1985.

Lederberg, J., Lederberg, E. M., Zinder, N. D., and Lively, E. R. (1951). "Recombination analysis of bacterial heredity," *Cold Spring Harbor Symposia on Quantitative Biology*, 16: 413–443.

Monod, J. (1971). *Chance and Necessity: An Essay on the Natural Philosophy of Modern Biology*. Knopf, New York.

Morgan, T. H. (1910). "Chromosome and heredity," *American Naturalist*, 44: 449–496.

Morgan, T. H., Bridges, C., and Sturtevant, A. H. (1925). "The genetics of drosophila," *Bibliographia Genetica*, 2: 1–262.

Newman, S. A., and Comper, W. D. (1990). "'Generic' physical mechanisms of morphogenesis and pattern formation," *Development*, 110: 1–18.

Pauling, L., and Corey, R. B. (1950). "Two hydrogen-bonded spiral configurations of the polypeptide chains," *Journal of the American Chemical Society*, 71: 5349.

Pauling, L., and Corey, R. B. (1951). "Atomic coordinates and structure factors for two helical configurations of polypeptide chains," *Proceedings of the National Academy of Sciences (USA)*, 37: 235–240.

Provine, W. B. (1971). *The Origins of Theoretical Population Genetics*. University of Chicago Press, Chicago.

Romashoff, D. D. (1925). "Über die Variabilität in der Manifestierung eines erblichen Merkmales (Abdomen abnormalis) bei *Drosophila funebris* F," *Journal für Psychologie und Neurologie*, 31: 323–325.

Sapp, J. (1987). *Beyond the Gene: Cytoplasmic Inheritance and the Struggle for Authority in Genetics*. Oxford University Press, New York.

Sarkar, S. (1989). *Reductionism and Molecular Biology: A Reappraisal*. Ph.D. Dissertation, Department of Philosophy, University of Chicago.

Sarkar, S. (1996a). "Biological information: a skeptical look at some central dogmas of molecular biology," in S. Sarkar (ed.), *The Philosophy and History of Molecular Biology: New Perspectives*. Kluwer, Dordrecht, pp. 187–231. Chapter 9 of this volume.

Sarkar, S. (1996b). "Decoding 'coding'—information and DNA," *BioScience*, 46: 857–864. Chapter 8 in this volume.

Sarkar, S. (1998). *Genetics and Reductionism*. Cambridge University Press, New York.

Sarkar, S. (1999). "From the *Reaktionsnorm* to the adaptive norm: The norm of reaction, 1909–1960," *Biology and Philosophy*, 14: 235–252.

Sarkar, S. (2001). "Reductionism in genetics and the Human Genome Project," in R. Singh, C. Krimbas, D. B. Paul, and J. Beatty (eds.), *Thinking about Evolution: Historical, Philosophical, and Political Perspectives*, vol. 2. New York: Cambridge University Press, pp. 235–252.

Sarkar, S. (2002). "From the Reaktionsnorm to the Evolution of Adaptive Plasticity: A Historical Sketch, 1909–1999," in T. DeWitt and S. M. Scheiner (eds.), *Phenotypic Plasticity: Functional and Conceptual Approaches*. Oxford University Press, New York.

Schlichting, C. D., and Pigliucci, M. (1998). *Phenotypic Evolution: A Reaction Norm Perspective*. Sinauer, Sunderland, Mass.

Timoféeff-Ressovsky, H. A., and Timoféeff-Ressovsky, N. W. (1926). "Über das phänotypische Manifestieren des Genotyps. II. Über idio-somatische Variationsgruppen bei *Drosophila funebris*," *Wilhelm Roux' Archiv für Entwicklungsmechanik der Organismen*, 108: 146–170.

Vogt, O. (1926). "Psychiatrisch wichtige Tatsachen der zoologisch-botanischen Systematik," *Journal für Psychologie und Neurologie*, 101: 805–832.

Wagner, G. P. (2000). "What is the promise of developmental evolution? Part I: Why is developmental biology necessary to explain evolutionary innovations?" *Journal of Experimental Zoology (Molecular Development and Evolution)*, 288: 95–98.

Watson, J. D., Gilman, M., Witkowski, J., and Zoller, M. (1992). *Recombinant DNA*, second ed. W. H. Freeman, New York.

Woltereck, R. (1909). "Weitere experimentelle Untersuchungen über Artveränderung, speziell über das Wesen quantitativer Artunterschiede bei Daphnien," *Verhandlungen der deutschen zoologischen Gesellschaft*, 19: 110–173.

4 Reduction: A Philosophical Analysis

4.1 Introduction

The mechanical philosophy of the seventeenth century, associated with figures such as Descartes, Boyle and Huygens, initiated a pattern of explanation that has persisted in many sciences into the twenty-first century. The mechanical philosophy sought to explain all physical and biological phenomena on the basis of contact interactions between impenetrable particles of matter of which all systems are composed. In the life sciences this led to the conceptualization of living organisms as machines; the most significant result that emerged from this point of view was Harvey's model of the circulatory system. Two assumptions are implicit in the pursuit of mechanical explanations: (1) that phenomena in one domain can be explained on the basis of principles from another domain; and (2) the system to be explained is viewed hierarchically with phenomena at one level being explained from principles operative at lower levels of organization. The philosophical term "reduction" refers primarily to a type of explanation that is directly descended from these mechanical explanations.

The mechanical philosophy ultimately proved inadequate as a foundation for physical theory. Primarily this happened because of Newton's law of gravitation which, by unifying celestial mechanics with terrestrial mechanics, remained a paradigm for natural law until the twentieth century. By insisting on contact interactions, the mechanical philosophy denied the possibility of action at a distance, which is incorporated in the Newtonian gravitational interaction. In the nineteenth century Helmholtz attempted to relax the mechanical philosophy to allow for all central forces, including gravitation. Even this could not save the mechanical philosophy because some electromagnetic phenomena proved to be recalcitrant.

Meanwhile, Helmholtz and many others also attempted to ground "biology" (a term that only goes back to 1800) on mechanistic explanations alone, by which they meant purely physicochemical explanation. This program became central to biology after Lamarck and others insisted on a purely materialistic interpretation of living organisms. What is called "reductionism" today—and the term will be defined more precisely below—is the direct descendant of this program. The main impetus for endorsing reductionism in biology has come from the spectacular successes of molecular biology during the last fifty years (see below).

Discussions of reductionism involve both ontological and epistemological questions, which should be kept distinct. The former include questions about what exists in a system, whether the laws and mechanisms at one level determine phenomena at another, etc. The latter include the questions about whether reductionist explanations of phenomena can be offered and whether research programs should be based on reductionist strategies. Even if one is an ontological reductionist, it is possible to deny epistemological reductionism, for example, if attempts at reductionist explanation of some phenomenon continue to fail. Such failures can arise because successful explanation depends on both the questions asked and what is known. For instance, a system may be too complex for a reductionist answer to be given for some question asked about it because of the current limitations of computational ability. In biology this is often the case: at present it would be foolhardy to suggest that all organismal behavior—let alone behavior at higher ecological levels of organization—can be given reductionist explanations in terms of molecular interactions. Nevertheless, it is probably true that these interactions determine the phenomena at all higher levels.

Restricting attention to epistemological issues, "reduction," as noted before, is a type of explanation. "Reductionism" is usually taken to be either or both of the theses that: (1) in a given domain reductions will be forthcoming; and (2) research programs should be based on the search for reductions. Here, reductionism will be construed as the conjunction of both these theses. The first makes reductionism an empirical claim: it may fail just as the mechanical philosophy ultimately failed. The second makes reductionism a prescriptive program for biological research.

Finally, it should be noted that there is one type of question that is ubiquitous in biology and for which there are no straightforward reductionist explanations. Such questions are those of origin: why certain features of an organism arose during its evolutionary history. These questions are typically answered functionally, appealing to natural selection, or historically, appealing to historical contingencies, or both. Whether such explanations can eventually be subsumed under the reductionist camp, and whether such a project is even biologically interesting, remains doubtful.

4.2 Formal Issues

Systematic philosophical analysis of reductionism began with Nagel in 1949 and almost all formal models remain variants of Nagel's model. These models assume that (1) reduction is a relation between theories; and (2) the reduced theory is logically deduced from the reducing one; after (3) terms in the two theories are connected to each other by bridge laws.

All three assumptions of this model have been controversial: (1) Theories are supposed to be sets of universal laws. Yet, in many areas of science where reductions are supposed to be occurring, from the kinetic theory of matter to molecular biology, there seem to be no such laws. Rather, at the level of the reducing discipline, there only are mechanisms. (2) Scientific derivations often involve approximative techniques and numerical calculations that do not satisfy the pristine requirements of logical deductions. (3) The form of the bridge laws has been subject to more philosophical controversy than any other aspect of reduction. At one extreme are those who hold that "types" in the reduced theory must be connected to "types" in the reduced one: no single entity at the former level can be connected to many at the lower level (sometimes called multiple realizability). This position is almost certainly too strong to be tenable. It excludes the standard type of situation where macroscopic phenomena are reduced to microscopic ones: to every macrophysical state (say that of a cylinder of gas) there is always an almost infinite number of microphysical states (molecular arrangements of the gas). Another issue has been whether these bridge laws must be biconditionals ("if and only if" claims) since, for explanations to go through, conditionals ("if then" statements connecting the reducing domain to the reduced one) are sufficient. Those who hold the former position are usually motivated by the ontological hope of establishing identities between entities at the two levels.

It is important to note that some of these formal issues are not about reduction *per se* but about the nature of explanation in general, for instance, the role of theories and whether approximations are permissible. It is probably helpful to ignore these issues in epistemological discussions of reduction, in order to concentrate on what additional requirements otherwise acceptable explanations must satisfy to qualify as reductions.

4.3 Substantive Issues

Quite independent of the form that reductions must take is the question as to what substantive assumptions are involved in reductions. Two intuitions lie behind attempts to describe certain explanations as reductions: that phenomena in one domain are being explained by—or reduced to—phenomena in another, and that there is a hierarchical representation in which explanations go from the bottom towards the top. These intuitions can be made precise using three criteria: (1) fundamentalism—the explanation of a feature of a system invokes only rules (the "fundamental rules") from a different, more fundamental realm; (2) abstract hierarchy—the system is represented as a hierarchy of entities with lower levels of the hierarchy being regarded as more fundamental than upper ones; and (3) spatial hierarchy—the hierarchy mentioned in (2) is one in physical (rather than some abstract) space.

The three most important types of reduction are: (a) weak reductions: those that only satisfy criterion (1); (b) abstract reductions: those that satisfy both criteria (1) and (2) but not (3); and (c) strong reductions: those that satisfy all three criteria. Genetics provides illustrations of all three of these types, classical genetics of the first two, and molecular genetics of the third.

Consider (the controversial) attempts to explain the variability of a phenotypic trait in a population from the variation at the genetic level on the basis of a high value for the statistic "broad heritability." Such explanations, to the extent that they work, are weak reductions. The genetic level is different from the phenotypic one and presumed to be more fundamental since genetic differences are assumed to explain phenotypic differences, but not conversely. The fundamental rule is that which says that genetic and environmental influences on a trait can be factored using the analysis of variance. That the genome has a hierarchical structure plays no role in these explanations. Thus, of the three substantive criteria for reduction, only criterion (1) is satisfied.

In classical genetics, two of the standard ways to attribute a genetic aetiology to a trait are through segregation and linkage analyses. As before, criterion (1) is satisfied. However, segregation and linkage analyses assume Mendel's laws (in the usual diploid case) which, in turn, postulate a hierarchical organization of the genome into linkage groups, individual loci and alleles at those loci. However, and this point should be more widely appreciated than it is, this hierarchy is an abstract one. Mendelian explanations are not committed to any particular physical composition of the genes. Mendel formulated his rules long before the physical basis of heredity was known. The reason why there is sometimes confusion on this issue is because the abstract Mendelian organization of the genome maps approximately on to the physical structure of chromosomes (only approximately, because there can be overlapping genes, genes within genes, etc., at the physical level). This type of reduction is sometimes called genetic reduction, and the associated reductionist thesis, genetic reductionism.

Finally, in molecular genetics, detailed physical models of genes are constructed and, as explanations proceed from the details of the physical structure and interactions of the parts, strong reductions result. These explanations rely on mechanisms governed chiefly by "lock-and-key" fits between interacting parts of molecules (for instance, the active sites of enzymes). To the extent that these molecular mechanisms explain the replication of chromosomes and thereby show why the Mendelian rules are true, this amounts to a reduction of Mendelian genetics to molecular genetics. However, because such explanations do not conform to the formal requirements of Nagel's model, this putative reduction of Mendelian to molecular genetics has been controversial.

The substantive issues treated so far are epistemological. Turning to ontological issues, vitalism was a doctrine that denied ontological reductionism because it postulated the existence of special forces in living systems. Since its demise, substantive ontological issues about reductionism have not generally been controversial in biology. Even those who deny the possibility of strong reductions in biology admit that the lower levels determine the upper; no change can occur at an upper level without a change at the lower level. This position is sometimes called "supervenience" in the philosophical literature.

4.4 Scientific Status

In this section the current status of reductionism in the various biological subdisciplines will be very briefly indicated. Attention will be restricted to substantive issues, which are the ones of most scientific interest.

4.4.1 Molecular Biology

It is in molecular biology that one finds the most significant triumphs of physical reductionism, of the mechanistic view of life that has guided a significant part of biological research since the nineteenth century. How reduction proceeds in molecular genetics has already been indicated in the last section. The same principles apply elsewhere in molecular biology, resulting in strong reductions. For instance, antibodies interact with antigens when there is a fit in shape between the antigen and the active site of the antibody. Two models have been particularly important in underscoring the success of reductionism in molecular biology by providing reductions of phenomena that have traditionally been part of the repertoire of antireductionists: (1) the allostery model shows how cooperative phenomena, as exemplified by the sigmoidal binding curve of haemoglobin and oxygen, can be give a reductionist explanation; and (2) the operon model shows how phenomena involving feedback can also be similarly explained. What deserves attention, though, is that reduction is to the laws of macromolecular physics and not physics or chemistry at some lower level of organization.

This should not be taken to suggest that all explanations in molecular biology involve strong reductions. In particular, explanations using the concept of information are not reductionist in this sense since rules about information are not lower-level physical rules.

4.4.2 Genetics

The types of reduction that are found in classical genetics have already been discussed in the previous section. Restricting attention to genetic reductions, it is clear that these explanations work in some cases, for instance, in explaining polydactyly or the sickle cell trait in humans. However, genetic reductionism becomes much more controversial when attempts are made to extend it to more complex traits, in particular, human behavioral traits including mental disease traits such as schizophrenia. Prospects for genetic reductionism in these contexts seem dim at present. Genetic reductionism only remains tenable because an adequate theory of development is yet to be formulated. If such a theory emerges, and is based on the primacy of genes, then genetic reductionism will be vindicated.

4.4.3 Evolutionary Biology

In the context of reductionism, the issue most in dispute in evolutionary biology is whether evolution at all levels of the taxonomic hierarchy can ultimately be explained by selection (or other evolutionary mechanisms) acting at the level of genes. If so, this leads to the view, popular among population geneticists, that evolution is only a change of gene frequencies. (This position may be viewed as an evolutionary extension of genetic reductionism.) There is little dispute that evolution can be tracked using gene frequencies. However, critics of gene selectionism maintain that this is only book-keeping; selection is supposed to be directly operating at higher taxonomic levels. Most of these critics endorse selection at the level of individual genotypes. However, even higher-level units such as groups or kin are probably necessary to explain certain social traits such as altruism.

4.4.4 Ecology

In population ecology, explanations of populational features such as growth rates, relative species compositions, etc., on the basis of the interactions between individuals satisfy the requirements of strong reductions. The fundamental rules are those that govern the interactions between individuals. In community ecology, systems are often represented using trophic or food webs. If these have a hierarchical structure, explanations using them are abstract hierarchical reductions; otherwise they are weak reductions. The fundamental rules are those that govern the interactions between species. Since ecosystem ecology is much less clearly defined than either population or community ecology, the status of explanations in it is also less clear. If the fundamental rules used are those involving energy and matter flows, the system usually does not have a hierarchical representation and only weak reductions are possible.

In recent years, an intriguing proposal has been that all ecological phenomena can be explained on the basis of individual-based models. In effect, this says that community and ecosystem ecology is to be reduced to population ecology as a strong reduction. In the past such a claim would have been implausible, given the complexity of most ecological systems. However, the advent of high-speed computation has made this proposal much more plausible. In consonance with usage in the philosophy of the social sciences, this type of reductionism has been called methodological individualism in ecology.

4.4.5 Neurobiology

Because of the traditional interest in the mind–body problem, more philosophical work on reductionism has been done in the context of neurobiology than in any other biological context. At one extreme are claims of strong reduction: mental properties are taken to be properties of brains. Explanation then proceeds by exploiting the anatomy of the brain and the biochemical interaction of parts. This proposal is supported by the fact that increasingly detailed neuroanatomical studies have led to the functional localization of many forms of mental behaviour. However, given the present status of knowledge in the neurosciences, any final commitment to strong reductionism seems unwarranted, even though the search for such reductions continues to constitute fertile research programmes.

At the other extreme are claims that even weak reduction of mental phenomena to neurobiological principles will be impossible. According to this view, mental phenomena only supervene on neurobiological facts; theories of mind must draw on traditions other than those of functional neuroanatomy to be successful, for instance, on assumptions about information storage and manipulation in the computer model of the mind. Somewhere in between these antireductionist claims and strong reductionist claims are models such as those of neural networks. Explanations using these models assume that neuroanatomy matters, but what really matters are not the detailed biochemistry and structure of the interacting parts but, rather, how the system is organized. In recent years, it has become clear that some mental functions, such as memory, are not localized to unique brain regions. This may indicate that the complexity of the relation between brain and mind may require topological or other forms of systemic nonreductionist explanation. In general, these nonreductionist claims do not deny the supervenience of the mental upon the physical. What is at stake is the epistemological issue of explanation.

4.5 Research Methodology

In each of the fields considered above, there is almost complete consensus that reductionist research strategies continue to be useful even if they are not the only ones that should be pursued. What remains in dispute among biologists is whether too much emphasis is being put on reductionist research programs. Philosophers have pointed out that such research programs are committed to the principle that the better explanations always emanate from lower levels of organization. One starts at the lowest level and searches for explanations at the next higher level iteratively only if explanations are not found at that level. This is a metaphysical principle that is incompatible with the empiricist interpretation of science, which urges a search for the best explanation—one that explains the most phenomena most accurately—irrespective of the level from which the explanation comes. At least in evolutionary biology, it is clear that a reductionist methodology has led to biases against groups and higher taxa as units of selection. Finally, it should be noted that biology is replete with methodologies that are not reductionist, for instance, the statistical analysis of traits in populations. It follows that a claim that biology operates successfully only with reductionist research programme is clearly untenable.

Further Reading

Churchland, P. S. (1986). *Neurophilosophy*. Cambridge, Mass.: The MIT Press.

Coleman, W. (1971). *Biology in the Nineteenth Century*. Cambridge: Cambridge University Press.

Hull, D. (1972). *Philosophy of Biological Science*. Englewood Cliffs, N.J.: Prentice-Hall.

Huston, M., DeAngelis, D., and Post, W. (1988). New computer models unify ecology. *BioScience* 38: 682–691.

Nagel, E. (1961). *The Structure of Science*. New York: Harcourt, Brace and World.

Rosenberg, A. (1985). *The Structure of Biological Science*. Chicago: University of Chicago Press.

Rosenberg, A. (1994). *Instrumental Biology*. Chicago: University of Chicago Press.

Sarkar, S. (1996). Biological information. In: Sarkar, S. (ed.), *The Philosophy and History of Molecular Biology: New Perspectives*. Dordrecht, The Netherlands: Kluwer. (Chapter 9 of this volume.)

Sarkar, S. (1998). *Genetics and Reductionism*. New York: Cambridge University Press.

Schaffner, K. (1993). *Discovery and Explanation in Biology and Medicine*. Chicago: University of Chicago Press.

Shimony, A. (1987). The methodology of synthesis: parts and wholes in low-energy physics. In: Kargon, R. and Achinstein, P. (eds.), *Kelvin's Baltimore Lectures and Modern Theoretical Physics*. Cambridge, Mass.: The MIT Press.

Sober, E. (1984). *The Nature of Selection*. Chicago: University of Chicago Press.

Wimsatt, W. C. (1976). Reductive explanation: A functional account. In: Cohen, R. S. and Sonné, R. (eds.), *PSA-1974*. Dordrecht: Reidel.

Wimsatt, W. C. (1980). Reductionist research strategies and their biases in the units of selection controversy. In: Nickles, T. (ed.), *Scientific Discovery: Case Studies*. Dordrecht: Reidel.

Part II Function

5 Reductionism and Functional Explanation in Molecular Biology

5.1 Introduction

Philosophical discussions of the issue of reductionism in molecular biology have routinely been remarkably confused over the use of the related terms, "reduction" and "reductionism."[1] For example, the main defender of the relevance of reductionism in molecular biology, Schaffner, has argued that though molecular biologists are not actively carrying out reductions, the result of their activities is a reduction of parts of biology to physics and chemistry (Schaffner 1974). Meanwhile, Wimsatt (1976) has defended the position that what molecular biologists are doing constitutes "reductions," but his explication of that notion has virtually no similarity to that of Schaffner. Further, the main critic of the relevance of reductionism in molecular biology, Hull, has argued that though what occurs in biology is not reduction in Schaffner's sense, there might be an "informal" sense in which reduction occurs (Hull 1976). At one point, Ruse (1971) even provided what amounts to a partial explication of a notion of "informal reduction" which was to be contrasted, presumably, with "formal reduction" which, he then claimed, was not occurring in molecular biology.

The roots of this confusion lie in the common failure to maintain two sets of distinctions that are quite critical to any clear analysis of the issues involved. The *first* of these consists of a quite simple distinction and is much less often missed than the second. This distinction is that between "reductionism" used to describe actual scientific practice (the research strategies of the scientists involved), and "reductionism" used to describe the structure of explanations afforded by molecular biology. Once this distinction is admitted, Schaffner's position, for example, becomes clear: "reductionism cannot be used to describe the former situation, but can be used to describe the latter."[2] Almost all critics of reductionism in molecular biology from Ruse (1971) and Hull (1972) to Kitcher (1984) do not clearly maintain this distinction perhaps because they want to argue against the role of reductionism in either context. However, this failure leads to a serious misunderstanding of the history of molecular biology, as will be indicated below.

The *second* set of distinctions arises from the various construals that have been given to "reduction" in either of the two contexts separated above. Some have construed "reduction" as a relation between theories (Schaffner 1967, 1976; Hull 1972; Balzer and Dawe 1985, 1986, *etc.*) while sometimes offering radically different explications of that relation. Schaffner (1967), for example, construes "reduction" as the existence of a deductive relation between two theories, codified in first-order logic, the entity and predicate terms of which have been appropriately connected using synthetic identities. The theory that is so deduced, the "reduced theory," is the biological one; the theory used to carry out the deduction, the "reducing theory," is the physical or chemical one. Balzer and Dawe (1985), however, adopt the method of reconstructing theories using informal set-theoretic predicates advocated by Sneed (1971) and construe "reduction" as the existence of a certain relation between prospective *models* of these theories. However, others such as Wimsatt (1976) prefer not to construe "reduction" as a relation between theories at all but, in the context of molecular biology, as a relation between some observed biological behavior and the physical or chemical mechanisms used to explain its occurrence. A similar construal is implicit in Kauffman (1972). Finally, Rosenberg (1978, 1985) has argued that all that is involved in the reduction of the behavior of biological organisms to physics and chemistry is their "supervenience" on the latter: there can be no alteration of biological behavior without a corresponding alteration of physical and chemical behavior of the entities involved. However the use of the latter behavior to explain the former is precluded because of the complexities involved.

The distinctions implicit in these conflicting construals of "reductionism" are best captured using an insight due to Mayr (1982). Mayr distinguishes between three broad categories of reductionism: *constitutive reductionism, explanatory reductionism*, and *theory reductionism* (1982, 59–63). Once these categories are distinguished, the various notions and explications of "reduction" can then be organized into these categories though

Mayr does not carry out this additional step of clarification. The category of constitutive reductionism simply consists of those explications or notions of reduction that require that all biological processes occur in such a way that they are consistent with physical law. In effect, all that this category excludes is any vestige of vitalism. The category of explanatory reductionism includes those explications or notions of reduction that require that biological processes are explained by underlying physical and chemical ones.[3] Finally, the category of theory reductionism includes those explications or notions of reduction that necessarily construe it as a relation between theories. If this relation involves explanation, as it almost always does, then this category can be taken to be even more restrictive, or "stronger," than the category of explanatory reductionism.[4] Notions or explications of reduction falling into any of these categories can potentially be used either to investigate the research strategies pursued in molecular biology or to examine the structure of explanations afforded by it. Thus the two sets of distinctions are independent of each other.

For the sake of convenience, in the following discussion, a notion or explication of reduction will be said to fall into the category of constitutive reductionism if it falls in that category and no other; to fall into the category of explanatory reductionism if it falls into that category but not into the category of theory reductionism; and to fall into the last category simply when it does so. Thus Rosenberg's notion of reduction by supervenience (1978, 1985) falls into the category of constitutive reductionism. The explications of Wimsatt (1976) and Kauffman (1972) both fall into the category of explanatory reductionism. Schaffner (1967) and Balzer and Dawe (1985, 1986) have provided radically different explications of reduction that fall into the category of theory reductionism.

A failure to maintain both sets of distinctions has led to a lack of appreciation of the complexity of the role of reductionism in the history of molecular biology. If the first distinction is admitted, the question whether the research strategies of molecular biology were reductionist is independent of whether the structure of explanations in it are reductionist simply because research strategies can fail. Schaffner (1974), who implicitly and consistently maintains this distinction, has argued that the research strategies in molecular biology are not reductionist but the resulting explanations are. However, Schaffner works only in the context of his own explication of reduction, ignoring the second set of distinctions introduced above. Thus, it is quite possible that whereas even if all explications of reduction falling under the category of theory reductionism fail to capture the research strategies followed by molecular biologists, some explication from the category of explanatory reduction might successfully do so. It turns out that, for some quite significant developments in the history of molecular biology, not even any explication of reduction falling in the category of explanatory reductionism can capture the research strategies involved (Sarkar 1989). One well-known example is that part of phage research that was guided by Delbrück and motivated by his search for complementarity in biological phenomena (Sarkar 1989, 90–167). The aim of this research was to discover biological phenomena that could not be explained by ordinary physics and chemistry but would require some "complementary" explanation.[5] Another less-known example is constituted by the theoretical attempts to decipher the genetic code in the 1950's which were guided by assumptions about the efficient storage and transmission of information far removed from physical or chemical considerations (Sarkar 1989). The results of these analyses, for either category of theory or explanatory reductionism, question the importance of reductionism in the research strategies historically followed by molecular biology.

However, both these research strategies were failures. Molecular biology is yet to come up with any phenomenon that cannot be explained by ordinary physics and chemistry, at least to the extent to which data are available. Further, the actual decipherment of the genetic code in the early 1960s showed the naïvete of the theoretical attempts of the 1950s. This raises the possibility that the structure of explanations afforded by molecular biology is still reductionist according to some explication or notion of reductionism. This possibility is enhanced by the observation that research in molecular biology continues to be quite successful in providing physical and chemical explanations of biological phenomena.

The exploration of this possibility becomes interesting because, once the second set of distinctions introduced above is carefully maintained, a curious fact about past philosophical discussions of reductionism in molecular biology emerges. There has been considerable criticism of the point of view that the structure of explanations in molecular biology is reductionist (Hull 1972, 1976; Ruse 1971; Kitcher 1984). Yet, with the exception of Rosenberg (1978, 1985), all these past criticisms have only considered explications of reduction falling within the category of theory reductionism. The most influential criticisms have been due to Hull (1972) and Kitcher (1984). Hull considers the relation between classical and molecular genetics and, in the context of the explication of reduction offered by Schaffner (1967), essentially points out that molecular biology provides explanatory mechanisms and not a reducing theory of the kind that Schaffner's explication requires. Kitcher discusses the same case and, unlike Hull, fails to find an adequate theory of classical genetics that would be a candidate for reduction.[6] Both of these arguments rely on one common objection: the absence of appropriate theories. At most these arguments can be extended to all explications of reduction that fall within the category of theory reductionism. They leave explications of reduction falling in the category of explanatory reductionism untouched.

Rosenberg (1985), by arguing for the supervenience of biological interactions on physical or chemical ones, does deny the possibility of any explication of reduction from the category of explanatory reductionism capturing the structure of explanations in molecular biology. The source of this argument is complexity. Since the same, or at least very similar, biological entities can have a very wide variety of physical or chemical constitutions, any attempt to explain their properties at the physical or chemical level would be incredibly complex. However, incredible complexity does not entail impossibility. More importantly, Rosenberg's argument goes against the justifiable intuition that everyday molecular biology is providing more and more examples of physical and chemical explanations of biological phenomena. Thus, this argument does not detract much from the possibility that there is some explication of reduction from the category of explanatory reductionism that captures the structure of explanations in molecular biology.

However, the thesis maintained in this paper is that there exists a class of explanations in molecular biology, namely, functional explanations, whose structure cannot be captured by any model of reduction that falls into the category of explanatory reductionism, let alone theory reductionism. If this thesis is true, it presents a more significant problem to the program of construing explanations in molecular biology as reductionist than the previous objections of Hull and Kitcher which only bring into question explications of reduction from the category of theory reductionism. However, the effect of this thesis on that program is diluted because (i) the scope of such explanations in molecular biology is quite limited and (ii) even at the present state of knowledge, it appears possible that such explanations might soon be captured by some explication of reduction from the category of explanatory reductionism. None of these points seem to have been previously noticed in the literature.

Section 5.2 begins with a discussion of an example of functional explanation in molecular biology that was introduced by Rosenberg (1985) who, however, failed to appreciate its full significance probably because of a failure to maintain the second set of distinctions introduced above. An explication of functional explanation due to Wimsatt (1972) is then adopted and this delegates to the theory of natural selection a critical role in ensuring the causal adequacy of such explanations. Since the theory of natural selection cannot be explained yet in physical or chemical terms, the structure of functional explanations cannot be captured by any explication of reduction from the categories of explanatory and theory reductionism. In section 5.3 it is argued, however, that functional explanations only occur in molecular biology as answers to questions of origin and often provide only partial answers to these questions. Thus the relevance of the considerations adduced in section 5.2 to the program of construing explanations in molecular biology as reductionist is severely limited. In section 5.4 it is shown that, using some of the work done by Eigen and coworkers (Küppers 1975) in a different context, it might be possible to construct a purely physical theory of natural selection at the molecular level which is the level at which functional explanations in molecular biology occur. Thus some of these functional explanations might be captured by an explication of reduction from the category of explanatory reductionism. Details of this part of the analysis appear in Sarkar (1988).

5.2 Functional Explanation, Natural Selection, and Explanatory Reductionism

The use of functional explanations in molecular biology is very well illustrated by the following example originally invoked by Rosenberg (1985, 38–43). The genetic material in all living organisms, except some viruses, is DNA, which consists of chains of four types of nucleotide bases, namely adenine (A), cytosine (C), guanine (G), and thymine (T). The order in which these bases appear in a DNA chain determines what sequence (if any) of amino acid residues in the polypeptide chain that DNA chain codes for. The process of producing a polypeptide chain from a DNA chain begins with a process called "transcription" during which the order of the nucleotides in the DNA chain is transcribed to a corresponding chain of RNA. For three of the four base types, the last-mentioned process takes place by the use of the standard base pairing scheme that occurs between complementary strands of DNA: the C, G, and T in the DNA corresponds to a G, C, and A in the RNA respectively. However, the A in the DNA does not correspond to a T in the RNA (as it would in a complementary strand of DNA), but to a new type of base, uracil (U). In other words, thymine does not occur in RNA at all, and wherever it might have been expected to occur by base pairing from its source DNA, uracil occurs in its place.

The question that now arises is that of the source of this difference between DNA and RNA. After all, the substitution of uracil for thymine does not alter any of the coding properties of the nucleic acids: there are still four types of bases, and three nucleotides code for each amino acid residue. Therefore, RNA with uracil instead of thymine would have the same coding properties as it would have had if thymine had occurred in it. Further, the difference in structure between thymine and uracil appears to be small. This similarity of structure ensures that were uracil present in DNA instead of thymine, the DNA double helix structure could still be formed by the same base pairing mechanism that normally exists. In other words, DNA with uracil instead of thymine would continue to possess two of its most biologically significant properties, namely, that of coding for the amino acid residues in polypeptide chains, and that of maintaining the stable structure of the DNA double helix.[7] Yet there is this difference between DNA and RNA. Furthermore, in all biological organisms, thymine is basically synthesized from uracil. The reaction is endothermic and, therefore, energetically expensive for the organism. While this last fact could potentially be used to explain the occurrence of uracil in RNA out of thermodynamic considerations, it makes it even more difficult to explain the occurrence of thymine in DNA.

The provisional explanation of the occurrence of thymine in DNA instead of uracil is actually a little complicated. Cytosine, one of the nucleotide base types in DNA, can easily convert to uracil by deamination. Such deamination, when it occurs, destroys the coding property of a strand of DNA since it introduces a nucleotide base type, uracil, which is not normally present in DNA. However, such a situation cannot usually occur since there are present in the cell a set of enzymes which, through a complicated process, remove uracil from a strand of DNA and replace it with cytosine. Now suppose that uracil were a normal component of DNA. Then the deamination of cytosine would not introduce into a strand of DNA a type of nucleotide base that did not normally occur in it. Then the sort of repair mechanism that excises uracil from DNA and replaces it with cytosine would be ineffective: it would end up excising a normal component of DNA. Further, if no such repair mechanism existed, the DNA would be extremely susceptible to having its coding property altered to that of a different strand of DNA by the deamination of cytosine assuming, of course, that a cytosine → uracil mutation would not always be silent.[8] Hence, the system as it exists now, including the incorporation of thymine rather than uracil in DNA, results in a greater stability of the code represented in that strand of DNA than what would have been obtained if uracil had occurred instead, even though the presence of thymine does lead to a greater expenditure of energy during the production of DNA.

These considerations, at best, account for the presence of thymine rather than uracil in DNA. They do not account for the presence of uracil in RNA. This is provisionally accounted for in the following manner. A strand of DNA has a long existence, usually for about the same time as the life of the cell in which it occurs. RNA, on the other hand, is short-lived: a strand of RNA merely has to retain its form for the time it takes it to move from its source DNA to a ribosome and for the time it requires for translation to occur at the ribosome. Further, from each strand of DNA numerous strands of RNA are transcribed for translation at the ribosome. Therefore, the lesser stability of the code, as represented in the RNA, which results from the occurrence of uracil instead of thymine, almost never results in the production of a polypeptide chain with an altered amino acid residue sequence. Therefore, the extra energy expenditure that would be necessary for the formation of thymine-containing RNA is not warranted. Given all of these considerations, the provisional explanation of the source of the difference between DNA and RNA is finally complete. The presence of thymine in DNA ensures greater stability of the code incorporated in it, which more than compensates for the additional energetic cost of the production of thymine from uracil. Uracil is present in RNA since the added stability that thymine would provide does not warrant the additional energy expenditure for the short-lived RNA.

There are five features of this explanation that need to be noted. *First*, the explanation appeals to *results, effects*, or *consequences* of the difference between RNA and DNA. The *effect* of the presence of thymine in DNA is the greater stability (as compared to RNA) of the code incorporated in that DNA. On the surface such an explanation appears not to be *causal*: the offered explanation does not appear to be only in terms of antecedent factors which is required by virtually any explication of causality.

Second, not all effects of the difference between DNA and RNA—for example, the difference in molecular weight—are relevant for the explanation. The kind of effect that enters into the explanation is called a *function* and explanations of this kind are called *functional explanations*. Functions, then, are certain effects of some feature of the organism (or part of the organism), and for functional explanations to be offered it therefore becomes incumbent to provide a method by which functions can be distinguished from other kinds of effects.

Third, it is important to note that the explanation offered depends on a wide variety of factors including, for example, the nature of the mechanism that repairs DNA by recognizing uracil and replacing it with cytosine. In general, factors such as these, which provide the context in which functional explanations can be offered, are critical to the adequacy of such explanations. All three of these points will be addressed by the explication of functional explanation adopted later in this section.

Fourth, what has been offered so far is only one example of a functional explanation in molecular biology. However, other examples abound. The most important of these refer to the various schemes that are offered to account for the properties of the genetic code. For instance, the nature of the degeneracy of the genetic code is such that even when substitution mutations are not silent, most of them will result in the replacement of a hydrophilic or hydrophobic amino acid residue by a residue of a similar type in the ensuing polypeptide chain. Moreover, since the genetic code is nonoverlapping, a mutation, even when it is not silent, affects only one residue in the corresponding polypeptide chain. Both of these observations are routinely explained by noting that the nature of the genetic code is such that it functions to maintain the fidelity of the relation between a particular strand of DNA and the polypeptide chain as much as possible.[9]

Fifth, in the example being discussed in detail, and in the other examples considered in the last paragraph, the question that was being posed was always one about the existence of some feature, whether it be of DNA and RNA or the genetic code. This suggests that functional explanations are offered in molecular biology only in response to such questions which will be called "questions of origin." In section 5.3, a general argument is given that attempts to show that this is so. For the rest of this section it will simply be assumed to be true. This does not entail, however, that attributions of function can only be made in the context of answering such questions. The attribution of functions and the use of such attributions in *functional explanations* are separate issues, though the former is necessary for the latter. Functional attributions can occur without attempting explanations: the issue of answering questions of origin, as will be argued in section 5.3, arises only when such attributions are made as part of the process of explanation.

Various explications of functional explanations have been offered in order to ensure their causal adequacy. The one adopted here (with notational modifications) is due to Wimsatt (1972) which has the dual advantage of treating functional attribution as having explanatory power and giving a very detailed treatment of the contextual requirements for the adequacy of a functional explanation.[10] Intuitively, the idea behind this explication is quite simple. What requires explanation is the existence of some feature of a biological entity that is being investigated in molecular biology. A causal chemical theory, which need not be any more than a description of chemical mechanisms (the theory T, in the characterization below), identifies the effects of the various properties of this feature in the relevant chemical or biological circumstances.[11] A second causal theory (T' in the characterization below), which is the theory of natural selection in the case of molecular biology, determines which of these effects count as functions. As is conventionally assumed, only those effects that enhance the fitness of the entity involved are considered as functions. Since fitness is correlated with the probability of survival, the identification of some effect of a feature of the entity in question as a function explains, at least partially, the existence of that feature. Thus functional attribution plays a critical role in such explanations.

More formally, Wimsatt's explication attempts to provide a causal warrant for a functional explanation by invoking true "function-statements" of the form: "According to theory T, a function of feature x, in having property Y, in system S, in environment E relative to purpose P is to bring about consequence C" (1972, 32). The theory T is required to be causal. Thus, it ensures that the consequence C, which can, for all practical purposes, be identified with the function under consideration, like all other effects of the feature x, in these circumstances, is causally "brought about." A second causal theory, T', usually some sort of a selection theory, whose role is explicitly acknowledged in this explication, must fulfill two aims. *First*, it must already have shown which effects or consequences of the feature x, having property Y, in environment E, are to be considered as functions with respect to purpose P. Thus it permits the choice of certain effects or consequences as functions which was earlier shown to be necessary for adequate functional explanation. *Second*, it must specify the purpose P satisfied by the function in question and the way in which it is satisfied, namely, the consequence C. When T' is a selection theory, as is the case in almost all biological contexts, the relevant purpose P is always to increase fitness. Note that the theory T permits the construction of sentences formally similar to function-statements for any effect of the feature x, having property Y, in system S, in environment E. Only when it works in conjunction with the theory T' do purposes and functions get identified and actual function-statements can be constructed. The function-statement answers the question of origin being raised simply because T' specifies the purpose P to be one of increasing fitness which is correlated with a higher probability of survival and, therefore, existence.

In the case of functional explanations in molecular biology, the theory T is the chemical theory that specifies the mechanisms which ensure that the feature x, in having the property Y, in system S, in environment E, causes the consequence C. T' is simply the theory of natural selection: it requires that some feature x, in having the property Y in these circumstances, is functional if the consequence C enhances the fitness of the organism in question. In the case of the difference between DNA and RNA, the feature x is the occurrence of thymine in DNA. The purpose P is, of course, to increase fitness. The property Y is the ability of thymine to resist excision by the enzyme that excises uracil from DNA. The system S is the strand of DNA. The environment E is the cell environment including all the various enzymes required for the excision of uracil and the incorporation of cytosine in its place in the DNA. Finally, the consequence C is the possibility of repair of DNA after deamination of cytosine. For the functional explanation in question to work, two conditions must hold. *First*, the function-statement characterized by these assignments must have empirical support. This is ensured by the empirical truth of the chemical theory T that was invoked. *Second*, the theory T', which, in this case, is the theory of natural selection, must also similarly be empirically true.

It is easy to see how this explication of functional explanation addresses the first three features of functional explanation that were noted above. *First*, this explanation is causal even though it explicitly invokes consequences or functions simply because the two theories that are involved are both causal. *Second*, it is explicit that the theory T' determines which effects of feature x, having the property Y, in the relevant circumstances constitute functions. *Third*, this explication invokes the environment E and the system S thereby emphasizing the various contextual factors that have to be considered in judging the adequacy of functional explanations.

The discussion in this section has demonstrated the existence of functional explanations in molecular biology and has outlined the requirements that must be satisfied in order to ensure their causal adequacy. Nothing has yet been said regarding the issue of reductionism. The question that is at stake here is whether the structure of functional explanations in molecular biology, as explicated above, can be captured by any explication of reduction from the category of explanatory reductionism.[12] There are two theories that play a critical role in this explication of functional explanation. The *first* of these, T, is the chemical theory which ensures that the particular feature in question, in the appropriate circumstances, causally produces the consequence which is identified as a function. Clearly, this should not present any problems for an explication of reduction from the category of explanatory reductionism. In fact, since this is a chemical theory, which explains at the chemical level, often through the description of appropriate mechanisms, a failure of a particular explication of reduction from the category of explanatory reductionism to capture such an explanation would only militate against that explication.

It is the *second* theory invoked in the explication of functional explanation given above, namely T', the theory of natural selection, that presents serious difficulties. The theory of natural selection is obviously neither a physical nor a chemical theory. There is considerable controversy over exactly what the structure of the theory of natural selection is.[13] It is uncontroversial, however, that this theory, broadly construed, requires that differential fitnesses of entities, in specified environments, causally produce differential probabilities of survival. It is in this form that the theory enters the explication above. However, the theory, as stated, is not capable yet of receiving physical or chemical explanation. Thus, no model of reduction from the category of explanatory reductionism can capture the structure of any explanation that critically involves the theory of natural selection.[14] Thus functional explanations pose a new sort of problem for the program of construing the structure of explanations in molecular biology as reductionist even in terms of an explication of reduction from the category of explanatory reductionism.

5.3 The Scope and Power of Functional Explanations in Molecular Biology

The last section has shown that there exists a class of explanations in molecular biology, namely, functional explanations, that cannot be captured by any explication of reduction from the category of explanatory reductionism. To judge the importance of this situation to the program of construing explanations in molecular biology as reductionist, it becomes incumbent to attempt to determine how large this class is. It has already been indicated in the last section that this class consists of attempts to answer questions of origin. However, no argument was offered in defence of this claim and this is the first task that is taken up in this section. It is later argued that functional explanations might only provide incomplete answers to questions of origin.

In order to argue that functional explanations occur in molecular biology only in response to questions of origin, it is not enough simply to list a number of functional explanations and show that they all occur in this way. Even if all known functional explanations were adduced, the possibility that there might yet be some functional explanation that serves some other purpose is left open. Thus, some more "transcendental" argument is necessary and the one to be offered here relies on a distinction between *questions of mechanism* and *questions of origin.*

Questions of mechanism arise when the behavior of some feature of an organism is probed with the purpose of discovering how it occurs or is accomplished. Thus the question how ions are transported across the cell membrane of bacteria or how the eye of some animal processes light signals is a question of mechanism. To begin to answer these questions, a mechanism for the transport of small charged particles across a bilipid layer and a mechanism showing how photons cause electric signals in optic nerves have to be elaborated. On the other hand, one can ask why fetal hemoglobin differs from hemoglobin A in humans or why a codon has three bases. These are *questions of origin:* they probe the source of a particular feature of an organism. Elaborating mechanisms that describe results of these features—how fetal hemoglobin and hemoglobin A have different affinities for oxygen and carbon dioxide in different ionic environments or how triplet assignments lead to the possibility of degeneracy in the genetic code—obviously do not, by themselves, answer the pertinent question: why these features are there in the first place. Even elaborating the process (or mechanism) by which such features are produced, such as listing the different genes that code for hemoglobin A and fetal hemoglobin, might not, depending on the context, be an adequate answer though it does appear, in some sense, to answer the question why there is a difference between these two molecules. For what might be being asked are the much harder questions, namely, why there are two hemoglobins and not just one and, assuming that this question receives a satisfactory answer, a second one, why the two hemoglobins differ exactly in the way they do. This point becomes even more explicit in the case of the triplet codon: clearly what is being asked for are not some mechanisms that produce DNA, RNA, ribosomes, or the enzymes involved in replication, transcription and translation. What is at stake is why each codon has three (and not, for instance, one, two, or four) bases. To answer that question is to speculate about the origin of the genetic code which is a much harder question.

Functional explanations occur in molecular biology only when attempts are made to answer questions of origin.[15] Consider any question of mechanism: what is being probed is how some particular behavior of an organism (or its part) occurs or is accomplished. To offer a functional argument towards that end would be either to elaborate some function of that behavior or to show that the behavior under question is the function of some other process. The first alternative obviously does not answer the question that is being asked. A function is a particular result or effect of a feature of an organism (or its part). Even a list of all such effects of the feature, that is, the behavior, which would automatically include all possible functions, would not by itself explain how such behavior is brought about which was the question being asked. The second alternative is more interesting, though equally inappropriate. Suppose the question asked concerned the transport of ions across cell membranes of bacteria and, for the sake of simplicity and exactness, the question was restricted to the case of *E. coli*. If it were asked how such transport occurred, the following sort of functional argument might be offered. There are several proteins, notably the colicins, whose function, when attached to cell membranes of *E. coli*, is to form channels for ion transport (Cleveland et al. 1983). These proteins do attach to cell membranes of *E. coli* and, because of their function, the transport of ions takes place. There are two problems with such a response: in one sense it goes too far, that is, the invocation of function is gratuitous and unnecessary, and in another sense it does not go far enough because, while it begins to give the required answer, much more can be said even at the present stage of biological knowledge. The attempted answer goes too far because it would be enough, as far as explaining how the behavior of ion transport is brought about is concerned, merely to state that the effect of the attachment of the colicins to the cell membranes of *E. coli* is to produce channels for ion transport without having to decide whether this effect is one that can be called a function. In fact, in this example, when colicins are present, the channels produced almost always lead to the death of *E. coli* because all the ions necessary for the functioning of the cell leave its interior through these channels—it would be odd to call such an effect a function. The sense in which the answer does not go far enough is that what might be being asked for is more detail, that is, how the colicins attach to the cell membrane and, exactly how, by their conformation or other interactions, they cause channels to be formed for ion transport.

The problem with the first alternative in the discussion above shows that the function of a particular behavior of an organism (or its part) does not address the question how it is brought about. The second alternative does address this question but the first problem with the answer to the question raised by it, demonstrates that functional considerations are unnecessary when attempts are made to answer questions of mechanism. Taken together they show that functional explanations cannot be adequate when offered as answers to questions of mechanism. Thus functional explanations can, at most, only be offered in response to questions of origin. Even here, however, as will be discussed below, they might provide only partial answers and, sometimes, no answer at all.

It is important to note that trying to decide whether some particular question that is asked is a question of mechanism or a question of origin depends critically on the context in which the question is asked and can be non-trivial. It might, for example, be tempting to think of questions of mechanism as *"how*-questions": *How* does an organism (or a part of an organism), O, perform behavior, B?"; and questions of origin as *"why*-questions": "*Why* does an organism (or a part of an organism), O, have feature, F?" The distinction between "feature" and "behavior" in these questions is not important: features can include behaviors as will be evident in the example discussed below. However, such an analysis of questions of mechanism and questions of origin is simplistic because it ignores the context in which the question is asked. Though *how*-questions are usually questions of mechanism, *why*-questions can routinely be either depending on the context. The last point is illustrated by the following example. Hemoglobin A is known to exhibit the Bohr effect. When the concentration of oxygen is high and carbon dioxide low, giving rise to a high pH (or low hydrogen ion concentration) in an aqueous environment, hemoglobin A has a lower affinity for carbon dioxide and a higher affinity for oxygen than in the reversed situation. Since oxygen concentration is high and carbon dioxide concentration low in the lungs, and the opposite is true in the capillaries, this property makes hemoglobin A admirably suited to transfer oxygen from the lungs to the tissues and carbon dioxide back from the tissues to the lungs. Now consider the question: "Why does hemoglobin A exhibit the Bohr effect?" It can either be a question of mechanism or a question of origin depending on the context. If it is the former, then the answer to the question involves, for example, elaborating the quaternary structure of hemoglobin A, that is, the joint conformation of the four amino-acid chains that compose hemoglobin A, and how this leads to allostery, that is, the assumption of different conformations in different environments, which in turn physically accounts for the changed oxygen affinities. If, on the other hand, the question being asked is a question of origin, then the answer to it involves the elaboration of the importance of this property of hemoglobin A in the functioning of the circulatory system and, therefore, why a molecule of this sort is suited to it. The context in which the question was asked determines which of these answers was being asked for, and unless that context is specified, there is no way of determining whether the question asked was one of mechanism or one of origin.

These considerations do not, of course, deny that the elaboration of mechanisms can be quite crucial in answering questions of origin. An exact and detailed knowledge of the interactions of some feature of an organism (or a part of an organism) can be necessary even to determine if that feature has a function. Sometimes the knowledge that is required is so detailed that it must be at a molecular level. In the example of the uracil-thymine difference between RNA and DNA, it was only when the details of DNA repair were discovered at the molecular level, that functional considerations could begin. The point, however, is that when functional explanations are offered to answer questions of origin, just a complete knowledge of the various mechanisms involved is not enough for adequate explanation.

The discussion so far in this section has limited the role of functional explanation to answering questions of origin. However, even in this role, there are three factors that make the success of functional explanations often open to question. *First*, since the theory of natural selection plays such a crucial role in providing a warrant for functional explanations in molecular biology, those features of the organism (or its part) that exhibit functional behavior must be adaptive. Moreover, since functional explanations are offered in molecular biology specifically to answer the questions of the origin of those features, it has actually to be shown that those features rose by random variation, and were immediately and consistently adaptive for that function, in order for such explanation to be adequate. Thus, these features have to be adaptive in a stronger sense than is normally used. This point is made clearer using a distinction initially made by Gould and Vrba (1982), that between *adaptations* and *exaptations*. A feature that is functional, at the present time, as indicated by the theory of natural selection, could have originated by random variation and subsequent selection for that function, or it could have originated in some other fashion though it might now be persisting because of selection for the current function.[16] The former are *adaptations*. The latter are *exaptations* and might have arisen in a variety of ways: they might have initially been created by random variation and selected for some other function than the present one, or they could have arisen out of some other kind of process—perhaps a purely physical process—and are only being subjected to selection constraints now. Functional explanations are adequate as answers to questions of origin in this stricter sense of the term. If a feature is functional in some specified way but arose through some other process, citing the current function clearly does not provide a complete answer to the question of its origin.

The last point leads to a *second* factor regarding the success of functional explanations in answering questions of origin: a particular functional explanation might only provide a partial answer to the question of origin that was posed. Gould and Vrba give the example of repeated segments of DNA, which might have arisen simply because they did not code for any polypeptide chain and were thus virtually invisible at the phenotypic level. Therefore, there was no selection constraint, except the obvious weak energetic ones, to prevent their accumulation. These segments might now have the function of providing multiple copies of genes that code for particularly important proteins. To answer the question of their origin, that is, why they now exist, the functional explanation with respect to their current function which explains their persistence *and* the account of their initial creation and accumulation must both be given. Thus, when only exaptations are involved, functional explanation with respect to their current function only provides an incomplete answer to the question of their origin.

Third, it could be the case that a question of origin of some feature might have no functional explanation as an answer. For example, the so-called mechanistic models for the origin of the genetic code, which claim that there is a specific physico-chemical relationship between a codon and the amino acid residue it codes for (perhaps mediated by a complicated pathway), account for the origin of the degeneracy of the genetic code without invoking any functions at all (Lewin 1974, 34–36). According to these models, since there is a physico-chemical that assigns an amino acid residue to a codon, the fact that there are six different codons for leucine, serine, or arginine, for example, is explained by the physico-chemical characteristics of those codons, these residues, and the expression pathways, and have nothing to do with the effects or consequences of this degeneracy.[17]

Reductionism and Functional Explanation

The last factors considered reinforce a point noted in section 5.2, namely, that functional attribution, by itself, need not have anything to do with answering a question of origin. If the theory T', in the explication adopted in section 5.2, which permits identification of "functions," is a *selection* theory, as is usually the case in *biological* contexts, then functional attribution can be connected at least to the persistence of some feature, if not the origin. If this theory is not a selection theory, this connection cannot necessarily be made. Further, a distinction has to be maintained between functional attribution and the use of such attribution as part of functional *explanation*. If the argument given earlier in this section is valid, then functional *explanation* arises in molecular biology only in the context of attempts to answer questions of origin. In such a circumstance, since what is being asked is a question of origin, the purpose that a functional explanation serves is to answer such a question. This does not, in any way, preclude functional attribution in other contexts.

5.4 Molecular Evolution and the Possible Recovery of Explanatory Reductionism

The argument given in section 5.2, against the possibility of capturing the structure of functional explanation in molecular biology by any explication of reduction from the category of explanatory reductionism, relied on the impossibility of a physical or chemical explanation of the theory of natural selection at the present time. There is a possibility that, at the *molecular level* at which molecular biology generally operates, this difficulty might eventually be overcome. However, it should be emphasized that this possibility is yet far from being actualized and is being offered here only as a very optimistic, and perhaps not equally realistic, promissory note. Since this possibility has already been discussed, with appropriate technical detail, in the philosophical literature (Sarkar 1988), only a very brief and informal discussion will be offered here for the sake of completeness. The relevance of this possibility to the program of construing the structure of explanations in molecular biology will then be discussed.

During the course of the development of a theory of the origin of life, Eigen and coworkers (Eigen 1971, 1983) have elaborated a theory of molecular evolution that can easily be separated from the particular initial conditions making it attractive as a theory of that origin (Sarkar 1988). Basically, this theory considers chemical systems where different nucleotide chains replicate, either by autocatalysis, or by the production of complementary intermediates as in the case of DNA replication in contemporary organisms. If the same number of different types of nucleotide chains are initially present, their relative frequency in subsequent generations need not be the same if available resources for replication, such as the number of nucleotide bases or available energy, is limited. In such circumstances, the relative frequency of a particular type of nucleotide chain will be proportional to the ratio of its formation rate to the average formation rate for all types, all other factors being constant.[18] Thus the "fitnesses" of these types can be defined using purely physical or chemical concepts. What results is a physical theory of natural selection of molecular species.[19]

Now imagine that strands (or species) of two different types of DNA, "DNA_t," containing thymine, and "DNA_u" containing uracil, were undergoing replication in a chemical environment containing the enzymes responsible for replacing uracil by cytosine. After several generations, the proportion of DNA_t to DNA_u would have greatly increased because the total rate of production of DNA_u would be much lower than that of DNA_t simply because strands of DNA containing uracil would constantly be transmuted to some other species by the repair mechanism. Therefore, the fitness of DNA_t species is greater than that of DNA_u species in this environment. Note, now, that the theory of natural selection that is involved in the explication of functional explanation becomes, in these circumstances, a purely physical theory. Thus the difficulty, mentioned in section 5.2, about the possibility of capturing the structure of functional explanations by some explication of reduction from the category of explanatory reductionism is removed. Further, since the chemical explanations involved in this account are so simple, virtually *any* plausible explication of reduction from that category would have to capture these explanations.

However attractive the sort of schema just given might seem, especially to committed reductionists, there are three very important reasons to be cautious about the success of such schema. *First*, the theory of molecular evolution that is being invoked here is still in its infancy and wanting in both theoretical development and, especially, experimental confirmation. Any use of it, especially to draw broad philosophical conclusions, must be tentative. *Second*, the particular example just discussed involves at least one very dubious assumption. This is that the two types of DNA mentioned were replicating in an environment that already contained the enzymes for the replacement of uracil by cytosine. It is hard—though not impossible—to imagine a situation in which such a system could have evolved at so early a stage of evolution that both types of DNA were present. *Third*, the physical explication of natural selection that has been utilized in this section is limited to the molecular level. There is no straightforward way to extend this to higher levels of organization, such as the cellular or tissue levels, let alone the organismic and even higher levels. There is no reason to suppose that the selection constraints operating to ensure the adequacy of functional explanations in molecular biology always involve only the molecular level. Thus, even if this theory of molecular evolution is fully successfully developed and confirmed, there might yet be cases of functional explanation in molecular biology that cannot be underwritten using only it and not invoking natural selection at some other level.

In any case, the purpose of this somewhat speculative section has been to show that *some* functional explanations in molecular biology *might eventually* be captured by explications of reduction within the category of explanatory reductionism. Perhaps the real moral to be drawn from this section is a simple and important one: the answer to some philosophical questions must necessarily await an answer to scientific ones.

5.5 Conclusion

The main purpose of this chapter has been to show that there exists a class of explanations in molecular biology, namely, functional explanations, whose structure cannot be captured even by explications of reduction from the category of explanatory reductionism, let alone theory reductionism. This has been argued for in section 5.2. There it has been observed that the theory of natural selection plays a critical role in assuring the causal adequacy of functional explanations. Since this theory of natural selection is itself incapable, at present, of physical or chemical explanation, its use precludes the possibility of capturing the structure of functional explanations in molecular biology in the manner just mentioned. In section 5.4, however, it has been argued that some such functional explanations might ultimately be captured by these explications of reduction though that will have to await the further development, and experimental tests, of theories of molecular evolution. Thus that section argues, in part, against the force that should be attributed the main conclusion of this paper.

However, a much more important limitation on the force of this conclusion comes from the considerations in section 5.3 which restrict functional explanations in molecular biology to answering questions of origin. Therefore, in the final analysis, the extent to which this paper brings into question the program of construing the structure of explanations in molecular biology as reductionist, under some explication of reduction from the category of explanatory reductionism, depends on the importance of questions of origin in molecular biology. This is an issue that is not precise enough for any definitive decision. On the one hand most of molecular biology concentrates on questions of mechanism, in answering which it has had some spectacular successes. On the other hand, questions of origin are of profound interest and have always been so in biology. It is likely, moreover, as knowledge of molecular mechanisms increases and it becomes possible to give more definitive answers to such questions; they will continue to be more and more fruitfully asked. Thus, it seems reasonable to conclude that any point of view regarding molecular biology that ignores these questions, and their answers, is seriously mistaken. Among these would be a point of view that is so strictly reductionist, in the senses invoked under either category of theory or explanatory reductionism, that would simply abandon these questions in order to preserve a reductionist metaphysics.

It is possible, of course, that ultimately the theory of natural selection, at all levels, will be underwritten by physical theory. In such a circumstance no functional explanation will pose a problem to the program of construing explanations in molecular biology as reductionist under some explication of reduction from the category of explanatory reductionism. However, such a situation is not imminent in the near future. In the meantime, the thrust of this paper must be taken to be another attack, though a limited one, to the notion that explanations in molecular biology are always "reductionist" even under the construal of that term by explications of reduction from the category of explanatory reductionism.

Acknowledgments

Thanks are due to W. Wimsatt, K. Schaffner, and R. McClamrock for their comments on an earlier versions of this paper.

Notes

1. The two discussions that have been relatively clear on the use of these terms are Wimsatt (1976) and Schaffner (1974). However, even they do not fully lay out the sets of distinctions proposed here and elaborated in Sarkar (1989).

2. This distinction is, in fact, implicitly made, and consistently used, by Schaffner (1974).

3. It might appear that the category of explanatory reductionism is more restrictive than the category of constitutive reductionism. However, such a claim involves an additional ontological commitment that the biological entity whose behavior is being explained and the physical and chemical systems whose properties are being used for the purpose of the explanation are the same "thing." While this might very well be true, it is not necessary to make this commitment for the sake of the arguments in this paper. It will, therefore, be avoided.

4. The qualification, "almost always," is made because in the case of at least one explication of reduction that falls under the category of theory reductionism, namely that of Balzer and Dawe (1985, 1986), it is not completely clear that explanation is involved.

5. This was the aspect of the history of molecular biology that was emphasized, perhaps too much, by Stent (1966, 1968).

6. The explication of reduction that Kitcher actually considers is due to Nagel (1961) which is a degenerate case of Schaffner's. The main difference between these two explications is that whereas Nagel envisions the deduction of the reduced theory from the reducing one, as the reduced theory currently exists, Schaffner admits the possibility that what can be deduced from the reducing theory is a corrected version of the reduced theory.

7. Of course, there can be some question whether the latter property should be considered particularly biologically significant. It might be held that some other structure could be just as efficacious. However, the double helix has some very useful properties that help maintain a stable structure. In particular, the helix provides a hollow cylindrical central core, protected from water, in which the hydrophobic purine and pyrimidine nucleotide bases are stacked.

8. A mutation is "silent" when it does not alter the particular amino acid residue which that codon (consisting of three nucleotide bases) codes for. This is possible because the genetic code is degenerate, that is, different codons often code for the same amino acid residue.

9. Explanations of this type lie at the core of the "selectionist" school of thought regarding the origin of the genetic code. More details of these explanations and regarding this school can be found in Lewin (1974, 34–36).

10. The first of these advantages is lost in the analysis of functions which sees functional behavior as "goal-directed" behavior used by Rosenberg (1985) following Nagel (1961). Functional attribution then becomes nothing more than a redescription of goal-directed behavior which is shown to be causally produced. Functions thus loose the explanatory force that is usually attributed to them in biology. The second of these advantages makes Wimsatt's explication preferable to a similar one due to Cummings (1975) which relies on the "wider context" to provide information about functional attribution but leaves the notion of the "wider context" embarrassingly vague.

11. The term, "theory" is being used somewhat loosely here. This point will be taken up in the following note.

12. It can, of course, also be asked whether the structure of such explanations can be captured by some explication of reduction from the category of theory reductionism. This seems impossible for basically the same problem cited by Hull (1972) and Kitcher (1984), that is, the absence of appropriate theories. First, what is being explained is the existence of some feature. I take it as uncontroversial that a statement asserting this existence is certainly not general enough, nor has the kind of organizing power, to be designated a "theory." Further, though Wimsatt's explication refers to a "theory" T, the chemical theory involved, the intent of his explication includes the possibility that this theory is merely a description of known mechanisms and thus not a "theory" in any significant sense. In any case, even if these claims prove controversial, they do not affect the main thrust of this paper, which is to question the ability even of explications of reduction from the category of explanatory reductionism to capture the structure of functional explanations in molecular biology.

13. Rosenberg (1985) and Ruse (1973, 1988), for example, present two radically different points of view. Rosenberg endorses an axiomatization due to Williams (1970) that finds no role for ordinary population genetics. Ruse holds that population genetics lies at the core of the theory of natural selection (and evolutionary theory in general).

14. The force of this conclusion will be mitigated somewhat, though only in the context of molecular biology, by the considerations adduced in section 5.4.

15. There is also a third type of question that might arise within the context of molecular biology and which is of some relevance. These are *questions of persistence:* they probe why an organism (or part of an organism) *continues* to have some feature that it does. A question of persistence is actually a part of a question of origin. The latter might be taken to consist of two separate questions, the first being one of the initiation of a certain feature, and the second being of its persistence. Sometimes, when questions of origin cannot be completely answered by a functional explanation, the question of persistence included in it might still admit such an answer. Thus func-

tional explanations might only provide partial answers to questions of origin. This point will be discussed in the text later. That questions of origin can be separated into questions of initiation and questions of persistence does not, in any way, affect the analysis of them that is attempted here.

16. In the latter circumstance, the question that might be asked and be answered by functional explanation is a question of persistence (see note 15).

17. Similarly even questions of persistence might not have adequate answers in terms of functional explanations.

18. Of course, all other factors need not be constant. The relative frequency will vary inversely with the ratio of its spontaneous decomposition rate to the average such rate, for instance. For details, see Sarkar (1988).

19. That what occurs is evolution by natural selection is shown by demonstrating that these systems satisfy the criteria for such evolution laid down by Lewontin (1970). This is done in Sarkar (1988).

References

Balzer W., Dawe C. M. (1985). Structure and comparison of genetic theories. Part 1. Character-factor genetics, *British Journal of the Philosophy of Science* v. 37, pp. 55–97.

Balzer W., Dawe C. M. (1986). Structure and comparison of genetic theories. Part 2. The reduction of character-factor genetics to molecular genetics, *British Journal of the Philosophy of Science* v. 37, pp. 177–191.

Cleveland M., Slatin S., Finkelstein A., Levinthal C. (1983). Structure-function relationships for a voltage-dependent ion channel: properties of COOH-terminal fragments of Colicin E1, *Proceedings of the National Academy of Sciences* v. 80, pp. 3706–3710.

Cummings R. (1985). Functional analysis, *Journal of Philosophy* v. 72, pp. 741–765.

Eigen M. (1971). Self-organization of matter and the evolution of biological macro-moleculae, *Naturwissenschaften* v. 58, pp. 465–523.

Eigen M. (1983). Self-replication and molecular evolution, in: D. S. Bendall (ed.), *Evolution from molecules to man*, Cambridge: Cambridge University Press, pp. 1–10.

Gould S., Vrba E. (1982). Exaptation. A missing term in the science of form, *Paleobiology* v. 8, pp. 4–15.

Hull D. (1972). Reduction in genetics—biology or philosophy, *Philosophy of Science* v. 15, pp. 135–175.

Hull D. (1976). Informal aspects of theory reduction, in: R. S. Cohen, A. Michalos (ed.), *PSA 1974*, Dordrecht: Reidel, pp. 653–670.

Kauffman S. A. (1972). Articulation of parts explanation and the rational search for them, in: R. C. Buck, R. S. Cohen (ed.), *PSA 1970*, Dordrecht: Reidel, pp. 257–272.

Kitcher P. S. (1984). 1953 and all that. A tale of two sciences, *Philosophical Review* v. 93, pp. 335–373.

Küppers B. (1975). The general principles of selection and evolution at the molecular level, *Progress in Biophysics and Molecular Biology* v. 30, pp. 1–22.

Lewontin R. (1970). The units of selection, *Annual Review of Ecology and Systematics* v. 1, pp. 1–18.

Lewin B. (1974). *Gene expression 1*, London: Wiley.

Mayr E. (1982). *The growth of biological thought*, Cambridge: Harvard University Press.

Nagel E. (1961). *The structure of science*, New York: Harcourt, Brace and World.

Rosenberg A. (1978). The supervenience of biological concepts, in: E. Sober (ed.), *Conceptual issues in evolutionary biology*, Cambridge 1984: MIT Press, pp. 99–115.

Rosenberg A. (1985). *The structure of biological science*, Cambridge: Cambridge University Press.

Ruse M. (1971). Reduction, replacement and molecular biology, *Dialectica* v. 25, pp. 39–72.

Ruse M. (1973). *The philosophy of biology*, London: Hutchinson.

Ruse M. (1988). *Philosophy of biology today*, Albany: SUNY Press.

Sarkar S. (1988). Natural selection, hypercycles and the origin of life, in: A. Fine, J. Leplin (ed.), *PSA 1988*, v. 1, East Lansing: Philosophy of Science Association, pp. 196–206. (Chapter 6 of this volume.)

Sarkar S. (1989). Reductionism and molecular biology: a reappraisal, Ph.D. dissertation, Chicago: University of Chicago, Department of Biology.

Schaffner K. (1967). Approaches to reduction, *Philosophy of Science* v. 34, pp. 137–147.

Schaffner K. (1974). The peripherality of reductionism in the development of molecular biology, *Journal of the History of Biology* v. 7, pp. 111–139.

Schaffner K. (1976). Reduction in biology: prospects and problems, in: E. Sober (ed.), *Conceptual issues in evolutionary biology*, Cambridge 1984: MIT Press, pp. 428–445.

Sneed J. D. (1971). *The logical structure of mathematical physics*, Dordrecht: Reidel.

Stent G. S. (1966). Introduction: waiting for the paradox, in: J. Cairns et al. (ed.), *Cold Spring Harbor: Cold Spring Harbor Laboratory of Quantitative Biology*, pp. 3–8.

Stent G. S. (1968). That was the molecular biology that was, *Science* v. 160, pp. 390–395.

Williams M. B. (1970). Deducing the consequences of evolution: a mathematical model, *Journal of Theoretical Biology* v. 29, pp. 343–385.

Wimsatt W. C. (1972). Teleology and the logical structure of function statements, *Studies in the History and Philosophy of Science* v. 3, pp. 1–80.

Wimsatt W. C. (1976). Reductive explanation: a functional account, in: R. S. Cohen et al. (ed.), *PSA 1974*, Dordrecht: Reidel, pp. 671–710.

6 Natural Selection, Hypercycles, and the Origin of Life

6.1 Introduction

Over the last eighteen years Manfred Eigen and his coworkers have postulated a new theory about the origin of life on earth that has presented a detailed account of how many of the features of extant living organisms (such as a universal genetic code and protein-nucleic acid interdependence) might have arisen from purely physical interactions.[1] This theory is critically based on the special dynamical properties of certain chemical cycles called "hypercycles" which cause some of them to exhibit hyperbolic growth over time while undergoing selection. The purpose of this chapter is to separate two aspects of this theory and then to study the first one in greater detail. The first aspect consists of a physical account of evolution by natural selection at the molecular level that depends only on the kinetics of certain chemical systems and can be applied to a variety of such systems. The second consists of the special dynamical properties that are exhibited when this account is applied, specifically, to hypercycles.

There are at least four reasons for making this separation. First, the separation of the various components of any theory which clarifies the relations between them is often of scientific and always of philosophical interest. Second, it is the first aspect of this theory alone that shows that, at the molecular level, purely physical interactions can cause the evolution of systems by natural selection in the sense that they satisfy the three criteria for such evolution laid down by Lewontin (1970). Third, it is the second aspect alone that makes this theory a theory of the origin of life. Fourth, the first aspect of this theory may well survive even if the general Eigen picture of the origin of life turns out to be unsatisfactory, that is, the second aspect of the theory turns out to be false, as has often been suggested.[2]

Besides the fourth reason above, there are at least three other reasons for concentrating on the first aspect of the theory as is done here. First, and this is an extension of the second reason listed above, explicating the mechanisms by which natural selection can occur, at any level, is of biological interest. Second, a physical explanation of natural selection, once again at any level, contributes to the program of finding physical explanation of biological phenomena, that is, explanatory reductionism, and is, therefore, of added philosophical interest.[3] Third, at the molecular level, the level of this discussion, physical explanation of natural selection provides a *physical* warrant for functional explanations in molecular biology, thereby also advancing the program of explanatory reductionism.

The separation between the two aspects of the Eigen theory of the origin of life is achieved here by the exposition, first, of a dynamical theory of chemical systems and then showing how this theory is applied to hypercycles. This is done by writing down phenomenological kinetic equations that make no assumption about the underlying processes of the system (section 6.2). Assumptions about the kinetics of the system are then introduced that result in evolution by natural selection. This is first done for an autocatalytic system which is, in many ways, the simplest system that exhibits such behavior (section 6.3). Next, systems that are catalytically coupled by having complementary replication are introduced and it is shown that the equations describing the dynamical evolution of the second system are formally identical to those of the first (section 6.4). Hypercycles are then briefly discussed to show where they fit in the scheme of things and why they are plausible candidates as systems by which life originated on earth (section 6.5). After this brief digression into hypercycles, a characteristic example of a functional explanation in molecular biology is introduced and the conditions for its adequacy explicated (section 6.6). The physical theory of evolution by natural selection for systems with complementary replication, developed in section 6.4, is then used to provide a tentative physical warrant for this functional explanation (section 6.7).

6.2 Phenomenological Equations and the Criteria for Evolution by Natural Selection

The following two general assumptions are being made for the chemical systems being considered here. First, it is assumed that the system consists of a population of n molecules, each of which is a polymer of length v built from λ types of monomers and an environment with a large supply of energy-rich monomers from which such polymers can be synthesized. Thus $N = \lambda^v$ types of molecules are possible in the population.[4] Second, it is assumed that the system is isolated in the following specific sense: (i) it is confined to a given volume V; and (ii) no polymers can enter the population from outside V though polymers can leave V, and thus leave the system. Now, let x_i be the number of polymers of the i-th type in the population. It is assumed that the population is large enough for all x_i to be treated as continuous variables.[5] Let R_i^B be the rate of increase of the i-th type of polymer due to formation, R_i^D its rate of decrease due to decomposition, and R_i^E its decrease due to emigration from V. Then:

$$\frac{dx_i(t)}{dt} = R_i^B - R_i^D - R_i^E \tag{6.1}$$

for $i = 1, 2, \ldots N$. This is the basic phenomenological equation. No assumption has yet been introduced about the form of any of the three rates occurring on the right-hand side. Specific forms for these rates determine the kinetics of the system as, for instance, autocatalysis discussed in section 6.3 and complementary replication discussed in section 6.4.

For each of these specific kinetic models, the task then becomes to show that the dynamical development of the system, simply because of the physical properties incorporated in the kinetics, obey Lewontin's three criteria of evolution by natural selection: (i) Phenotypic Variation: Different types of individuals, termed "phenotypes," in the population have different structure and behavior; (ii) Differential Fitness: Different phenotypes have different rates of survival and formation in the environment; and (iii) Heritability of Fitness: There is a correlation between the fitnesses of an individual and those individuals formed as a result of interaction between that individual and other parts of the system (Lewontin 1970).

6.3 Autocatalytic Replication with Natural Selection

An autocatalytic process is one in which a molecular type catalyzes its own formation. It is assumed here that this catalytic process admits occasional errors (due to quantum-mechanical uncertainties) which lead to the production of molecules of other types. It will be assumed that R_i^B has the form $A_i Q_i x_i + \sum_{j \neq i} \phi_{ij} x_j$. The first term represents correct autocatalysis. The nature of autocatalytic processes requires that this term be a monotonically increasing function of x_i. It is being assumed to be linear here as a first-order approximation that enhances formal simplicity.[6] A_i is the rate constant for this process: it is being assumed to be time-independent for the sake of simplicity. These two simplifying assumptions, linear dependence of the various rates on the x_i and the time-independence of the various rate constants, will be made throughout this section and the next. Q_i is a quality factor with a value between 0 and 1 that determines what fraction of the catalyzed x_i consists of accurate copies. The second term represents the increase of x_i due to errors in the autocatalysis of molecules of other types: $\phi_{ij} x_j$ is the number, x_j, that form x_i by erroneous catalysis. Similarly, R_i^D is assumed to be of the form $D_i x_i$. Finally, the emigration of all molecular types from V is assumed to be the same linear function of their respective numbers, that is, R_i^E is equal to $E x_i$. With these assumptions, equation 6.1 becomes:

$$\frac{dx_i(t)}{dt} = (A_i Q_i - D_i \quad E)x_i + \sum_{j \neq i} \phi_{ij} x_j \qquad (6.2)$$

for $i = 1, 2, \ldots N$.

Selection occurs in such a system under a variety of conditions that constrain its evolution. The constraint imposed here will be that the population in V remains constant, that is, n is a constant.[7] This constraint makes the system of differential equations represented by equations 6.2 nonlinear. A general solution of that system consists of:

$$x_i(t) = \frac{n \varepsilon_i(t)}{\int_\alpha^t \sum_k \left[(A_k - D_k) \frac{x_k(t')}{x_i(t')} \varepsilon_i(t') \right] dt'} \qquad (6.3)$$

where:

$$\varepsilon_i(t) = \exp\left[\int_0^t \left(A_i Q_i - D_i + \sum_{j \neq i} \phi_{ij} \frac{x_i(t')}{x_j(t')} \right) dt' \right] \qquad (6.4)$$

for $i = 1, 2, \ldots N$, and the lower limit of integration, α, is determined by the initial conditions. Construct the matrix $M_{ij} + (A_j Q_j - D_j)\delta_{ij} + \phi_{ij}(1 - \delta_{ij})$ where $\delta_{ij} = 1$ if $i = j$ and 0 if not. If M_{ij} is nonsingular and has no degenerate eigenvalue, then the solution can be written in the closed form:

$$x_i(t) = n \frac{\sum_j C_j q_{ij} e^{\lambda_j t}}{\sum_i \sum_j C_j q_{ij} e^{\lambda_j t}} \qquad (6.5)$$

for $i = 1, 2, \ldots N$, where q_{ij} is the i-th component of the eigenvector corresponding to the eigenvalue, λ_j, of M_{ij}, and the C_j are determined by the initial conditions.[8]

An examination of this solution shows that the molecular species with the largest value for $A_i Q_i - D_i$ among those present will dominate the population provided that the ϕ_{ij} are small. However, the other species will not completely die out, and, further, a first-order perturbation treatment shows that, to first order, their numbers will be proportional to the ϕ_{ij} (where the i-th species is the dominant one, and the index j refers to the other species) (Thompson and McBride 1974). Further, if, due to a mutation (that is, incorrect catalysis), a new type with an even higher value of $A_i Q_i - D_i$ is formed, it will eventually come to dominate the population. This means that, ultimately, the molecular type with the highest possible value of $A_i Q_i - D_i$ will dominate the population.

It is easily shown, now, that this system satisfies Lewontin's three criteria for evolution by natural selection. The criterion of Phenotypic Variation is satisfied because there are different molecular types in the population and because, since not all the Q_i are equal to 1, new types may be created from old ones. The criterion of Differential Fitness is satisfied because the molecular type with the highest value of $A_i Q_i - D_i$ will dominate and the proportion of the others will be determined by the rate at which they may be formed from this one. The criterion of Heritability of Fitness is also trivially satisfied because, in general, each molecule catalyzes the formation of another of its own type.

6.4 Complementary Replication with Natural Selection

Systems with complementary replication are those in which one polymer type catalyzes the formation of a second which, in turn, catalyzes the formation of the first. Such systems are particularly interesting because this is the model of replication in contemporary DNA and presumably was so in primordial RNA if, indeed, RNA was the original carrier of genetic information. In such a system the rate of production of a molecular type is a function of the number of molecules of its complement. A model of such a system can be represented using the same notation as in section 6.3. However, a slight change makes the properties of such systems much more perspicuous. Let σ take the values $+1$ or -1. Let the subscript σ, i characterize parameters corresponding to the molecular type complementary to those characterized by $-\sigma, i$. Other than this difference in the subscripts, let A, Q, D, E (which has no subscript), and ϕ have the same form and interpretation as in section 6.3. Given these assumptions and this notation, for this model, equation (6.1) becomes:

$$\frac{dx_i(t)}{dt} = A_{\sigma,i} Q_{\sigma,i} x_{-\sigma,i} - (D_{\sigma,i} - E) x_{\sigma,i} + \sum_{j \neq i} \sum_{\sigma = +1, -1} \phi_{\sigma, i; \sigma', i} x_{\sigma', j} \qquad (6.6)$$

for $\sigma = +1, -1$; $i = 1, 2, \ldots N$. This represents a system of differential equations just like equation 6.2 with one difference: there are now $2N$ equations rather than N. Nonlinearity is introduced here, as in section 6.3, by the selection constraint that the total population remains constant.

Formally, therefore, the system represented by equation 6.6 is the same as that represented by equation 6.2 and can be analyzed in the same way. If the following simplifying assumptions are made about the relations between each molecular type and its complement, then some particularly simple conclusions can be drawn (Thompson and McBride 1974). First, it is assumed that the production rate is the same for complementary types, that is, $A_{\sigma,i} = A_{-\sigma,i} = A_i$. This is reasonable, for instance, if molecules of complementary types serve as physical templates for the formation of each other. Second, it is assumed that the quality factor is the same for complementary types, $Q_{\sigma,i} = Q_{-\sigma,i} = Q_i$. Third, it is assumed that the decay rate is the same for complementary types, $D_{\sigma,i} = D_{-\sigma,i} = D_i$. This is reasonable, for instance, for all polymers with identical bonds between subunits if decay occurs by the disruption of such bonds. Given these assumptions it can be shown that the number of molecules of each type is a monotonically increasing function of $A_i Q_i - D_i$ provided that the mutation rates are small. The relative numbers of the various types ultimately present depends on the mutation rates as in the model discussed in section 6.3. This model satisfies Lewontin's criteria for the same reasons the last one did.

6.5 Hypercycles and the Origin of Life

The complementary replication model just discussed can be considered as representing a catalytic cyclic process involving just two molecular types (see figure 6.1a). Similarly, there can be cyclic processes with m members in which each type x_i catalyzes x_{i+1} for $i = 1, 2, \ldots m - 1$, and x_m catalyzes x_1

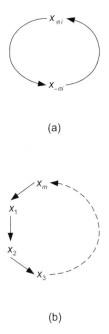

Figure 6.1
(a) A 2-member cyclic reaction like the complementary replication model discussed is section 6.4; (b) an *n*-member cyclic reaction. If the formation rate is nonlinear, either cycle is a hypercycle.

(see figure 6.1b). If at least one R_i^B in a cyclic process is of higher order than one in the x_is, then the system is called a hypercycle.[9] Hypercycles can be of several types. An *m*-member hypercycle is *simple* if: (i) for each i, R_i^B depends on x_i (giving rise to partial autocatalysis); each R_i^B is a homogeneous polynomial of the k-th degree in the x_is ($k \leq m$); and (iii) any x_i (of the members of the hypercycle in question) occurs only to the first order, if at all, in any R_i^B (see figure 6.2). In a *catalytic* hypercycle, some of the R_i^B need not contain x_i. The last situation holds, for instance, when some of the members cannot catalyze their own formation. An example of a catalytic hypercycle in which nucleic acids reproduce by complementary replication and also code for proteins which catalyze the former reaction (and others) is shown in figure 6.3.

Phenomenological equations (represented by equation 6.1) can be constructed for hypercycles. Eigen and coworkers have shown, by computer

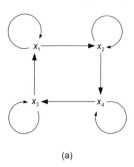

(a)

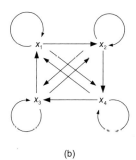

(b)

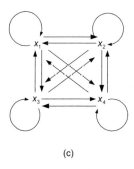

(c)

Figure 6.2
A 4-member hypercycle. (a) Degree = 2. $R_1^B = A_1Q_1x_1x_4$, $R_2^B = A_1Q_2x_2x_1$, etc.; (b) Degree = 3. $R_1^B = A_1Q_1x_1x_4x_3$, $R_2^B = A_1Q_2x_2x_1x_4$, etc.; (c) Degree = 4. $R_1^B = A_1Q_1x_1x_4x_3x_2$, $R_2^B = A_1Q_2x_2x_1x_4x_3$, etc.

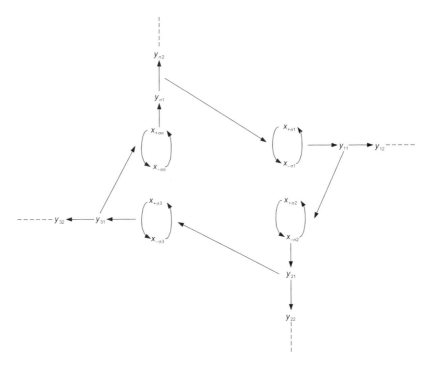

Figure 6.3
A catalytic hypercycle. The $x_{\sigma i} - x_{-\sigma i}$ pairs represent nucleic acids undergoing complementary replication. The $x_{\sigma i} - x_{-\sigma i}$ nucleic acids code for a protein y_{i1} which catalyzes the replication of $x_{\sigma, i+1} - x_{-\sigma, i+1}$, etc. The proteins can also catalyze other proteins in chain reactions of the form $y_{ij} \to y_{i,j+1} \to y_{i,j+2}$, etc. Such a hypercycle represents, in highly simplified and idealized form, the type of protein—nucleic acid coupling in living systems. Such a hypercycle, in the prebiotic environment, according to the Eigen theory, is the most likely candidate for the origin of life.

simulations (these systems are almost always too complicated for analytic solution) for a variety of situations that "once-for-all" selection occurs in such systems: the hypercycle type selected undergoes hyperbolic growth and all other types rapidly become negligible. Such a system presents an interesting model for the origin of life on earth for two reasons: (i) it can incorporate protein—nucleic acid coupling; and (ii), most important, it explains the universality of the genetic code: the genetic code that exists today is the one incorporated in the hypercycle selected in prebiotic times.[10]

6.6 Functional Explanation in Molecular Biology: DNA and RNA

The ability to provide a physical explanation of evolution by natural selection becomes particularly important when attempts are made to provide a physical warrant for functional explanations in molecular biology. A characteristic example of such explanations, originally invoked by Rosenberg (1985, pp. 38–43), is the following.[11] In virtually all living organisms there is a difference between DNA and RNA. While DNA consists of nucleotides of the four base types, adenine (A), cytosine (C), guanine (G), and thymine (T), RNA consists of A, C, G, and uracil (U) instead of T. The question that arises is the source of this difference. It does not alter the coding properties of either of the acids. U is so structurally similar to T that the possible structural roles are also unaltered. Further, T is synthesized from a uracil deoxyribonucleotide by an endothermic reaction. That suggests that whereas energetic considerations might partly account for the occurrence of U in RNA, they cannot account for the occurrence of T in DNA.

The provisional explanation of this difference between DNA and RNA is somewhat complicated. C in DNA can easily convert to U by deamination. When this occurs, the coding property of a DNA chain is destroyed because of the presence of a foreign base type, U. However, such a situation almost never pertains in a living cell because of the presence of a number of enzymes which, through a complicated process, remove U from DNA and replace it with C. Had U ordinarily occurred in DNA, this repair mechanism would be ineffective since it would excise a normal base type from DNA and replace it with something else. Further, in the absence of any repair mechanism, the coded information contained in the DNA chain would be quite unstable because not all C → U mutations are silent. Thus, the incorporation of T in DNA instead of U enhances the stability of the coded information contained in a DNA segment which more than makes up for the additional energy expenditure. Now DNA, as the genetic material, has a long existence (of the order of the lifetime of the cell) during which stable coded information needs to be maintained, whereas RNA has a relatively short life (of the order of the time it takes for translation at the ribosome). In the case of RNA, therefore, stability of the coded information is not so important, and is not worth the energy expenditure. An explanation of this kind is a functional explanation.[12] T occurs in DNA because it is the function of DNA to carry stable coded information over long periods of time and it is a function of T to enhance the stability of this coded information. No similar functional role could be assigned to T had it occurred in RNA.

In order to guarantee the adequacy of the explanation offered, two separate conditions must be satisfied, and this is generally true of functional explanations in molecular biology. First, the various factors that were adduced must be known to be empirically correct. It will be assumed here that they are; some reservations will be noted in section 6.8. Second, and much more important, it must be shown that the explanation is causal and, on the surface, there is at least one good reason to doubt that it is: it invokes a consequence of the incorporation of T in DNA rather than antecedent factors. The last problem is resolved by invoking "function statements": of the form: "According to theory T, a function of feature x, in having property Y, in system S, in environment E, relative to purpose P is to bring about consequence C" (Wimsatt 1972, p. 32). T is a causal theory that requires that, if x has Y in S in E, C obtains. Some other causal theory, T', normally an explicit selection theory, specifies what constitutes a purpose and when a function statement is true. Since both these theories are causal and function statements hold antecedent to the result in question actually occurring, the explanations become causal. In the example being discussed, x is the occurrence of T in DNA; Y is the ability of T to resist excision by the mechanisms that replace U with C; S is a DNA polymer; E is the environment containing the various enzymes responsible for DNA repair (as discussed above); P is the stability of the DNA sequence; and C is the repair of DNA. T is the chemical theory that specifies the various interactions that cause such repair: obviously it is causal. Finally, T' is the theory of natural selection at the molecular level applied to systems undergoing complementary replication as modeled in section 6.4 with all the simplifying assumptions along with the following definition and stipulation. Define as a "purpose" any property of a molecular type the augmentation of which increases its rate of growth relative to other types, all other factors remaining constant.[13] Next, stipulate that, a function statement is true if C causes a system to have some such property. T' is evidently causal since the theory in section 6.4 was manifestly developed as a causal theory and the addition of this definition and stipulation does not alter that fact. Given these considerations, if it can now be shown that the function statement obtained after these specifications is true, the functional explanation under question will have been shown to be adequate. It is obvious that C causes S to have P. What remains to be shown is that P satisfies the definition of purpose and it suffices to do this using any of the laws adduced in T'. In order to avoid repetition this is only done in the next section where it is simultaneously shown that the truth of the functional statement can be guaranteed on purely physical grounds.

6.7 A Tentative Physical Warrant for a Functional Explanation

The causal adequacy of functional explanation does not imply, of course, that this adequacy can be guaranteed on purely physical grounds. In order to accomplish the latter task two other criteria must be shown to be satisfied. First it must be shown that T and T' are theories that are true for physical reasons. Second, it must be shown that the truth of the function statement being invoked can be certified on purely physical grounds. In the example being discussed here T is simply the theory of chemical interactions that are responsible for DNA repair, and these obviously occur because of physical law. T' is the theory of selection at the molecular level as noted and augmented in section 6.6. This was manifestly constructed using not only causal, but only physical, assumptions. Thus the first criterion is trivially satisfied.

In order to satisfy the second criterion, it first needs to be shown that P satisfies the definition of purpose on purely physical grounds. In order to do this, assume, first, that S, one DNA polymer reproducing by complementary replication competes with a pair of complementary RNA polymers. Second, assume that E is the environment specified in section 6.6, that is, it contains some set of molecules that excise U from DNA and replace it with C and further require that no similar mechanism exists for RNA. Third, assume that $A_{DNA} \approx A_{RNA}$ and $D_{DNA} \approx D_{RNA}$. Now $Q_{DNA} > Q_{RNA}$ because random deamination causes the number of molecules of an RNA type to decrease by converting to new types during the course of replication. Hence $A_{DNA}Q_{DNA} - D_{DNA} > A_{RNA}Q_{RNA} - D_{RNA}$. Using the conclusions of section 6.4, the DNA pair will grow more rapidly than the RNA. Thus the greater stability of the DNA sequence causes the DNA polymer to grow more rapidly than the RNA polymers present. In other words, P satisfies the definition of a purpose on purely physical grounds.[14] Thus C causes S to have P is also true on physical grounds: in fact, it is trivially true because

DNA repair physically ensures that the DNA sequence is stable. Hence the second condition is also satisfied.

6.8 Conclusion

The purposes of this chapter were, first, to show that a physical theory of natural selection at the molecular level can be developed for a variety of systems of which that consisting of hypercycles is just one and, second, to show that this theory can be used to provide physical warrants for functional explanations in molecular biology. The first purpose has been successfully achieved.[15] Some reservations, however, need to be expressed about the success of the achievement of the second although, if the discussion in sections 6.6 and 6.7 has any merit, some success has indeed been achieved. First, it has been assumed without argument that the DNA–RNA example is typical. However, there is one good reason to make this assumption: no functional explanation that does not fit this pattern has yet gained prevalence in the literature.[16] Second, and more important, all three assumptions about the DNA–RNA competition made in section 6.7 are open to question on empirical grounds. In particular, the assumption that the DNA repair mechanism, which plays such a crucial role there, existed in the prebiotic environment or only shortly afterward is problematic although there is yet no evidence either way. Further, the assumption that there was an isolated region on earth in which primitive systems, of both DNA and RNA, reproduced themselves by complementary replication, might also be inadequate. On the other hand, if one or more of these assumptions is false, it might be the case that the functional explanation in question is itself incorrect: the correctness of functional explanations is proverbially difficult to gauge. As far as functional explanations go, all that this chapter does, perhaps, is to show the *type* of argument that would be necessary to provide physical warrants for them. To the extent that the second purpose is achieved, this chapter furthers the program of explanatory reductionism in molecular biology, that is, the program of showing that all phenomena in molecular biology can be explained on physical grounds. As far as that program is concerned this chapter also achieves another end that is of some interest to it: from a physical scientist's point of view it demystifies both evolution by natural selection and functional explanation.

Notes

1. Eigen (1971) is the source of these developments; the most complete account of the results so far obtained are Eigen and Schuster (1979) and Küppers (1983).

2. For a penetrating discussion of some of the most important objections to the Eigen theory of the origin of life, see Dyson (1985), chapter 2.

3. By "explanatory reductionism" is meant the explanation of upper-level phenomena (not necessarily laws) by means of phenomena and mechanisms at a lower level. For the distinctions between "theory," "explanatory," and "constitutive reductionisms," see Mayr (1982), pp. 59–63, and Sarkar (1989).

4. The assumption of equal length for all polymers in the population is made for simplicity. It can be disposed of with no loss of generality for the conclusions reached here.

5. Note that since V is a constant, the x_i can also be regarded as concentrations.

6. This assumption would be true if, for instance, each molecule served as a physical template for the formation of another molecule of the same type.

7. For a discussion of other constraints that result in selection, see Küppers (1983) and Bernstein et al. (1983).

8. For a derivation of equations 6.3, 6.4, and 6.5, see Sarkar (1989), chapter 4, and the references therein.

9. Note that the critical property of a hypercycle in this definition is the nonlinear dependence of the formation rate terms on the x_is and *not* the coupling together of different cycles. No precise definition of a hypercycle has been offered by Eigen and his coworkers. The one adopted here seems to capture the spirit of the discussion in Küppers (1983) in which the description closest to a definition is found.

10. See Dyson (1985) for a critical assessment of this model for the origin of life.

11. For more details of this example, and for other examples of functional explanations in molecular biology, see Sarkar (1991).

12. The account of functional explanation used here is due to Wimsatt (1972); for a defense of this approach in the general context of functional explanations in molecular biology, see Sarkar (1991).

13. Note that this definition of "purpose" is consistent with but not identical to that given by Wimsatt (1972). On Wimsatt's account, many of these properties will be lower level functions.

14. Note, therefore, that it does so on causal grounds, thus showing also that the adequacy of the functional explanation adduced in section 6.6 is guaranteed.

15. It should be emphasized, though, that natural selection occurs at many other levels: the individual, for instance. The considerations adduced here do not, in any immediately suggestive fashion, carry over to those levels.

16. See Sarkar (1989), chapter 4, for a defense of this claim.

References

Bernstein, H., Byerly, H. C., Hopf, F. A., Michod, R. A., and Vemulapalli, G. K. 1983. "The Darwinian Dynamic." *Quarterly Review of Biology* 58: 185–207.

Dyson, F. 1985. *Origins of Life*. Cambridge: Cambridge University Press.

Eigen, M., and Schuster, P. 1979. *The Hypercycle*. Berlin: Springer-Verlag.

Küppers, B.-O. 1985. *Molecular Theory of Evolution: Outline of a Physico-Chemical Theory of the Origin of Life*. Berlin: Springer-Verlag.

Lewontin, R. C. 1970. "The Units of Selection." *Annual Review of Ecology and Systematics* 1: 1–18.

Mayr, E. 1982. *The Growth of Biological Thought*. Cambridge, Mass.: Harvard University Press.

Rosenberg, A. 1985. *The Structure of Biological Science*. Cambridge: Cambridge University Press.

Sarkar, S. 1989. "Reductionism and Molecular Biology: A Reappraisal." Ph.D. Dissertation, Department of Philosophy, University of Chicago.

Sarkar, S. 1991. "Reductionism and Functional Explanation in Molecular Biology." *Uroboros* 1: 67–94. (Chapter 5 of this volume.)

Thompson, C. J., and McBride, J. L. 1974. "On Eigen's Theory of Self-Organization of Matter and the Evolution of Biological Macromolecules." *Mathematical Biosciences* 21: 127–142.

Wimsatt, W. C. 1972. "Teleology and the Logical Structure of Function Statements." *Studies in the History and Philosophy of Science* 3: 1–80.

7 Form and Function in the Molecularization of Biology

7.1 Formalism

During the first few decades of this century European art finally discovered the power of formalism which had long been known to many other societies, particularly those of central and southern Africa, which Europe, in its colonial frenzy, had scornfully designated as "primitive."[1] The new formalism came to dominate, however briefly, one medium after another, from painting through sculpture to photography and architecture.[2] No medium except those that, by convention, necessarily had "a story to tell" was immune to formalism's invasion and even these, including the novel and the new medium of cinema, did not go completely unscathed. Formalism did not play itself out in exactly the same way in all media. However, what is remarkable is that, from Wassily Kandinsky and Paul Klee in painting, to Le Corbusier and Mies van der Rohe in architecture, early twentieth-century formalism transcended the differences between media and can be characterized by a few basic ideas.

"Formalism," as I construe it, is *the pursuit of forms for their own sake*. In those media, such as painting, sculpture and photography, that can potentially admit a subject, the form becomes that subject. In media such as architecture, where there is no question of subject, form comes to dominate other pursuits such as immediate function. In any medium, there can be a wide variety of forms which can be distributed, if not precisely partitioned, into categories. What these categories can be depends on the medium in question and they cannot all be known a priori: a tradition of reflection, by both artists and critics, defines these categories.[3] In color photography, for instance, useful categories include color, line, tone, light, and balance. In contrast to black-and-white photography, texture *probably* is not. The particular forms of color are obvious. The forms that lines may take include intersections, parallels, repeats and so on. Their articulation can help achieve or destroy balance—the categories are not independent of each other. Perhaps because of the relative youth of color photography (compared to other media), and because there has been surprisingly little critical reflection on it, individual descriptive (rather than evaluative) terms for the individual forms of tone or light are not in common use. In literature, in very sharp contrast, given its long history of reflection, no such terminological poverty befuddles critical attention.

In all media, the use of "form" to refer both to particular forms and to what I have been calling "categories" is commonplace. I do not think that such a catholic use of "form" does any harm. It does not matter whether we say that Edward Weston, in later life, was pursuing color and tone as forms or that he was pursuing blues and reds, and tonal concentrations as forms (see Buchsteiner 1989, pp. 65–67). What is important is that form was being pursued, whether it be a particular one or a set of forms. What is also important is that the ability to distinguish different forms from each other—for instance, to distinguish color from tone or regularity from symmetry—is an important part of the skill of a critic. To actualize such potential differences is part of the skill of an artist.

Those who prefer desert ontologies designate a few forms—say, the triangle, square, and circle (see, e.g., Kandinsky 1979), or blue, red, and yellow (the "primary" colors), or the cube, sphere, and cylinder (e.g., Corbusier)—as fundamental. Other forms are to be reduced to, and constructed from, these fundamentals. However, fundamentalism is no more necessary in art than in religion: nothing prevents a formalist from preferring the ontology of a rain forest. For a formalist, forms are to be manipulated during the construction of a work of art and, above all, they are to be directly, that is, *sensually* appreciated. Forms can be symbols for other things. However, and this is critical to formalism, *that need not be the case.* Formalism cannot be reduced to a form of symbolism. The search for "meaning"—or, even worse, "truth"—before appreciation of a form is little more than yet another unfortunate incursion into aesthetics of the dubious linguistic turn of twentieth-century philosophy.[4] It makes a mockery out of modern art: we need not undertake an inquiry into meaning and truth before we pass from apprehension to appreciation of a work by Kandinsky or Weston.[5]

It should be clear, then, that I do not construe "formalism" as one of the two parts of a "form–content" dichotomy, usually (and, unfortunately) explicated by philosophers in analogy with the syntax–semantics distinction of logic. Even in those media that admit subjects, the contrast, here, is with "representation" rather than with subject or content. In a formalist work, whether there is a one-to-one correspondence between the parts and their relations in the work and the putative "reality" it refers to is no desideratum for its success. It is not that representation in art is itself a particularly lucid concept. Short of an at least implicit appeal to convention, it is hard to see why even the most "perspectival" of paintings should appear as "real" to anyone except a one-eyed midget. But a formalist work is nonrepresentational not because of a failure of correspondence or of "realism"; it is nonrepresentational because its subjects are the forms that are, in a sense, within itself.

In all media, formalism can involve a variety of strategies, from abstraction (or analysis) to construction (or synthesis). The forms that can (though need not be) designated as the most basic may be obtained from actual objects through abstraction: a circle may be abstracted from any reasonably symmetric closed curve, a right angle from an upright tree, as "irrelevant" detail is ignored during the pursuit of what are taken to be more fundamental forms.[6] If abstraction provides one part of formalism, construction provides its counterpoint. Either of these parts can be emphasized over the other. If Klee is taken to have abstracted geometry from nature, the de Stijl group put the emphasis on pure construction.[7] But, perhaps, the happiest products emerge when the dialectic between abstraction and construction remains dynamic but stable enough to prevent the dominance of one over the other—witness the greater critical success of Kandinsky and Mondrian.

Nevertheless, in most cases, abstraction and analysis have to precede, conceptually, if not temporally, construction and synthesis. To construct or synthesize, units are necessary, and these units, whether they are the primary colors or shapes or volumes, have to be available before they can be put together. Should these units ultimately owe their origin to some actual entity, they are at least indirectly the products of abstraction. Whether all units must have their ultimate genesis in this manner is an old question. Full-blown empiricists insist on an affirmative answer; most others remain noncommittal even if they do not indulge in explicit denial. Of course, any of these units may well stand in need of further refinement after construction but, though the dialectic has no end, it has a beginning of sorts.

During abstraction, what makes one form more "fundamental" than another? Stated in this fashion, this question appears to be ontological. But such concerns can be avoided. Shorn of irrelevant ontological connotations, one can—and, perhaps, must—ask why one form is to be preferred over another. Why prefer a circle to an ellipse? Why prefer yellow over green? There is no simple scientific answer. An ellipse is a more general conic section than a circle. In fact a circle is a degenerate ellipse. In the theory of additive color mixing, which goes back to Newton, red, green and blue, rather than red, yellow and blue, are the primary (or fundamental) colors.

Another form of the same question is to ask why certain details are irrelevant to the characterization of an underlying form. Or, to put it in a form sanctioned by the ancient Greeks, how is accident to be distinguished from essence? In either form, this question is fundamental to the analysis of formalism. Nevertheless, I will not attempt to begin to answer this question. There is no easy answer, and it suffices for my present purposes simply to note that some theory, implicit or explicit, has to provide the criteria by which a recursive consistent procedure of abstraction can be effected. In the Bauhaus in the 1920s, attempts were made to determine such criteria through psychological experimentation. Efforts of this kind have not found much sympathy among professional philosophers, presumably in an attempt to avoid a potential naturalistic fallacy in the context of aesthetics. But to the extent that a naturalistic perspective is of value even in obviously normative disciplines, and there is little reason to object to naturalism so long as it does not replace science with fantasy (as in evolutionary ethics or human sociobiology), investigations of this sort are philosophically important. At the very least they provide constraints that can serve to eliminate some of the putative candidates for such criteria that philosophy might put forward. If indigo and blue cannot routinely be distinguished by the human eye, there is little insight to be gleaned from a suggestion that these be two of at most three primary colors.

As I have already noted, the unit forms from which construction can proceed need not have been obtained through abstraction. Formalism that puts an extreme emphasis on construction from preordained units has been particularly influential in twentieth-century architecture. Extreme examples include the constructivism of de Stijl and, barely more successfully, the geometric constructions of Corbusier in the 1920s. In fact much of early modernism—of the so-called International Style—was a thoroughly formalist enterprise.[8] The case of architectural modernism is particularly relevant for my purposes because of its similarities to some of the developments in molecular biology that I will discuss below.

In architecture, formalism cannot be straightforwardly contrasted to "representation" since, except perhaps metaphorically, buildings do not represent. The useful contrast to form is function and attempts to resolve the tension between the two have led to such slogans as "form follows function." Of the most influential of the founders of early architectural modernism—Walter Gropius, Corbusier, and Mies—only the first paid any serious attention to function, though his successor at the Bauhaus, Hannes Meyer, attempted to transform it into dogma (Whitford 1984, p. 180). Corbusier wrote extensively—indeed, it is at least arguable that he wrote more extensively than he built. However, no matter whether his writings can be interpreted as endorsing the primacy of function over form, there is little evidence of that in his geometrical constructions—see for example the famous *Villa Savoye* (1930) at Poissy-sur-Seine.[9]

That Mies was no functionalist is scarcely controversial. Moreover, few architects have pursued formal elements, whether it be material and balance (the *Barcelona Pavilion*, 1929) or shape and light (the *Farnsworth House*, 1946-1951), as systematically as Mies. That the latter building is also a functional nightmare is hardly open to question (see Schulze 1985, pp. 252-259). Its original crimes included poor ventilation, inadequate temperature control and a failure to guard against the invasion of seasonal insects! Nevertheless, even such vociferous critics of architectural modernism as Jencks (1985, pp. 104-105) have felt compelled to pay tribute to the striking form of the *Farnsworth House*. If imitation is the best form of flattery, Philip Johnson's *Glass House* (1949) further underscores the importance of that building.

I do not have space, here, to pursue further how extensively Mies, Corbusier or, for that matter, other formalists such as Alvar Aalto, J. J. P. Oud, or the architects of de Stijl, influenced early modern architecture, though an investigation of this sort, and how formalism is related to traditional theories of architecture, would be a very welcome contribution to aesthetics.[10] However, I hope that the example of the Farnsworth House and the invocation of the names of Mies and Corbusier, together suffice to demonstrate that one significant part of modern architecture was guided by the pursuit of form at the expense of function.

7.2 The Physical Sciences

I wish to suggest that the pursuit of formal elements forms a significant part of research, especially theoretical research, in the empirical sciences. I do not wish to criticize or defend this feature of science in general. I merely wish, for the time being, to demonstrate its existence. There are circumstances in which, I think, it can be useful. Formal investigations might, for instance, prepare the way for new theories and models, contrive examples to illuminate some unexpected characteristic of a theory, or perhaps even generate mathematical or logical investigations that are interesting in their own right. I am going to set aside a proper discussion of these possibilities for some other occasion. In other instances, the pursuit of formal elements can lead to less desirable consequences. The examples I give below will serve to illustrate the latter point.

Meanwhile, note two consequences of the claim that I am advocating: (i) the development of science cannot be understood solely by concentrating on the relation of experiment to theory; and (ii) the pursuit of formal elements might well lead to what is usually called the underdetermination of theory by evidence. However, I wish to emphasize that this does not entail that no other influences, such as social or political influences, might also be important in the development of science or that the pursuit of formal elements necessarily leads to evidential underdetermination of theories.

Besides the general claim that the pursuit of formal elements is a significant part of science, I also wish to advocate a subsidiary one: that this pursuit can often be seen in the way in which certain parts of a scientific discipline get designated as "fundamental," whereby other parts implicitly get moved to its periphery. Given what I have said in the last section, this designation, in turn, shows that there is a similarity between formalist scientific and artistic pursuits: both are highly concerned with what ought to be considered "fundamental."[11] I will try to illustrate this latter point by a very brief excursion into contemporary physics and the sort of philosophical reflection that it has so far generated.

Form and Function 167

What is considered to be fundamental physics today is largely limited to just two areas: the study of spacetime, that is, general relativity, and the study of the smallest known constituents of matter and their interactions, including the framework theory that governs them, that is, quantum field theory. Both of these studies come together in modern cosmology. I do not wish to criticize the attention that is paid to these areas. However, I do wish to suggest that attention to these areas, along with an implicit concomitant decision not to regard other areas—the physics of viscous fluids or of large molecules or what has come to be called "physics on a human scale"—as equally "fundamental" is based, in part, on aesthetic preferences.[12]

When attention is fixed on particle physics, the usual defence of its fundamental importance takes the form what might be called "ontological fundamentalism": after all, all other bodies in the universe are "composed of" these fundamental entities. But, in a world of indistinguishable particles, transient resonances, virtual particles, and entangled states, the notion of "composed of" is far from clear.[13] Certainly, we cannot fall back upon precisely locating these entities at specific spatial points within composite systems. When, so far, nobody has even found indubitable evidence of the existence of free quarks, it should be clear that to say that a proton is "composed of" quarks is to make a rather different claim about composition than when I say that my body is "composed of" certain organs or that a bacterial cell is "composed of" certain organelles.[14] I wish to suggest that the models that the particle physicist constructs are the result of a process akin to the method of analysis in formalist art that I described earlier. There is little synthesis to provide a counterpoint. That is why particle physicists do not seem to be able to come up with any other experimental design than collisions of particles at higher and higher energies. I am not at all denying that there are differences, too, between artistic and scientific pursuits of form. In particular, empirical adequacy, as understood in the sciences is an important constraint that cannot be easily violated. All I am trying to demonstrate is the existence of ultimately aesthetic considerations, along with evidential ones, in science.[15]

A more sophisticated defence of particle physics would forego too much concentration on ontology of this sort and would emphasize, instead, deep "symmetries" of nature that the laws of high energy physics allegedly reveal. A similar argument can also easily be made for spacetime physics. But, note, here, that even the language is aesthetic. The pursuit of symmetry, especially the rather peculiar kind of symmetry revealed by the invariance of equations under transformations, is something that any formalist in the arts would feel happy with. Falling back on symmetry emphasizes the aesthetic dimension of science rather than denies it. A physicist could, at this point, once again attempt to invoke ontology and argue that these symmetries are there to be discovered in nature. But, as usual, the ontological move falls afoul of the usual empiricist objections that the symmetries are usually approximate, are properties of the models, and are subject to revision. Spacetime, we now know, may well be neither homogeneous nor isotropic. Indeed it need not even admit a single symmetry (in technical language, a Killing field).

There is yet another, and potentially quite powerful, defence of the ostensible fundamental nature of the lowest levels of organization in physics (that is, that of the particles of high energy physics). Casting ontology aside, the argument now becomes epistemological: all behavior of all bodies in the universe is to be explained in terms of their constituents. This is what philosophers call explanatory reductionism (Sarkar 1992a). This looks good on paper, but is little more than a dream in practice. No one understands quantum mechanics on a macroscopic scale yet, let alone quantum field theory. There are many important and startling connections between microscopic physics and macroscopic physics, especially in phenomena such as superfluidity and superconductivity. But all of these involve post hoc approximations. And they exhaust little of the known macroscopic phenomena of the physical world.

What does physics on a human scale look like? Physicists have already begun to discover surprising kinds of "universality" where certain macroscopic behaviors do not depend on the finer details of the behavior of the microscopic parts (see Leggett 1987). The new physics has begun to be able to account for the shapes of clouds. From the discovery of chaotical dynamical systems, it has already shown why weather prediction is so difficult. Moreover, chaotical phenomena of the same sort unite as apparently diverse subjects as physics and ecology. At the human scale, complexity and diversity are more apparent than the simplicity and elegance of the symmetries of particle interactions. The physicist is left with the daunting task of discovering general principles in the maze of things and behaviors that surround us as we go from day to day.

A sophisticated physicist would, no doubt, argue at this point that physicists do not deny that there are other areas of physics as fundamental as the study of high energy particles or of spacetime. They would point out, for example, that today's physicists are fascinated by scaling between levels of organization and the issues raised by complexity. This is true, but such sophisticated physicists are still relatively rare. That is why the physics community can present an almost monolithic front against critics of the supercollider project, the particle physicists' dream (and guarantee of job security, at least in the short term). That is also why philosophers of physics often rederive old results in new notation, especially in the quantum measurement problem, but pay little attention to areas of physics other than quantum mechanics, particle physics and spacetime.[16]

7.3 Biology

Let me turn to biology. For the last generation, research in biology has been dominated by molecular biology. Not only has molecular biology illuminated a wide variety of fields from genetics to immunology, some of its models, especially the DNA double helix, have become cultural icons of importance. On the surface, molecular biology appears as a natural, and perhaps even inevitable, development from biochemistry, as the chemical characterization and explanation of biological phenomena was pursued systematically as more and more experimental techniques became available. But there is more to this development than first meets the eye.

Central to molecular biology is molecular genetics. It is from this subfield that most (though not all) of the dominating ideas of molecular biology have been framed. In principle, molecular biology includes the study of all those molecules that comprise biological systems, whether they be lipids, nucleic acids or proteins. Indeed, in practice, molecular biology does include all of these in its domain. However, the study of DNA has come to dominate molecular biology to a greater and greater extent. Moreover, central to the conceptual structure of molecular biology is a concept of "information" which is construed exclusively as sequence information ultimately contained in the DNA. The "genetic code," from this point of view, mediates between the DNA sequence and everything else there is to life.

It is important to note how myopic this construal of biological information is. Even the chemical specificities of the proteins and, especially, other molecules that may initiate or terminate gene expression, are not ultimately deemed to carry information. This is not the place to pursue the possibilities generated if this restrictive notion of "information" is relaxed; suffice it to observe that this notion has been critical to conventional interpretations of molecular biology (see, e.g., Yockey 1992). For example, it is only because of it that the central dogma of molecular biology, that information flows from nucleic acid to protein, but never in the reverse direction, can be maintained.[17]

One consequence of maintaining this assumption is that the DNA sequence and the genetic code come to occupy central positions in the conceptual structure of molecular biology. It then makes sense to pursue the behavior of DNA, and even just the sequence, for its own sake. Moreover, the genetic code has a property that is comparatively unique in biology, namely, its relative *universality*: the code is almost exactly the same in virtually all species.

The relative universality of the genetic code was only demonstrated in the late 1960s. However, the fascination with the genetic code began long before the demonstration of its near universality. The idea of a "hereditary codescript" goes back to Erwin Schrödinger (1944) and the term "information" was explicitly introduced in a genetic context in 1953 (Ephrussi et al. 1953).[18] As soon as the double helix model of DNA was announced by James D. Watson and Francis Crick (1953), George Gamow (e.g., 1954) designated the relation between DNA and protein as being one of "coding". His own attempts to crack the code were futile but, in an important conceptual contribution, he distinguished between the problem of determining the mechanisms of gene expression and the abstract coding problem, that is, of finding the translation table between the DNA "language" written in a 4-letter alphabet and protein "language" with a 20-letter repertoire.[19]

In a striking formalist move, Crick, Griffith, and Orgel (1957) attempted to solve the abstract coding problem without reference to that of finding the mechanisms of gene expression. The formal theory that provided the background were assumptions about the desirable properties of biological information. The translation procedure, they argued, presented "two difficulties: (1) Since there are $4 \times 4 \times 4 = 64$ different triplets of four nucleotides, why are there not 64 kinds of amino acids? (2) In reading the code, how does one know how to choose groups of three?" (Crick, Griffith, and

Orgel, p. 417). The first problem was that of potential degeneracy. If 64 triplets only coded for 20 residues, some triplets would have to code for more than one residue. There was no experimental reason that precluded a degenerate code, but they obviously felt that it was undesirable. The second problem was that of synchronization: do we read the sequence ACCGTAGT as ACC, GTA, ... or CCG, TAG, ... or CGT, AGT, ... ?

Their solution to both problems was the ingenious "comma-free code." They assumed

> that there are certain sequences of three nucleotides with which an amino acid can be associated and certain others for which this is not possible. Using the metaphors of coding, we say that some of the 64 triplets make sense and some make nonsense. We further assume that all possible sequences of the amino acids may occur (that is, can be coded) and that at every point in the string of letters one can only read "sense" in the correct way.... It is obvious that with these restrictions one will be unable to code 64 different amino acids. The mathematical problem is to find the maximum number that can be coded. We shall show (1) that the maximum number cannot be greater than 20 and (2) that a solution for 20 can be given. (Crick, Griffith, and Orgel 1957, pp. 417–418)

Thus, purely formal considerations "solved" what was, arguably, the first theoretical problem in molecular biology. The particular solution they presented is shown in figure 7.1. 407 other such schemes are possible.

Probably, few ideas entirely bereft of experimental evidence have captured the imagination of working scientists as much as the comma-free code in the later 1950s. It spawned an entire industry of mathematicians and biologists who presented variants, attempted to generalize it, or simply referred to it without good reason in their publications. Among them were

X	Y	Z
A	A A	A A
A B	C B	B D B
B	B C	C C
		D

Figure 7.1
The comma-free code. Read cyclically in triplets in each of the three columns of the three classes (X, Y, and Z). A, B, C, and D can each be any of the four nucleotide base types. Exactly 20 triplets emerge up to cyclic permutation (after Crick, Griffith, and Orgel 1957).

the mathematician, S. W. Golomb, and such stalwarts of the new molecular biology as Max Delbrück and André Lwoff (Sarkar 1989). I have argued, elsewhere, that had the comma-free code or any of its variants turned out to be true, it would have been a remarkable success of a nonreductionist research strategy in molecular biology (Sarkar 1989, 1996). But history was not kind to the formalists.

In fact, when the code was finally deciphered in the early 1960s, it had none of the "desirable" properties envisioned for it in the 1950s. Some experimentalists, long skeptical of the mathematical models of the theorists, could barely contain their glee. They were also not hesitant to point out that, ultimately, the considerations that led to it were aesthetic, and to denigrate it on those grounds. "The early period [of research on the genetic code]," wrote Woese,

from 1954 until the discovery of the *in vitro* system [i.e., 1961] was dominated, of necessity, by theoretical speculations, for which the few facts then available served as seasoning rather than substance. Consequently, this early period was rather unfortunate, scientifically speaking, in that theories were judged almost exclusively along Platonic lines—by their internal consistencies, aesthetic qualities, and numerological appeal. (1963, p. 212)

Gamow's stereochemical codes were bad enough, but Crick's comma-free code, especially as mathematicized by Golomb, particularly irked Woese: that code "serves in retrospect as a good example of how, unchecked by fact, attractive hypotheses can become elevated to the level of dogma, and further fabricated into impressive, beautiful but illusory dream worlds. Sadly, such worlds have been the repository of all theories of the biological code to date" (1963, p. 216).

Nevertheless, the metaphor of the code continued to be used as an important organizing principle within molecular biology. One corollary was the exclusive attribution of ultimate or fundamental importance to the DNA sequence of an organism. Even though, at the level of DNA, information, was not stored uniformly, due to a variety of factors, including differing codon usages in different organisms, the degeneracy of the code, the existence of introns, *etc.*, the universality of the code could be used to argue that there was at least one level at which all living organisms could be viewed as being the same.

Note that there are many other levels of functional behavior where equally universal phenomena exist, for instance, in the dynamic redefinition of self and the distinction of nonself from self that forms part of the immune response of all living organisms. However, these phenomena do not—as yet—have the same simplicity and elegance that is so endearing about the genetic code. So far, they have failed to capture the biological imagination in the way that the code has. It is arguable that the Human Genome Project, the mechanical pursuit of DNA sequences at the expense of everything else, with little concern for the actual explanatory or cognitive value of the sequence, is the ultimate result of the deification of the code.

I have argued in detail, elsewhere, that the DNA sequence of any organism (except a virus) would be of little explanatory value at present.[20] I am not claiming that, a generation or so down the road, these sequences might well become part of the daily repertoire of molecular biologists. The Human Genome Project, however, wants to sequence something that is supposed to be "the" entire human genome by 2005. However, at present, thanks to the absence of any potential solution for a variety of other problems, especially the protein folding problem, we cannot get from a DNA sequence even to the structure of a protein, let alone biology at higher levels of organization. I am more than skeptical that these problems will have been sorted out by the time when the HGP's technicians begin churning out larger and larger DNA sequences. This is formalism, through and through, and almost entirely abstraction devoid of any concern for subsequent constructive possibilities. A certain set of entities, in this case the DNA sequences, has been identified as fundamental by criteria that are not purely epistemological and has since been relentlessly pursued for its own sake.

This is not to suggest that the aesthetic criteria are the only ones that have contributed to the pursuit of DNA. In a society where economic power is paramount, and wealth is inherited, the "stuff of inheritance" is obviously going to be interesting. This sociological factor certainly forms part of the explanation of the origin of eugenics and human sociobiology; it is at least quite likely to be important in the explanation of the rise of genetics and, in particular, of the Human Genome Project. Nevertheless, the possible role of aesthetic factors, which would have worked in harmony with the socioeconomic ones in this case, should not be ignored in attempts to understand the origins of contemporary molecular biology. Moreover, in the case of physics, at least, such socioeconomic explanations have never been particularly convincing and, ultimately, aesthetic concerns may well be more important to science than the ideological roots of scientific practices.

7.4 Diversity and Complexity

Formalism is by no means all that there is to any medium of art. Even in the twentieth century, Hans Hollein's *Städitches Museum Abteiberg* (Mönchengladbach, 1976–1982) provides the counterpoint to Mies's *Farnsworth House* without foregoing a modernist emphasis on material. Yousuf Karsh's portraits or Robert Capa's meticulous record of the desolation and heroism of war are as intriguing as the formalist nudes of Weston.

In physics the situation is no different. The diversity and complexity of physics at the human scale are ignored in the pursuit of "fundamentals" at the level of leptons and quarks. Nevertheless, the formation of minerals and mountains, or the patterns of clouds and waves are all governed by physical principles. Moreover, nothing more than an unproved assumption of explanatory reductionism justifies the designation of such phenomena as incidental, rather than central, to the pursuit of physics. When such phenomena have been pursued, interesting explanations have often been forthcoming, at least up to a first level of approximation. During the last decade, moreover, studies of complexity have made almost as many inroads into physics as anywhere else (Ford 1989). Ultimately, the choice of fundamental particles, rather than middle-sized objects, as the frontier of physics, is largely an aesthetic choice.

If diversity and complexity have finally breached the defences of physics, biology has traditionally reveled in their pursuit. Moreover, molecular biology has not been able to avoid complexity within its own ranks. The early universals of the field: the single genetic code, the operon model of gene regulation, the linear contiguous relation between DNA, RNA, and protein, have all fallen afoul of the unexpected complexity of eukaryotic genetics. If the comma-free code was an instance of the pursuit of form over representation, the Human Genome Project is the pursuit of form over function. In retrospect, it is hard to criticize the comma-free code. In the absence of any experimental evidence to be represented, it was not unreasonable—and intellectually exciting—to pursue form for its own sake. The Human Genome Project, however, is quite another story. Whatever excitement it generates is technological. How most of a sequence is to be related to biological behavior or function is simply mysterious at present, and unlikely to be much clarified within the next few years while the genome gets sequenced for its own sake. Nevertheless, however dimly we foresee, we charge ahead.

But what is most intriguing, for my present purposes, is that the same pattern of choices apparent in the pursuit of the arts also manifests itself in the sciences. Formalism in the arts is mimicked by the relentless drive to the smallest particles in physics, with the hope that the principles found there can be used to explain physics at all other levels of organization. A formal universalism is pushed in biology at the level of DNA sequence and code. The similarities go even further. In the arts or in the sciences, the skills generally required by the formalist do not completely coincide with those that are required by those pursuing diversity and complexity. The skills of the formalist are often technical, whether it be an architect's competence in structural technology, a physicist's grasp of differential equations or a biologist's knowledge of a polymerase chain reaction. If abstraction is pursued for its own sake, the attention to technique can become of paramount importance. "God is in the details" was Mies's famous dictum. A mathematical physicist must perforce study the divergence of series. A biologist must perfect DNA sequencing techniques.

Finally, just as formalism is but one of the many modes of artistic practice, with many other modes coexisting with it, physics has the option of pursuing everyday objects and processes, and biology that of exploring the diversity and complexity of organic life. In fact, to the extent that Theodosius Dobzhansky's famous dictum, that nothing in biology makes sense except in the light of evolution, is true, biology has no option other than to move beyond the formalism of the genetic code. This is not to suggest that the formalist mode of inquiry in biology (or, for that matter, physics) is inappropriate or sterile. Any such judgment would make a mockery of the excitement and fascination of molecular biology during the last generation. It is merely to note that this is not all that there is to the subject.

Acknowledgments

This paper is the transcript of a talk given at the Boston Colloquium for the Philosophy of Science, November 17, 1992.

Notes

1. My use of "formalism" should not be confused with any of the customary—and mutually inconsistent—uses in the literature of art criticism. I will explain my use in detail in the next few paragraphs. I am fully aware that the idea about formalism that I am advocating requires much more detailed elaboration and defence than what I provide here. My excuse is a lack of space and the constraint that I have to address the aesthetics of science, not just art-forms, in the context of this anthology. There are many similarities between what I am arguing here and what the French art historian, Henri Focillon, argued long ago (Focillon 1934; a highly interpretive English translation has recently been issued (Focillon 1989)). In between came two sets of debilitating developments, one in art history starting especially with Erwin Panofsky's iconographic analysis of forms (see Panofsky 1955), which is a collection of his most influential essays, and the other in philosophy, with Rudolf Carnap and the logical empiricists' interpretation of all form as syntax to be distinguished from semantics as subject (see note 4 below). Both of these have discolored subsequent philosophical accounts of artistic form (see Whyte 1951) with the full cooperation of the twentieth century's unfortunate linguistic turn in philosophy.

2. I am not using "medium" in any precise sense, accepting as a medium anything that commonly gets referred to as such.

3. It follows that what the appropriate categories are for a medium—"appropriate" in the sense of being useful in critical reflection—only gets discovered as a critical tradition develops.

4. The ultimate responsibility for this lies primarily with Carnap (1937). Philosophy is the syntax (later, also the semantics) of a language. Physicalism was taken to be obvious (though, in all fairness, Carnap was sensitive to the possibility of genuine introspective life unlike much more narrow-minded behavioristic physicalists such as Quine).

5. Nevertheless, I would like to suggest that the standard artists' claims of trying to reach "truth" in a work should be taken literally. What that truth ostensibly refers to is not necessarily anything about the state of the "physical" world (where "physical" only refers to what current physics allows) but, presumably, some other realm to be described and understood. That no such truth can be referred to is a dogma that has resulted from yet another dubious move in twentieth-century philosophy, namely, physicalism. Physicalism denies the possible "real" existence or explanatory value of these other realms but, once mental phenomena are recognized to exist—and only philosophers would deny the unequivocal existence of minds and mental phenomena—and once we reflect on how little behaviorist (or any nonmental) psychology tells us, I am somewhat mystified as to why we should believe in physicalism. Even if only as an aside, I should like to note that a hard-headed empiricism—trust only your experiences and avoid metaphysical (especially

ontological) commitments—argues against physicalism rather than for it. Such hard-headed empiricism would worry about the regularities of artistic experience, including the introspections of artists. In fact, it might even be a first step towards a "naturalistic" understanding of aesthetics but, about that, neither I nor, as far as I know, anybody else has so far had much to say.

6. See, e.g., the practices encouraged by Kandinsky (1979) while ignoring the dubious pseudo-psychological theories elaborated there which confer an illegitimate veneer of naturalism over what is otherwise a very interesting exercise of rule-governed art construction.

7. For Klee, see Lazaro (1957); for de Stijl, Overy (1991) is particularly useful.

8. I am only concerned with modernism in the 1920s and early 1930s. A much more complicated story would have to be told about its later disparate parts, including the obsessive minimalism of Mies and the brutalism of Corbusier, which contributed much to modernism's unpopularity in the late 1960s and 1970s (see Jencks 1985).

9. Indeed, it is doubtful that any tangible relation subsists between Corbusier's voluminous writings and much more modest building accomplishments during this period.

10. Unfortunately, the only book-length contribution devoted to the aesthetics of architecture from a philosophical orientation (Scruton 1979) shows little familiarity with the transformation of architectural practice *and theory* brought about by the technological and political experiments of the first half of the twentieth century. Worse, the book is devoted to developing the rather sophomoric thesis that architecture should be evaluated in continuity with sculpture.

11. It is also likely that the dialectic of abstraction and construction that I outlined there, in the context of modern art, might be equally applicable to the construction of scientific models. An exploration of this point, however, is far beyond the scope of this discussion, though it might be more important than any of the points being made here.

12. For "physics on a human scale" see chapter 4 of Leggett (1987).

13. See Shimony (1989), Close (1989), and Georgi (1989) for a survey of the current state of quantum mechanics and particle physics.

14. The latter notion of "composition," which is akin to what was used in classical physics does, however, influence modern particle physics. I have previously argued that the quark model constitutes an attempt to capture, to the extent that is possible, the classical notion of composition in modern high-energy physics (Sarkar 1980).

15. Note, moreover, the pursuit of forms in the arts is also not unconstrained. Depending on the medium, representation and function can well be regarded as constraints analogous to the evidential concerns of science.

16. See Shimony (1987), however, for a very important and welcome exception.

17. For a development of this point, see Sarkar (1996).

18. I am grateful to Joshua Lederberg for this reference.

19. For details of this history, see Sarkar (1989).

20. Part of this work was done in collaboration with A. I. Tauber. See Tauber and Sarkar (1992) as well as Sarkar (1992b).

References

Buchsteiner, T., *Weston* (Schaffhausen: Edition Stemmie, 1989).

Carnap, R., *Logical Syntax of Language* (London: Kegan Paul, Trench, Trubner, 1937).

Close, F., "The quark structure of matter," in P. Davies (ed.), *The New Physics* (Cambridge: Cambridge University Press, 1989), pp. 396–424.

Crick, F. H. C., Griffith, J. S., and Orgel, L. E., "Codes without commas," *Proc. Natl. Acad. of Sci. (USA)* 43: 416–421, 1957.

Ephrussi, B., Leopold, U., Watson, J. D., and Weigle, J. J., "Terminology in bacterial genetics," *Nature* 171: 701, 1953.

Focillon, H., *La vie des formes* (Paris: Presses Universitaires de France, 1934).

Focillon, H., *The Life of Forms in Art* (New York: Zone Books, 1989).

Ford, J., "What is chaos, that we should be mindful of it?" in P. Davies (ed.), *The New Physics* (Cambridge: Cambridge University Press, 1989), pp. 348–372.

Gamow, G., "Possible relation between deoxyribonucleic acid and protein structures," *Nature* 173: 318, 1954.

Georgi, H. M., "Grand unified theories," in P. Davies (ed.), *The New Physics* (Cambridge: Cambridge University Press, 1989), pp. 425–445.

Jencks, C., *Modern Movements in Architecture* (Harmondsworth: Penguin, 1985).

Kandinsky, W., *Point and Line to Plane* (New York: Dover, 1979).

Lazaro, G. D. S., *Klee* (New York: Praeger, 1957).

Leggett, A. J., *The Problems of Physics* (Oxford: Oxford University Press, 1987).

Overy, P., *De Stijl* (London: Thames and Hudson, 1991).

Panofsky, E., *Meaning in the Visual Arts* (Chicago: University of Chicago Press, 1955).

Sarkar, S., "On the concept of elementarity in particle physics," *Columbia Journal of Ideas* 5(3): 93–129, 1980.

Sarkar, S., "Reductionism and molecular biology: A reappraisal," Ph.D. dissertation, Department of Philosophy, University of Chicago, 1989.

Sarkar, S., "Models of reduction and categories of reductionism," *Synthese* 91: 167–194, 1992a. (Chapter 2 of this volume.)

Sarkar, S., "Para qué sirve el proyecto Genoma Humano," *La Jornade Semanal* 180: 29–39, 1992b.

Sarkar, S., "Biological information," *Boston Studies in the Philosophy of Science* 183: 187–231, 1996. (Chapter 9 of this volume.)

Schrödinger, E., *What Is Life? The Physical Aspect of the Living Cell* (Cambridge: Cambridge University Press, 1944).

Schulze, F., *Mies van der Rohe: A Critical Biography* (Chicago: University of Chicago Press, 1959).

Scruton, R., *The Aesthetics of Architecture* (Princeton: Princeton University Press, 1979).

Shimony, A., "The methodology of synthesis: Parts and wholes in low-energy physics," in R. Kargon and P. Achinstein (ed.), *Kelvin's Baltimore Lectures and Modern Theoretical Physics* (Cambridge, Mass.: MIT Press, 1987), pp. 399–423.

Shimony, A., "Conceptual foundations of quantum mechanics," in P. Davies (ed.), *The New Physics* (Cambridge, Cambridge University Press, 1989), pp. 373–395.

Tauber, A. I., and Sarkar, S., "The human genome project: Has blind reductionism gone too far?" *Perspectives on Biology and Medicine* 35(2): 220–235, 1992.

Watson, J. D., and Crick, F. H. C., "Molecular structure of nucleic acids: A structure for Deoxyribose Nucleic Acid," *Nature* 171: 737–738, 1953.

Whitford, F., *Bauhaus* (London: Thames and Hudson, 1984).

Whyte, L. L. (ed.), *Aspects of Form* (London: Lund Humphries, 1951).

Woese, C. R., "The genetic code—1963," *ICSU Review of World Science* 5: 210–252, 1963.

Yockey, H. P., *Information Theory and Molecular Biology* (Cambridge: Cambridge University Press, 1992).

Part III Information

8 Decoding "Coding": Information and DNA

Most biologists would probably be more bored by than surprised at, and certainly not impressed by the originality of, the claim that the central core of the conceptual structure of contemporary molecular biology can be encapsulated in the following three precepts:

- All hereditary information resides in the DNA sequences of organisms.
- This information is transferred from DNA to RNA through the process of transcription, and from RNA to protein through translation.
- This information is never transferred from protein to nucleic acid sequences.

Following Crick (1958), the last precept is usually called the "Central Dogma" of molecular biology.

These three precepts are so universally accepted that they usually find their way into introductory biology texts. Nevertheless, they are at best misleading, and at worst simply vacuous, in the sense that they can play no significant explanatory role in molecular biology. The reason for this is that there is no clear technical notion of "information" in molecular biology. "Information" is little more than a metaphor that masquerades as a technical concept and leads to a misleading picture of the conceptual structure of molecular biology. Metaphors are ubiquitous in science. When they provide succinct or easily comprehensible accounts of complex technical concepts, they are particularly useful for communicative or didactic purposes. However, when they serve only as surrogates for nonexistent technical concepts, their influence is less than benign. In molecular biology, "information" and associated terms (especially "coding") belong to the latter class of metaphors. This claim is, no doubt, a bold one—to develop the arguments in its favor requires an excursion into the history of molecular biology, a discussion of how the term "information" came to be introduced, and why, for all its initial plausibility as a useful theoretical concept, there is little to recommend its continued use today.

8.1 Information and Molecular Biology

Historically, the term "information" entered molecular biology as part of a putative theory of biological specificity. By around 1930, it had become clear that molecular interactions in living organisms are highly specific in the sense that particular molecules interact with exactly one, or at most a few, reagents. Enzymes act on specific substrates. Living organisms produce antibodies that are highly specific, not only to naturally occurring antigens, but also to artificial antigens (Landsteiner 1936). Even the action of genes was sometimes described using "specificity": genes specified precise phenotypes to different degrees of accuracy called their "specificity" (Timoféeff-Ressovsky and Timoféeff-Ressovsky 1926). In genetics, the ultimate exemplar of specificity became the gene–enzyme relationship (Beadle and Tatum 1941): "one gene–one enzyme" was perhaps the most important organizing hypothesis of early molecular biology.

By the end of the 1930s, a highly successful theory of specificity (and one that remains central to molecular biology) emerged. Due primarily to Linus Pauling (e.g., Pauling 1940), although with many antecedents, this theory claimed that the behavior of a macromolecule is determined by its conformation, and that what mediates biological interactions is a precise lock-and-key fit between the shapes of molecules. In the 1940s, when no three-dimensional structure of a biological macromolecule had yet been determined, the conformational theory of specificity was speculative. The demonstration of its approximate truth for a wide variety of interactions, which came in the late 1950s and 1960s, was one of molecular biology's most significant triumphs. Just as "one gene–one enzyme" was the archetypal slogan of early molecular biology, "structure determines function" came to be the dominating principle of the field during its triumphant 1960s.

Meanwhile, back in 1944, in *What Is Life?*, Erwin Schrödinger had introduced a conceptual scheme that raised the possibility of a startlingly different source of specificity. Schrödinger asked how so tiny an object as the nucleus of a fertilized cell could contain all of the specifications (i.e., instructions) necessary for normal development of an adult organism. He stated that there existed in the nucleus some structure whose organization was interpreted as "an elaborate code-script," which he compared to the Morse code (Schrödinger 1944). Although he was willing to countenance codes in more than one dimension, even a linear code based on a 5-letter alphabet and words of up to 25 letters could generate more than 10^{17} patterns. Thus the arrangement of the units rather than their physical shape became the source of specificity in Schrödinger's model. In the postwar era, when many scientists who were initially trained in physics turned their attention to biology for the first time, *What Is Life?* was influential in setting the agenda of a new biology (Sarkar 1991).

The 1940s also saw an explosive growth of microbial genetics, starting with Luria and Delbrück's (1943) demonstration of spontaneous mutagenesis in bacteria; continuing, especially, with Avery and colleagues' (1944) demonstration of DNA as the likely genetic material; and culminating with Joshua Lederberg's discovery of recombination in bacteria (Lederberg and Tatum 1946a,b). "Transformation," "induction," and "transduction" were some of the new terms introduced to describe these phenomena (Ephrussi et al. 1953). In an attempt to navigate through this terminological morass, Ephrussi et al. (1953) suggested that the term "interbacterial information" replace them all. This was the first modern use of "information" in genetics. Ephrussi et al. (1953, p. 701) emphasized that the use of this term "does not necessarily imply the transfer of material substances, and [that they] recognize the possible future importance of cybernetics at the bacterial level."

Immediately after the publication of Ephrussi et al. (1953)—in fact, in the next issue of *Nature*—Watson and Crick (1953a) published the double helix model of DNA. The base pairing—A:T and C:G—that they proposed showed a possible way in which the specificities between the two helices could be involved in the formation of exact replicas. Moreover, in their second paper on the model (Watson and Crick 1953b, p. 964), they went on to use "information" explicitly, and defined it implicitly as what the "code" carried: "The phosphate-sugar backbone in our model is completely regular but any sequence of the pairs of bases can fit into the structure. It follows that in a long molecule many different permutations are possible, and it therefore seems likely that the precise sequence of bases is the code which carries the genetic information."

"Information" was finally defined explicitly by Crick in 1958, who identified it with the specification of a protein sequence. Crick's (1958) concern was the synthesis of proteins. There were three separate factors involved, he argued: "the flow of energy, the flow of matter, and the flow of information" (Crick 1958). The former two exhausted the physics and chemistry of the situation, whereas information was peculiar to biological systems. Crick (1958, p. 144) defined "information" with more care than ever before in this context: "By information I mean the specification of the amino acid sequence of the protein." He took it for granted that the genetic information was encoded in a DNA sequence. The physics and chemistry of folding of proteins, Crick hypothesized, were purely a result of its amino acid sequence. This is the well-known "sequence hypothesis" (which remains unproved because of the continued insolvability of the protein folding problem). Finally, he put this formalized notion of information to additional use:

The Central Dogma ... states that once "information" has passed into protein *it cannot get out again.* In more detail, the transfer of information from nucleic acid to nucleic acid, or from nucleic acid to protein may be possible, but transfer from protein to protein, or from protein to nucleic acid, is impossible. Information means here the *precise* determination of sequence, either of bases in the nucleic acid or on amino acid residues in the protein. (Crick 1958, p. 153; italics in the original)

This assumption about the one-way transfer of information did not arise from physical considerations. Rather, it was Crick's way to give a molecular characterization of neo-Darwinism. "[I]t can be argued," he explicitly observed, "that [the protein] sequences are the most delicate expression possible of the phenotype of an organism" (Crick 1958, p. 142). Therefore, the one-way transfer of information ensured that changes that occur initially at the phenotypic level cannot induce genotypic changes and be inherited.

Crick implicitly distinguished two different types of specificity: that of each DNA sequence for its complementary strand, as modulated through base pairing; and that of the relationship between DNA and protein. The latter was modulated by genetic information. This notion of information was combinatorial: all that was required was that the code perform its function from the sequence of bases in a DNA segment. Schrödinger's (1944) "arrangement" came to "encode information," resulting in a new theory of specificity, distinct from the conformational theory.

8.2 The Theory of the Genetic Code

Even before Crick's (1958) explicit identification of "information" with "specificity," the same idea was being used systematically. For instance, in 1955, during a symposium on enzymes, Mazia (1956) argued that the role of RNA was to carry "information" from the nuclear DNA to the cytoplasm for the synthesis of proteins. At the same conference, Spiegelman (1956) argued that the "informational complexity" required for the formation of proteins made RNA and DNA the only two plausible candidates for being templates for protein formation. Lederberg (1956) noted that "information" was what "specificity" was "called nowadays."

However, what fully embedded "information" into the conceptual framework of molecular biology was the idea of a genetic code, that is, the idea that the relationship between DNA and protein should be conceived of as one of coding. After the formulation of the double helix model in 1953, this idea entered molecular biology largely through the work of George Gamow and his collaborators, who attempted to deduce properties of the genetic code from plausible theoretical assumptions about information. (See Gamow et al. 1955 for a review, and Judson 1979 and Sarkar 1989, 1996, for critical discussions.) Broadly speaking, these assumptions were about the efficiency of information transfer and storage using DNA sequences. The basic problem was conceived of as that of determining, first, what sets of DNA nucleotides coded for individual amino acid residues and, second, whether these sets had any overlap when a long stretch of DNA coded for an amino acid residue sequence. Gamow formulated a variety of coding schemes from different sets of assumptions. By the late 1950s, these schemes were all shown to be incorrect.

The most interesting and, in some ways, influential of the theoretical attempts to establish the nature of the genetic code was the "comma-free" coding scheme of Crick et al. (1957), who introduced relatively sophisticated ideas about information storage and transmission. Their only experimental criterion for judging success was the "magic number," 20, of amino acid residues that occurred naturally in proteins and, therefore, must be coded for by DNA. They rejected overlapping codes on experimental grounds and argued that it was natural to restrict attention to triplet codes because doublet ones would allow only 16 coding units, whereas triplet ones, allowing 64, were clearly sufficient.

From their point of view, there were two problems to be solved. First, there was the problem of potential degeneracy—if 64 triplets code for only 20 residues, then in some cases several triplets would code for a single residue. Although there was no experimental ground to reject a degenerate code (such as the one that is now accepted), they believed that it was undesirable. Second, there was the problem of synchronization—if A, C, G, and T are the four nucleotide base types, is the sequence ACCGTAGT to be read as ACC, GTA, ..., as CCG, TAG, ..., or as CGT, AGT, ...? Their solution to both problems was ingenious. By attempting to solve only the synchronization problem, they also removed degeneracy and, incidentally, obtained the magic number, 20.

Their solution to the synchronization problem began with the assumption that only some triplets had "sense," that is, could code for residues. Of the 64 possible triplets, the 4 with one base type, such as AAA, had to be rejected immediately: otherwise a sequence such as AAAACGA could potentially be read ambiguously as AAA, ACG, ... or as AAA, CGA, ... This left 60 possibly meaningful triplets. These segregate into 20 sets of three, each set consisting of a triplet and its two cyclic permutations (e.g., ACG, CGA, and GAC). If the possibility of ambiguity is to be avoided, only one triplet from each set can be meaningful. For instance, if ACG and CGA were both meaningful, ACGACGT could potentially be read as ACG, ACG, ... or as CGA, CGT, ... Thus, at most 20 meaningful triplets were possible. Crick et al. (1957) went on to show that a solution with exactly 20 triplets was possible by explicitly constructing such a set. That solution was not unique. Crick et al. managed to find 288 different solutions, and there are in fact as many as 408 (Golomb 1962).

Crick et al. (1957) went on to discuss a possible physical interpretation of the code that is historically important because it suggested that intermediate molecular complexes are involved in the translation of a nucleic acid sequence to a polypeptide one. This suggestion was one of the first statements of the "adaptor hypothesis," the adaptors eventually being identified with transfer RNA (tRNA). However, these physical considerations did not form the basis for their coding scheme. That basis was provided by the two desiderata of solving the degeneracy and synchronization problems. There was no experimental ground for these; the only experimental constraint was the magic number of 20. The desiderata were based on two claims about the nature of biological information. The first was a claim of a certain kind of simplicity: a degenerate code is not as simple as a nondegenerate one. The second was an assumption about efficiency: if synchronization is not automatically determined by the nature of the code, then ambiguous translation can occur by a shift of the reading frame, resulting in errors. The comma-free code satisfied both desiderata.

But why settle for synchronization alone? In 1958, Max Delbrück pointed out that genetic information ultimately resides in double-stranded DNA rather than in only one strand (Golomb et al. 1958). Consequently, synchronization must simultaneously hold for both strands. The "dictionary," Delbrück insisted, would be better off if this additional constraint on synchronization were also satisfied. This constraint was a natural extension of comma-freedom for a single strand and was called "transposability." Once it is imposed, triplet codes can no longer code for the 20 amino acid residue types known to occur in proteins. Triplet codes can now code for at most 16 residue types because the "converse" of each meaningful triplet, as determined by base pairing, must, when read from right to left, also belong to the set of coding triplets. This rules out many of the meaningful triplets of the original comma-free codes. Attention, therefore, shifted to quadruplet codes, and in 1962 Samuel Golomb reported from computer searches that there can be at least 57 quadruplet codons.

Pursuing coding as a mathematical idea, Golomb (1962) introduced another scheme, that of "biorthogonal codes," based on the mathematical theory of Hadamard matrices, which had six nucleotides code for each amino acid residue. Its biological motivation remains mysterious. Golomb provided no reason for its introduction but was apparently sufficiently convinced of the biological potential of all of these new formal schemes to observe: "It will be interesting to see how much of the final solution [of the coding problem] will be proposed by the mathematicians before the experimentalists find it, and how much the experimenters will be ahead of the mathematicians" (Golomb 1962, p. 106). Subsequent developments, as it turned out, did not show much respect for the mathematicians.

8.3 Theory and Experiment

What Golomb was apparently unaware of is that by 1961 it had become clear that the code was triplet (Crick et al. 1961), showing that his speculations—and those of Delbrück—had, for all their analytic sophistication, little relevance to biology. That same year, the first codon was experimentally deciphered by Matthaei and Nirenberg (1961a,b), using a cell-free system and RNA sequences. They determined that UUU, a triplet not permitted to be meaningful by the comma-free code, codes for phenylananine. As this result was verified, and other results began to come in, it became clear that the genetic code was not remotely comma-free. It was, in fact, highly degenerate. Colinearity of the code was demonstrated in 1964 (Yanofsky et al. 1964), and by 1966 the entire genetic code was established (Woese 1963; Ycas 1969). Synchronization turned out to be controlled by a variety of mechanisms, none of which could reasonably have been predicted from a consideration of constraints on the flow or storage of information. Failures in synchronization, as exemplified by the expression of frame-shifted sequences, were still possible. What the mechanisms controlling synchronization usually did was to prevent frame-shifted polypeptide formation, although this did not become clear until the 1980s (Atkins et al. 1991).

What had gone wrong with the comma-free code is that none of the elegant properties that were imposed on the code from considerations about information are revered in the living organism. The attempts to decipher the code in the 1950s using these ideas were an unmitigated failure. Even the ideas that the code was fully synchronized and fully sequential eventually came to be modified through the discovery of frameshift mutations and noncoding regions of the genome. The only idea to be validated, besides the uniformity of codon length, was Schrödinger's original one—merely that of the existence of a genetic code. But even the simplest form of this idea comes with two attendant metaphors: that of information and, perhaps even more perniciously, that of DNA as language, with the code used to interpret DNA symbols into meaningful proteins. The latter metaphor deserves more systematic analysis than is possible here. It is particularly harmful because it disregards the fact that, ultimately, DNA is a molecule interacting with other molecules through a complex set of mechanisms. DNA is not just some text to be interpreted, and to regard it as such is an inaccurate simplification.

The theoretical schemes for the genetic code are particularly important because the success of any of them would have provided at least a rudimentary theory of information for molecular biology. However, despite their failure, the ideas of coding and information in molecular biology persisted. The reason for this is the unusual simplicity of prokaryotic genomes. Much of molecular biology, especially in the 1960s, was established through the study of only a single species, *Escherichia coli*, the genome of which is particularly well behaved. Every DNA segment has either a coding or a regulatory function. For coding regions, transcription results in a complementary mRNA molecule that, with no further modification, is translated at the ribosome. If, following Crick, the specificity of a DNA sequence is identified with the information encoded in it, then the three precepts stated at the beginning of this article become plausible.

However, Crick's definition of information remains quirky: for instance, longer segments of DNA, unless they encode more genes, cannot be said to carry more information than shorter ones. Regulatory sequences do not have any coding role: they cannot be said to contain information in any straightforward sense. Worse, because the code is so arbitrary (i.e., there are no theories that explain why a particular codon codes for its amino acid residue), the concept of information cannot be invoked in any explanatory role: there is no potential for explaining novel biological phenomena by appeals to some property of information. Nevertheless, these problems become insignificant once attention shifts to eukaryotic genetics.

8.4 The Unexpected Complexity of Eukaryotic Genetics

Monod (1971) is responsible for what should perhaps be called the "Central Myth" of molecular genetics: that what is true for *E. coli* is true for elephants. Were this correct, then the notion of information formulated by Crick (1958), and the framework of coding that emerged from it, would not only make the three precepts at the beginning of this article true, but would also make the linguistic view of genetics palatable. However, developments since the early 1970s have shown Monod's claim of absolute universality to be incorrect. When molecular biologists turned to eukaryotic genetics in the late 1960s, they were in for surprise after surprise, so much so that Watson et al. (1983) entitled a chapter on eukaryotic genetics, "The Unexpected Complexity of Eukaryotic Genes." Precisely because of its bewildering complexity, eukaryotic genetics has no simple precepts like those that apply to prokaryotic genetics. Four sets of developments show

how the simple information/coding picture inherited from *E. coli* begins to fall apart:

- The genetic code is not universal, although the amount of known variation is not great (see Fox 1987 for a review). At present, the most extensive variations have been found in mitochondrial DNA, in which, for instance, across all the major kingdoms UGA codes for tryptophan rather than causing translation to terminate as it does in the usual code. It can be argued that mitochondrial DNA is "special" because mitochondria probably arose as independent organisms that were subsequently incorporated into eukaryotic cells. However, in the nuclear DNA of at least four species of protozoa, UAA and UAG can code for glutamine rather than terminating translation. Moreover, in many species, UGA codes for amino acid residues that do not belong to the standard set of 20. In some viral DNA sequences UGA and UAG are sometimes, but not always, read through, that is, ignored both as termination signals and as codons (Fox 1987). Even in the same RNA sequence, these codons sometimes result in termination and are sometimes ignored. For example, the virus $Q\beta$ has a coat protein that is usually produced by having UGA read as a termination codon. However, 2 percent of the time it is ignored, resulting in a longer functional protein (Fox 1987).

- The discovery of frameshift mutations has destroyed any residual belief in a natural synchronization of the genetic code. The extent to which frame shifts are present in organisms is largely a matter of conjecture (Atkins et al. 1991). Sometimes, frame shifts at the DNA level are used to transcribe an RNA segment that is translated into a different protein than the standard one (Fox 1987).

- Not all DNA segments have a coding or regulatory function. In human genomes, as much as 95 percent of the DNA may have no function. Inside a segment that, as a whole, has a coding function, coding regions, called "exons," are interspersed with noncoding regions, called "introns." In almost all eukaryotes, the portions of RNA corresponding to the introns are spliced out after transcription. Moreover, alternative splicing (the production from the same transcript of different RNA segments, coding for different proteins) has also been found (Smith et al. 1989). Moreover, there are large segments of nonfunctional DNA between genes (i.e., between segments with known coding or regulatory roles). The existence of introns and other nonfunctional DNA segments makes it impossible to simply read off a DNA sequence and predict an amino acid sequence (even when all regulatory regions are known).

• Besides splicing, several types of mRNA editing are also now known (Cattaneo 1991). DNA segments producing transcripts that are subsequently so edited are called "cryptic genes." For instance, in mammalian intestinal cells, a certain C nucleotide in the mRNA for apolipoprotein becomes deaminated, converting it to a U and creating a stop codon. Deamination of C to U, and the reverse process (U → C amination), occur in several plant mitochondrial mRNA transcripts as well. Moreover, even more unusual behaviors have been observed with mitochondrial RNAs in which bases can be deleted or inserted. The latter, especially, leads to a situation that can be interpreted as the formation of proteins for which there are no genes. In an extreme case, in the human parasite *Trypanosoma brucei*, as many as 551 Us are inserted throughout the transcript coding for NADH dehydrogenase subunit 7, and 88 are deleted (Koslowski et al. 1990). In this case, the DNA segment encoding the primary transcript can hardly be considered a gene for NADH dehydrogenase subunit 7. By looking at the DNA sequence it would be impossible to predict beforehand that this was the protein that would eventually be produced. Moreover, in almost all eukaryotes bases are added as "tails" and "caps" to the RNA after transcription.

In the present context, RNA editing is the most interesting facet of eukaryotic genetics: it already shows how the first precept described at the outset of this article (that all information resides in the DNA sequences of the genome) cannot be universally true. Acceptance of the idea that not all information resides in DNA sequences implies the acceptance of the idea that not all information proceeds as a transfer from DNA to RNA to protein, at least through the conventional coding relationship. Then the Central Dogma becomes dubious.

It would be unreasonable to criticize molecular biologists for not predicting these complexities in the 1960s, long before there was any experimental evidence for them. Nevertheless, these complexities show that the information/coding picture inherited from *E. coli* should no longer be regarded as the conceptual core of molecular biology. In particular, one should wonder whether the idea of a genetic code captures something important about biological systems or whether it is simply a metaphor that has epiphenomenally emerged from the accident that both nucleic acids and proteins happen physically to be linear molecules.

8.5 Cybernetics and Information Theory

The argument that has been developed so far assumes that "information" should be basically understood as Crick (1958) construed it, that is, what is specified by a DNA sequence through the genetic code. The problems with that interpretation nevertheless leave open the possibility that there is some other interpretation of that term that will allow its recovery as an interesting theoretical concept of molecular biology. Historically, there have been two such interpretations, and these have to be disposed of to complete the argument of this chapter.

The first of these was already alluded to by Ephrussi et al.'s (1953) comment that understanding information transfer may involve exploring "cybernetics at the bacterial level." Cybernetics, from their and other similar points of view, might provide the theory for the use of information in molecular biology. However, nobody seems to be certain about what constitutes cybernetics (see Pierce 1962). The term "cybernetics" was popularized with messianic fervor by Norbert Wiener (1948). However, all that is clear from Wiener's work is that cybernetics is a theory of regulated and, especially, self-regulating systems. Regulation was posited to occur through "feedback," a concept that had entered biology long before the invention of cybernetics, but was co-opted into the cybernetic framework (Keller 1995; Sarkar 1996). Feedback provides the information for regulation. In cybernetics, the concept of information gets no more explicit than the indication that "information" is that which enables regulation.

The value of cybernetics in molecular biology is doubtful, although putative cybernetic interpretations of genetics began as early as 1950 (see Kalmus 1950). If the published record is taken as evidence, these interpretations had negligible impact during the 1950s and early 1960s, when the conceptual framework of molecular biology came to be established. They were given a new lease of life by Monod (1971) in *Chance and Necessity*, when he reinterpreted much of his earlier work, including the model of allosteric regulation of proteins and the operon model of bacterial cell regulation, as examples of cybernetic systems (Sarkar 1996).

Whatever plausibility this interpretation may have had in 1971, it falls apart, once again because of the unexpected complexity of the eukaryotic genome. Eukaryotic gene regulation is not well understood even today, but it is clear that no model similar to the operon can account for the regulation of eukaryotic genes (see Sarkar 1996 for details). Cybernetics appears to have been little more than a diversion in the development of molecular biology, but even if it is somehow reinstated (although it is hard to see how), its associated concept of information (that which enables regulation) cannot enable a recovery of the three precepts mentioned at the beginning of this article: "information" as "feedback" is hardly what resides in DNA, passes from DNA to RNA to protein, or, for that matter, can make the Central Dogma true.

A second alternative interpretation of information emerged from the mathematical theory of communication, which eventually came to be called information theory (Shannon 1948). In information theory, the amount of information is measured by the logarithm of the relative number of choices available during a communication process. Information connotes uncertainty; formally, its numerical value is determined by an entropy function that is similar to the usual entropy of statistical mechanics. In the 1950s, there were many attempts to apply this notion of information to molecular biology. Branson (1953) calculated the information content of polypeptide sequences using empirical frequencies of the various residues to calculate the uncertainty at each position of a sequence. In a similar manner, Linschitz (1953) attempted to calculate the information content of a bacterial cell. However, by 1956, even the staunchest proponent of using information theory, Quastler (1958, p. 399) conceded at least temporary defeat:

Information theory is very strong on the negative side, i.e. in demonstrating what cannot be done; on the positive side its application to the study of living things has not produced many results so far; it has not led to the discovery of new facts, nor has its application to known facts been tested in critical experiments. To date, a definitive judgment of the value of information theory in biology is not possible.

Sporadic attempts to apply information theory directly to molecular biology continue, but the results are less than exciting. For instance, a major result of Yockey's (1992) attempt to apply information theory to molecular biology is that polypeptides may not code for DNA sequences (which is Yockey's version of the Central Dogma). The basis for this "theorem" is the degeneracy of the genetic code: a given polypeptide sequence can be encoded by different DNA sequences. The conclusion is correct. What is mysterious is why information theory—or any abstract theoretical framework—has to be invoked to make so trivial a point. It is a trivial combinatorial fact that was known by Gamow, Crick, or anyone else who had ever thought about the relation between DNA and protein.

Recently, Thomas Schneider and his collaborators (starting with Schneider et al. 1986) have made promising use of information theory to find the most functionally relevant parts of long DNA sequences when these are all that are available. The basic idea, which goes back to Kimura (1961), is that functional portions of sequences are most likely to be conserved through natural selection. These will therefore have low information content (in Shannon's sense). Whether Schneider's methods will live up to their initial promise remains to be seen. Nevertheless, for conceptual reasons alone, this notion of "information" (i.e., Shannon information) is irrelevant in the present context. According to this notion, for DNA sequences the "information" content is a property of a set of sequences: the more varied a set, the greater the "information" content at individual positions of the DNA sequence. But "information" in this scheme is not actually what an individual DNA sequence contains, that is, not what would be decoded by the cellular organelles. Worse, what Kimura's (1961) argument suggests is that what should be regarded as biologically informative—functional sequences—are exactly those that have low "information" content.

8.6 Conclusions

Thus, neither cybernetics nor formal information theory can rescue the concept of information for molecular biology in such a way as to permit the recovery of the conventional picture of DNA sequences encoding information to be decoded using the genetic code. The natural conclusion is that the conventional picture should be abandoned. However, because the concepts of information and coding have been central to how molecular biology is currently understood, abandoning these concepts will have important consequences. There are at least five such consequences, and a little reflection shows that they are, in fact, desirable:

- If biological "information" is not DNA sequence alone, other features of an organism can also contain information. This is precisely what recent discoveries indicate. In particular, the developmental fate of a cell might be largely a result of features such as methylation patterns of DNA, which are not even ultimately determined only by DNA sequences (see Jablonka and Lamb 1995). These "epigenetic" patterns can be inherited for several cell generations. Different cells in the same organism, presumably with identical DNA sequences, can have different epigenetic patterns. These differences can result in cell specialization and differentiation, the usual prelude to developmental changes. Epigenetic specifications are also critical in generating differences in offspring (of sexually reproducing organisms), depending on whether an allele is inherited from the mother or the father. Epigenetic specifications are sometimes transmitted across organismic generations. If "information" is to have any plausible biological significance, it would be odd not to regard the transfer of these specifications as transfers of information. The conventional DNA-based concept of information precludes this possibility.

- The Central Dogma of molecular biology is false if it is construed as a universal biological law. However, a less grandiose claim, that protein sequences do not directly specify nucleic acid sequences in the way in which the latter specify the former, remains true. This humbler claim does not have the majestic rhetorical power of the Central Dogma, but does this retreat really undermine some putative insight that is enshrined in that dogma? The usual defense of the general biological importance of the Central Dogma is that it is a statement at the molecular level of the noninheritance of acquired characteristics (see, for example, Crick 1958 and Maynard Smith 1989). However, this interpretation of the Central Dogma is entirely unjustified. Acquired characteristics are occasionally inherited, although usually not (Jablonka and Lamb 1995; Landman 1991). What ensures that even those acquired characteristics that involve changes in DNA are not inherited in higher animals is the segregation of the germline from the soma. But plants have no germline, and the extent of its segregation in animals varies greatly across phyla (Buss 1987). Nevertheless, whatever the relation between nucleic acid and protein, that relation shows no such variability across phyla: ipso facto the Central Dogma, even if it were true, could not be either an explanation or an alternative synonymous statement of the alleged noninheritance of acquired characteristics. There is certainly something peculiar, and extremely interesting, about how DNA resists easy change across phyla. But this observation is something to be studied and understood, not something to be explained away on the basis of some alleged law about some incoherent notion of information.

• Many influential contemporary discussions of the origin of life have concentrated on the origin of information, in which information is construed simply to be nucleic acid sequences (e.g., Eigen 1992). Implicit in these discussions is the assumption that nucleic acid sequences ultimately encode all that is necessary for the genesis of living forms and, therefore, that a solution to the problem of the initial generation of these sequences will solve the problem of the origin of life. The move away from sequences would put these efforts in proper perspective: to explain the possible origin of persistent segments of DNA does not suffice as an explanation of the origin of living cells.

• The emphasis on DNA sequences that marks contemporary molecular biology is misplaced. Therefore, the sorts of arguments that were mustered to initiate the Human Genome Project (HGP—a crash program to sequence DNA blindly, that is, without first determining the functional roles of the segments to be sequenced) are less than compelling. This is not a new point. It has previously been made, on the basis of other considerations, by many critics of the HGP (Davis 1992; Lederberg 1993; Lewontin 1992; Sarkar 1992). These arguments, taken together, strongly suggest that the HGP should be limited to the mapping of all known genetic loci to specific positions on chromosomes and the sequencing of only those segments that are found to have some functional interest. There is little scientific rationale for the blind sequencing of DNA, and the shift of scarce resources to it is unjustified. Human and other genome sequences will ultimately be sequenced, with or without the HGP, but should such sequencing proceed at a normal pace, not only would such a shift of resources not occur but there would be more time to prepare for the well-known social and ethical problems that the HGP raises (see, for example, Holtzman 1989).

• Abandoning the coding metaphor will also do much to liberate biology from the unfortunate linguistic metaphor of an organism's (or a cell's) DNA sequence being a message in some language to be decoded. Despite the immense popularity of this metaphor (see, for example, Wills 1991 and Pollack 1994), at the technical level the linguistic metaphor is at best only as helpful in understanding biology as the concept of coding. The complexities of eukaryotic genetics show that the code is of only limited use in the transition from DNA to an organism's biology. Given a DNA sequence, simply to read off an amino acid sequence requires that it be known whether any nonstandard coding is being used, what reading frame is to be used, that all gene–nongene and intron–exon boundaries are known, and what kinds of RNA editing will take place. Even at the metaphorical level, it is unlikely that these complexities can all be treated as questions of language: after all, natural languages do not contain large segments of meaningless signs interspersed with occasional bits of meaningful symbols. Of course, even with an amino acid sequence, biology has barely begun: one then faces the problem of going to higher levels of organization, and in the absence of a solution to the protein folding problem there is little prospect for doing that if one really starts from a DNA "text." In any case, the sterility of the informational picture of molecular biology is a much-needed reminder that DNA is, ultimately, a molecule and not a language.

Acknowledgments

Parts of this article also appear in Sarkar (1996), which provides a more detailed treatment of many of the issues discussed here. Thanks are due to Angela Creager, Larry Holmes, Manfred Laubichler, Lily Kay, Evelyn Fox Keller, Joshua Lederberg, Richard Lewontin, William C. Wimsatt, and two anonymous referees for extensive discussions and comments on an earlier version of this article. Work on this article was partly funded by a fellowship from the Dibner Institute at the Massachusetts Institute of Technology.

References

Atkins, J. F., Weiss, R. B., Thompson, S., Gesteland, R. F. 1991. Towards a genetic dissection of the basis of triplet decoding, and its natural subversion: programmed reading frame shifts and hops. *Annual Review of Genetics* 25: 201–228.

Avery, O. T., MacLeod, C. M., McCarty, M. 1944. Studies on the chemical nature of the substance inducing transformation of pneumococcal types: induction of transformation by a deoxyribonucleic acid fraction isolated from pneumococcus III. *Journal of Experimental Medicine* 79: 137–157.

Beadle, G. W., Tatum, E. 1941. Genetic control of biochemical reactions in *Neurospora*. Proceedings of the National Academy of Sciences of the United States of America 27: 499–506.

Branson, H. R. 1953. A definition of information from the thermodynamics of irreversible processes. Pages 25–40 in Quastler, H., ed., *Essays on the use of information theory in biology*. Urbana (IL): University of Illinois Press.

Buss, L. 1987. *The evolution of individuality*. Princeton (NJ): Princeton University Press.

Cattaneo, R. 1991. Different types of messenger RNA editing. *Annual Review of Genetics* 25: 71–88.

Crick, F. H. C. 1958. On protein synthesis. *Symposium of the Society for Experimental Biology* 12: 138–163.

Crick, F. H. C., Griffith, J. S., Orgel, L. E. 1957. Codes without commas. *Proceedings of the National Academy of Sciences of the United States of America* 43: 416–421.

Crick, F. H. C., Barnett, L., Brenner, S., Watts-Tobin, R. J. 1961. General nature of the genetic code for proteins. *Nature* 192: 1227–1232.

Davis, B. D. 1992. Sequencing the human genome: a faded goal. *Bulletin of the New York Academy of Medicine* 68: 115–125.

Eigen, M. 1992. *Steps towards life: a perspective on evolution*. Oxford (UK): Oxford University Press.

Ephrussi, B., Leopold, U., Watson, J. D., Weigle, J. J. 1953. Terminology in bacterial genetics. *Nature* 171: 701.

Fox, T. D. 1987. Natural variation in the genetic code. *Annual Review of Genetics* 21: 67–91.

Gamow, G., Rich, A., Ycas, M. 1955. The problem of information transfer from the nucleic acids to proteins. *Advances in Biological and Medical Physics* 4: 23–68.

Golomb, S. W. 1962. Efficient coding for the desoxyribonucleic acid channel. *Proceedings of the Symposium for Applied Mathematics* 14: 87–100.

Golomb, S. W., Welch, L. R., Delbrück, M. 1958. Construction and properties of comma-free codes. *Biologiske Meddelelser Kongelige Danske Videnskabernes Selskab* 23(9): 1–34.

Holtzman, N. 1989. *Proceed with caution*. Baltimore (MD): The Johns Hopkins University Press.

Jablonka, E., Lamb, M. J. 1995. *Epigenetic inheritance and evolution: the Lamarckian dimension*. Oxford (UK): Oxford University Press.

Judson, H. F. 1979. *The eighth day of creation.* New York: Simon and Schuster.

Kalmus, H. 1950. A cybernetical aspect of genetics. *Journal of Heredity* 41: 19–22.

Keller, E. F. 1995. *Refiguring life: metaphors of twentieth century biology.* New York: Columbia University Press.

Kimura, M. 1961. Natural selection as a process of accumulating genetic information in adaptive evolution. *Genetical Research* 2: 127–140.

Koslowski, D. J., Bhat, G. J., Preollaz, A. L., Feagin, J. E., Stuart, K. 1990. The MURF3 gene of *T. brucei* contains multiple domains of extensive editing and is homologous to a subunit of NADH dehydrogenase. *Cell* 62: 901–911.

Landman, O. E. 1991. The inheritance of acquired characteristics. *Annual Review of Genetics* 25: 1–20.

Landsteiner, K. 1936. *The specificity of serological reactions.* Springfield (IL): C. C. Thomas.

Lederberg, J. 1956. Comments on the gene–enzyme relationship. Pages 161–169 in Gaebler, O. H., ed., *Enzymes: units of biological structure and function.* New York: Academic Publishers.

———. 1993. What the double helix has meant for basic biomedical science: a personal commentary. *Journal of the American Medical Association* 269: 1981–1985.

Lederberg, J., Tatum, E. L. 1946a. Gene recombination in *Escherichia coli. Nature* 158: 558.

———. 1946b. Novel genotypes in mixed cultures of biochemical mutants of bacteria. *Cold Spring Harbor Symposia on Quantitative Biology* 11: 113–114.

Lewontin, R. C. 1992. *Biology as ideology: the doctrine of DNA.* New York: HarperPerennial.

Linschitz, H. 1953. The information content of a bacterial cell. Pages 251–262 in Quastler H., ed., *Essays on the use of information theory biology.* Urbana (IL): University of Illinois Press.

Luria, S. E., Delbrück, M. 1943. Mutations of bacteria from virus sensitivity to virus resistance. *Genetics* 28: 491–511.

Matthaei, J. H., Nirenberg, M. W. 1961a. Characterization and stability of DNAse-sensitive protein synthesis in *E. coli* extracts. *Proceedings of the National Academy of Sciences of the United States of America* 47: 1580–1588.

———. 1961b. The dependence of cell-free protein synthesis in *E. coli* upon naturally occurring or synthetic polyribonucleotides. *Proceedings of the National Academy of Sciences of the United States of America* 47: 1588–1594.

Maynard Smith, J. 1989. *Evolutionary genetics*. Oxford (UK): Oxford University Press.

Mazia, D. 1956. Nuclear products and nuclear reproduction. Pages 261–278 in Gaebler, O. H., ed., *Enzymes: units of biological structure and function*. New York: Academic Publishers.

Monod, J. 1971. *Chance and necessity: an essay on the natural philosophy of modern biology*. New York: Knopf.

Pauling, L. 1940. A theory of the structure and process of formation of antibodies. *Journal of the American Chemical Society* 62: 2643–2657.

Pierce, J. R. 1962. *Symbols, signals and noise*. New York: Harper and Brothers.

Pollack, R. 1994. *Signs of life: the language and meanings of DNA*. Boston: Houghton Mifflin.

Quastler, H., ed. 1958. The status of information theory in biology: a round-table discussion. Pages 399–402 in Yockey HP, ed., *Symposium on information theory in biology*. New York: Pergamon Press.

Sarkar, S. 1989. Reductionism and molecular biology: a reappraisal. [Ph.D. dissertation.] Department of Philosophy, University of Chicago, Chicago, IL.

———. 1991. What is life? Revisited. *BioScience* 41: 631–634.

———. 1992. Para qué sirve el proyecto Genoma Humano. *La Jornade Semanal* 180: 29–39.

———. 1996. Biological information: a skeptical look at some central dogmas of molecular biology. Pages 187–231 in Sarkar S, ed., *The philosophy and history of molecular biology: new perspectives*. Dordrecht (the Netherlands): Kluwer. (Chapter 9 of this volume.)

Schneider, T. D., Stormo, G. D., Gold, L., Ehrenfeucht, A. 1986. Information content of binding sites on nucleotide sequences. *Journal of Molecular Biology* 188: 415–431.

Schrödinger, E. 1944. *What is life? The physical aspect of the living cell*. Cambridge (UK): Cambridge University Press.

Shannon, C. E. 1948. A mathematical theory of communication. *Bell System Technical Journal* 27: 379–423, 623–656.

Smith, C. W., Patton, J. G., Nadal-Ginard, B. 1989. Alternative splicing in the control of gene expression. *Annual Review of Genetics* 23: 527–577.

Spiegelman, S. 1956. On the nature of the enzyme-formation system. Pages 67–92 in Gaebler, O. H., ed., *Enzymes: units of biological structure and function*. New York: Academic Publishers.

Timoféeff-Ressovsky, H. A., Timoféeff-Ressovsky, N. W. 1926. Über das phänotypische manifestieren des genotyps. II. Über idiosomatische variationsgruppen bei Drosophila funebris. *Roux Archiv für Entwicklungsmechanik der Organismen* 108: 146–170.

Watson, J. D., Crick, F. H. C. 1953a. Molecular structure of nucleic acids—a structure for deoxyribose nucleic acid. *Nature* 171: 737–738.

———. 1953b. Genetical implications of the structure of deoxyribonucleic acid. *Nature* 171: 964–967.

Watson, J. D., Tooze, J., Kurtz, D. T. 1983. *Recombinant DNA: a short course*. New York: W. H. Freeman and Co.

Wiener, N. 1948. *Cybernetics*. Cambridge (MA): MIT Press.

Wills, C. 1991. *Exons, introns, and talking genes: the science behind the human genome project*. New York: Basic Books.

Woese, C. R. 1963. The genetic code—1963. *ICSU Review of World Science* 5: 210–252.

Yanofsky, C., Carlton, B. C., Guest, J. R., Helsinki, D. R., Henning, U. 1964. On the colinearity of gene structure and protein structure. *Proceedings of the National Academy of Sciences of the United States of America* 51: 266–272.

Ycas, M. 1969. *The biological code*. Amsterdam (the Netherlands): North-Holland.

Yockey, H. P. 1992. *Information theory and molecular biology*. Cambridge (UK): Cambridge University Press.

9 Biological Information: A Skeptical Look at Some Central Dogmas of Molecular Biology

9.1 Introduction

Biologists would probably be bored rather than surprised if the most important part of the conceptual structure of contemporary molecular biology is encapsulated in the following three precepts; (i) all hereditary information resides in the DNA sequence of organisms; (ii) this information is transferred from DNA to RNA through the process of transcription, and from RNA to protein through translation; (iii) this information is never transferred from protein to nucleic acid sequence. All these claims are so universally accepted that they usually find their way into introductory biology texts and may well be taken to form the theoretical core of molecular biology (to the extent—and this is a matter of controversy—that molecular biology has any theory).

Most biologists would probably find the following two additional, and somewhat more interpretive, claims only very slightly more controversial: (iv) precept (iii), which is usually called the "central dogma of molecular biology" is an explication, at the molecular level, of the well-known biological fact that acquired characteristics cannot be inherited; and (v) most, if not all, of the behavior of organisms is ultimately determined by their DNA sequences. These last two claims have long had influential critics, and though what will be said here will argue explicitly against claim (iv), and implicitly against claim (v), I will largely leave them aside for some other occasion.

Meanwhile, what I will try to do here is suggest that the first three precepts (i–iii) are at best misleading and more likely, simply vacuous in the sense that they do not perform any significant explanatory or predictive role at all. The reason for this is that there is no clear technical notion of "information" in molecular biology. It is little more than a metaphor that masquerades as a theoretical concept and, as I shall argue in detail, leads to a misleading picture of the nature of possible explanations in molecular biology. I will do so by describing how "information" came to be introduced in molecular biology, look at three different ways in which it has been construed, examine the biological theories in which these construals are supposed to be embedded, and argue that these theories are either not theories about "information" as it is customarily used in molecular biology (as for instance in the precepts [i], [ii] and [iii] above), or are of little predictive or explanatory value.[1] I will also point out that this failure of information-based reasoning is philosophically interesting because had it not been so, that is, had "information" managed to play a significant explanatory role, then that would have provided a striking example of nonreductionist explanation in biology.

Since several of these arguments will hinge on what constitutes a "significant explanation," it should be incumbent upon me to indicate what I mean by that term. I will not attempt to provide an explication of "explanation" here. Suffice it to note that there is no philosophical consensus at present about the nature of scientific explanation and there is no good reason even to believe that a single explication will capture the various modes of explanation—statistical or deductive, historical or predictive, functional or mechanistic, and so on—that have come into vogue. All I will assume are two "adequacy criteria," one whose violation would preclude a purported explanation from being an *explanation* and another whose violation would preclude an otherwise adequate explanation from being *significant*. Thus, these criteria are intended as necessary but not sufficient conditions, and I am not even assuming that they constitute all such necessary conditions that may be formulated. Since I will be arguing that these criteria will be violated in certain putative significant explanations in molecular biology that I shall analyze, it will suffice for my present purposes not to pursue "significant explanation" any further.

The first of the two criteria is easy enough to formulate: the factors invoked in any explanation must help codify a body of knowledge so that it may be viewed as conforming to some general pattern. The intuition behind this criterion is that new explanations, whatever else they do, must at least add some new structure into the framework of what we already understand. If an explanation is anything like a deductive-nomological one, this criterion would be easily satisfied provided that the universal statement assumed in such an explanation is truly nomological. If an explanation is statistical, the relevant statistical factors must have some broad applicability. Otherwise these factors cannot aid the process of codification that I am assuming to be a necessary feature of an explanation. The emphasis, here, is on that *substantive* issue rather than on the formal structure of the explanation. The point, here, is that no matter how "explanation" is formally explicated, this adequacy criterion must be respected.

The second criterion is also simple: a significant explanation must answer new questions that are recognized to be important. Once again, this is a substantive rather than a formal criterion. Moreover, it is context-dependent. The stage of scientific enquiry determines what is important. The questions that were important to molecular biology in the 1950s are not those that are important now. However, loss of significance of a pattern of explanation is not always simply a function of age: Newtonian or Darwinian explanations are often significant even today. Nevertheless, should the putative scope of a pattern of explanation become circumscribed to smaller and smaller domains over time, its significance is likely to decrease along the way. I will argue, later, that this is precisely what happened to information-based explanations in molecular biology, though this kind of development is by no means the only mechanism for the loss of significance.

These two criteria are undoubtedly very weak. They do not exclude the possibility that there might be more than one explanation, or even more than one significant explanation, of the same phenomenon. They do not demand physicalism, that is, that explanations ultimately be based on physical principles.[2] They do not demand prediction. Nevertheless, I will argue, that explanations in contemporary biology which invoke "information" fail to meet even these criteria.

Similarly, since I will be making claims about reductionist explanations, it is also incumbent upon me to explain what I mean by "reduction" especially since this concept has been construed and explicated in a wide variety of ways by philosophers. I will attempt to do so succinctly but fully in this paper, but not here. Rather, I will take it up when I get to that stage of my argument (section 9.6). I will there distinguish five different senses of "reduction," all of which are sometimes relevant in scientific contexts, and argue that whereas a large and interesting class of explanations in molecular biology satisfy the strictures of the strongest of those senses, those that would invoke "information" do not.

9.2 Background: Information Enters Molecular Biology

Let me turn to the introduction of "information" into molecular biology. That term apparently had not been used until 1953, though the concept was certainly implicitly used in earlier discussions, particularly in Schrödinger's (1944) *What Is Life?* By 1957 it had become part of the standard conceptual apparatus of molecular biologists. Arguably—though I will not pursue this argument here because my purpose is to criticize current uses of "information" rather than to justify its initial introduction—the explicit introduction and systematic use of that concept was the most important part of the reconceptualization involved in the emergence of molecular biology as a conceptual enterprise that was identifiably distinct from its immediate ancestors, notably biochemistry.

This is not the place to attempt to give an adequate history of these conceptual changes, though, given the importance of the "molecularization of biology," it is particularly unfortunate that we do not have an adequate sketch of these developments.[3] I will simply list, here, the steps that are most important for the elaboration of the argument of this paper. I hope that this will at least indicate the conceptual background against which the post-1953 developments took place:

(i) During the first three decades of this century it became clear that the molecular interactions which occurred in living organisms were highly "specific" in the sense that particular molecules interacted with exactly one, or at most a very few, reagents. Enzymes acted specifically on their substrates. Living organisms produced antibodies that were highly specific not only to naturally occurring antigens but also to artificial antigens to which neither they nor their ancestors had ever been exposed (see, e.g., the comprehensive review in Landsteiner (1936)). From the mid-1920s, even the action of genes began to be described using "specificity" (Timoféeff-Ressovsky and Timoféeff-Ressovsky, 1926). In genetics, the ultimate exemplar of specificity became the gene–enzyme relationship in the 1940s: "one gene–one enzyme" was perhaps the most important organizing hypothesis of early molecular biology.

(ii) By the end of the 1930s, a highly successful theory of specificity, one which remains central to molecular biology today, had also emerged. Due primarily to Pauling (e.g., 1940) and his collaborators, though with many antecedents, this theory claimed (1) that the behavior of a biological macromolecule was determined by its conformation, and (2) what mediated biological interactions was a precise "lock-and-key" fit between the shapes of the molecules. Implicit in the second assumption is the rather striking idea that biological interactions are mediated by very weak interactions such as hydrophobic bonds rather than the stronger covalent and ionic interactions which are the staples of inorganic reactions. In the 1940s, when the three-dimensional structure of not even a single protein had been experimentally determined, the conformational theory of specificity was still speculative. The demonstration that it was at least approximately true for a wide variety of biological macromolecules in the late 1950s and 1960s has been one of molecular biology's most significant triumphs.

(iii) In 1944, meanwhile, in *What Is Life?*, Schrödinger introduced a conceptual scheme that raised the possibility of a startlingly different source of specificity. Schrödinger asked how so tiny an object as the nucleus of a fertilized cell could contain all the specifications necessary for the normal development of an individual organism. His answer was "an elaborate code-script" which he compared to the Morse code (1944, p. 61). Though he was willing to countenance codes in more than one dimension, even a linear code based on a 5-letter alphabet and up to 25-letter words could generate over 10^{17} patterns. Thus the arrangement of the units rather than their physical shape became the source of specificity in Schrödinger's model. In the postwar era, *What Is Life?* was remarkably influential in orienting a generation of physically minded researchers to biology (though the extent of its influence is usually overstated [see Olby 1971; Yoxen 1979]).

(iv) The 1940s also saw significant growth in our knowledge of microbial genetics, starting with Luria and Delbrück's (1943) demonstration of spontaneous mutagenesis in bacteria and continuing, especially, with Avery et al.'s (1944) demonstration that DNA is the likely genetic material and culminating, in a sense, with Lederberg's discovery of genetic recombination in bacteria (Lederberg and Tatum 1946a,b). "Transformation" and "transduction" were just some of the new terms introduced to describe these phenomena (Ephrussi et al. 1953). In an attempt to navigate through this terminological morass, Ephrussi et al. suggested that "the term 'interbacterial information' ... replace those above" (1953, p. 701). It was the first modern use of "information" in genetics, and they went on to emphasize that it "does not necessarily imply the transfer of material substances, and recognize the possible future importance of cybernetics at the bacterial level" (p. 701).

(v) Immediately afterwards—in fact, in the next issue of *Nature*—came the double helix model of DNA (Watson and Crick 1953a). It showed exactly how DNA was a linear molecule. By base pairing—A:T and C:G—it also showed a possible way in which the specificities between the two helices could be involved in the formation of exact replicas. Moreover, in their second paper on the model, Watson and Crick (1953b) went on to use "information" explicitly, and implicitly defined it as what the "code" carried:

> The phosphate-sugar backbone in our model is completely regular but any sequence of the pairs of bases can fit into the structure. It follows that in a long molecule many different permutations are possible, and it therefore seems likely that the precise sequence of bases is the code which carries the genetic information. (1953b, p. 964)

"Information," however, would not be explicitly defined for another five years, until Crick (1958) identified it with the specification of a protein sequence. Crick thus clearly distinguished two different types of specificity: (i) the specificity of each DNA sequence for its complementary strand, as modulated through base-pairing; and (ii) the specificity of the relation between DNA and protein. The latter was modulated by "genetic information." This notion of information was combinatorial since, as will be described below, by this point, all that was required of the code for it to perform its function was a precise sequence of bases. Since Crick also assumed that the relation between DNA and protein strands was sequential, the amount of information carried by a DNA sequence was directly proportional to its length. Schrödinger's (1944) "arrangement" thus came to encode "information" and a new theory of specificity, distinct from Pauling's conformational theory, resulted.[4]

9.3 Codes, Templates, and the Central Dogma

What molecular biologists customarily mean by "information" emerged from these developments.[5] The route, however, was somewhat circuitous or, at least, it involved what in retrospect appears to be a very curious digression. That digression was the work of Gamow who was the first to identify a "coding problem." Slowly, but ultimately unequivocally, he proceeded to separate its formal (and, with hindsight, "informational") part

Central Dogmas of Molecular Biology 211

from its physical basis and, perhaps most important, enticed actual molecular biologists, including Delbrück and Crick, to play the game of solving the coding problem *formally* using only very general facts about the DNA–protein relationship as experimental constraints. Gamow's program was an unmitigated failure, if success is judged by the satisfaction of explicit goals. Nevertheless, in it lies the origin of the standard information-based picture of contemporary molecular biology. I will outline—and criticize—this program next. Other possibilities for "information" will be taken up in subsequent sections.

9.3.1 Template Codes

As just noted, attempts to crack what came to be called the "genetic code" were systematically initiated by Gamow (1954a,b) who developed several detailed (and hinted at many more) general models of the code in the 1950s. Though he explicitly envisioned the formation of a linear amino acid residue sequence of a protein from a DNA sequence as a process of "translation," Gamow's initial attempts to decipher the genetic code were based on the standard stereochemical model of specificity. In Gamow's first scheme, an amino acid residue fit exactly into the diamond-shaped "hole" formed by four DNA bases, two on each helix, with complementary bases forming the lateral diagonal of the diamond. The resultant code was triplet since only three independent bases specified the residue: the two on the top and bottom of the diamond and either one of the complementary base pair along the middle. By an ingenious argument, which already went beyond stereochemistry, Gamow showed that there were only 20 possible diamonds, provided that lateral symmetry (i.e., parity) was irrelevant.[6]

This experimental constraint that there are exactly 20 amino acid residue types, rapidly became what all early models of the code tried to predict. In 1954, Gamow's list of 20 residues that were universal to proteins was faulty, though Crick and Watson soon guessed what has since become the canonical list. Moreover, Gamow (1954b) was willing to countenance an additional five residues that, he thought, also naturally occurred in proteins. His model for the code could easily be extended to these (and, if necessary, up to 32 residue types) provided that the parity of the middle pair was made relevant. Gamow's stereochemical assumptions were unjustifiable as Crick, Watson and others soon realized. Somewhat strangely,

Gamow never constructed a physical model to test the putative lock-and-key fit between the residues and the DNA double helix though he routinely referred to such a strategy for testing his model. Moreover, he seems to have been unaware of the possibilities that RNA mediated the translation of DNA to protein, and that protein synthesis took place in the cytoplasm, not at the DNA, both facts which were gradually established during that period.

Nevertheless, even after casting stereochemistry temporarily aside, there were additional empirical problems with the diamond code. As a code, it was overlapping, that is, the middle and top vertices of one diamond formed the lower and middle vertices of the next.[7] This property of the code puts restrictions on the possible adjacencies of different diamonds. For example, six of the diamonds could never follow themselves. Twelve could occur at most in repeats of two, while only the remaining two could repeat themselves indefinitely. These restrictions provided quite stringent experimental tests for the code, especially as the amino acid residue sequence of insulin was beginning to become available from the sequencing work of Sanger and his associates (Sanger and Tuppy 1951a,b; Sanger and Thompson 1953a,b). In 1954 Gamow had two sequences available, one 21, and the other 30 residues long. He proceeded to attempt to attribute particular diamonds to residues and immediately found out that the adjacency restrictions on the diamonds were inconsistent with the known sequences.

Gamow (1954b) tried to resolve this problem with *ad hoc* suggestions. Perhaps the adjacency restrictions on the diamonds could occasionally be violated. Perhaps insulin was not a good test since, according to Gamow, it was not a "hereditary protein," that is, one that is coded for by an inherited gene. The first of these possibilities could be tested on the basis of Gamow's stereochemical mechanism, using an atomic model kit but, once again, Gamow made no effort in that direction. That would have doomed the diamond code right from the start. Instead, he turned to statistical studies, essentially abandoning the diamond code. Gamow and Metropolis (1954) tried to correlate residue pairs known from sequencing to those that were predicted from various coding schemes. Their initial analyses found no evidence of adjacency restrictions.

By 1955, Gamow's enthusiasm for the coding problem had sparked efforts towards its solution by many others. By this point Gamow had clearly distinguished the abstract coding problem, "that of translating a four letter code to a twenty letter code" (Gamow et al. 1995, p. 24), from that of finding the mechanism of translation. The latter problem, he continued to insist, had to be solved for a full solution of the coding problem. In practice, however, Gamow and his collaborators altogether ignored the latter problem. In 1955, when Gamow, Rich and Ycas published a comprehensive review of attempts to decipher the code theoretically, it had become clear that the original diamond code was experimentally inadmissible. In its place Gamow et al. (1955) proposed a "triangular code" which came in two versions: compact and loose. Showing increased biological sophistication, this code was based on RNA rather than DNA as the template. RNA was assumed to be helical, which was not unreasonable at that time. Depending on the pitch of the helix, any three adjoining nucleotides on an RNA chain formed either an isosceles or an equilateral triangle, and any four formed two such triangles with two shared vertices. In the compact triangular code each of these triangles corresponded to an amino acid residue. In the loose triangular code, only the last nucleotide was shared as a vertex by two successive triangles. Thus, the compact triangular code was triplet with an overlap of two, whereas the loose was triplet with an overlap of one. Both presented adjacency restrictions. Those posed by the compact version were even stronger than those of the diamond code and could be promptly ruled out using the available amino acid residue sequences. The loose version presented weaker restrictions, but these were so weak that Gamow et al. gave up trying to correlate such codes with sequences. Instead, they turned to statistical considerations. Once again, their analyses and those by Gamow and Ycas (1955) found no such restrictions.

These later statistical arguments of Gamow were probably not fully convincing to those few who considered them. They were almost universally ignored primarily because Brenner (1957) soon demonstrated that the available amino acid residue sequences ruled out all overlapping codes. Brenner assumed that the code was triplet, that the overlap was two, and that the code could be degenerate. Since any two adjacent triplets shared two nucleotides, he pointed out, any given triplet can be preceded or succeeded by at most four different triplets. Let the former be called its "N-neighbors" and the latter its "C-neighbors." Now, assume that each triplet codes for one amino acid residue. Then, given a residue, for every different set of four of its N-neighbors and C-neighbors, one triplet must be assigned to it. Note that more than one triplet can be assigned to a residue in this fashion—the code was allowed to be degenerate. However, using the best-known amino acid residue sequences in 1957, Brenner found that at least 70 triplets were necessary, more than the 64 that were combinatorially possible from a triplet code. Already in 1957, between Gamow's statistics and Brenner's *reductio ad absurdum* demonstration, overlapping triplet codes were dead, and the possibility of deducing the code on formal principles alone must have seemed hopeless to many of its former proponents.

9.3.2   From Comma-free Codes to the Central Dogma

Crick et al. (1957) thought otherwise. They introduced somewhat more sophisticated ideas implicitly based on assumptions about the efficiency of information storage and transmission. In sharp contrast to Gamow, for them, the only problem to be solved was the "formal" coding problem. The desideratum for success was the ability to obtain the "magic number," 20. The stereochemical mechanism of coding was, no doubt interesting, but only peripheral to the formal problem of coding. Rejecting overlapping codes on experimental grounds, Crick et al. also chose to ignore partially overlapping ones as being improbable. Moreover, they argued that it was "natural" to restrict attention to triplet codes since doublet ones would only allow 16 coding units whereas triplet ones, allowing 64, were clearly sufficient.

From their point of view there were two problems that had to be resolved. The first was that of the potential degeneracy. If 64 triplets only coded for 20 residues, in some cases, several different triplets would have to code for a single residue. There was no experimental reason that precluded a degenerate code, but Crick et al. felt that it was undesirable. The second problem was that of synchronization: if A, C, G, and T represent the four nucleotide base types, is the sequence ACCGTAGT read as ACC, GTA, ... or CCG, TAG, ... or CGT, AGT, ... ? Their solution to both problems was ingenious. By explicitly attempting only to avoid the problem of synchronization, they also removed degeneracy and obtained the magic number, 20.

They assumed that only some triplets had "sense," that is, could code for residues, while others could not. Of the 64 possible triplets, the four with one base type, such as AAA, had to be immediately rejected: otherwise a sequence such as AAAACGA could potentially be ambiguously read as AAA, ACG, ... or AAA, CGA, ... This left 60 possibly meaningful triplets. These segregate into 20 sets of three, each set consisting of a triplet and its two cyclic permutations (e.g., ACG, CGA and GAC). Now, from each such set, only one triplet can be meaningful if the possibility of ambiguity is to be avoided. For instance, if ACG and CGA are both meaningful, ACGAGGT could potentially be read as ACG, ACG,... or CGA, GCT,... Thus, *at most* 20 meaningful triplets were possible. Crick et al. (1957) went on to show that a solution with *exactly* 20 triplets was possible by exhibiting such a set. That solution was not unique. They managed to find 288 different solutions; there can be as many as 408 (Golomb 1962).

They went on to discuss a possible physical interpretation of the code which is historically important because it suggested that intermediate molecular complexes mediated the translation of a nucleic acid sequence to a polypeptide one. It was one of the first statements of the "adaptor hypothesis," the adaptors eventually being identified with transfer RNA (tRNA). However, these physical considerations did not form the basis for their coding scheme. That basis was provided by the two desiderata: (i) that there should be no degeneracy; and (ii) the synchronization problem should be avoided altogether by the very nature of the code. There was no experimental justification for the desiderata; the only experimental constraint in their analysis was the "magic number" 20. Moreover, these desiderata were not based on any physical considerations. Instead, what was implicitly being used were two claims about the nature of biological information. One was a claim of a certain kind of *simplicity*: a degenerate code was not as simple as a nondegenerate one. The other was, at least approximately, an assumption about *efficiency*: if synchronization was not automatically determined by the nature of the code, ambiguous translation could occur by a shift in the position at which reading began, that is, from a shift in reading frame. Errors could occur and would have, somehow, to be corrected. The comma-free code solved these problems.

Crick et al. had truly freed the coding problem from Gamow's persistent stereochemical worries. But once stereochemistry was cast aside, what remained? In practice, Gamow and his collaborators had never actually built any physical models. They had simply claimed that they had stereochemistry in mind while the data they used—adjacency restrictions, the statistics of adjacencies—refuted their assumptions without ever being based on any physical (or chemical) argument. Once they had failed, in retrospect, it appears that they had no other theoretical framework to fall back upon. Brenner (1957), too, had only negative results to offer but had used the concept of "information" systematically. Crick et al. had not used "information" explicitly but, presumably because they were moving towards it, they had implicitly attributed properties such as simplicity and efficiency to the code. But simplicity and efficiency in doing what? Could it have been "simplicity and efficiency of specificity"? Perhaps, but only at the cost of radically altering the concept of specificity.

Instead, explicitly acknowledging an emergent conceptual framework, no longer centered on physical models of specificity, Crick (1958) turned to "information." His immediate concern was the synthesis of proteins. There were three separate factors involved, he argued: "the flow of energy, the flow of matter, and the flow of information" (p. 144). The former two exhausted anything that physics considered; "information" had finally been liberated for a life of its own. Crick defined it with more care than ever before in a biological context: "By information I mean the specification of the amino acid sequence of the protein" (p. 144). He took it for granted that the genetic information was encoded in a DNA sequence. The physics of the folding of a protein, Crick hypothesized, was taken care of by its amino acid sequence. This became the "sequence hypothesis." Finally, this formalized notion of information was put to additional use:

The Central Dogma
This states that once "information" has passed into protein *it cannot get out again*. In more detail, the transfer of information from nucleic acid to nucleic acid, or from nucleic acid to protein may be possible, but transfer from protein to protein, or from protein to nucleic acid is impossible. Information means here the *precise* determination of sequence, either of bases in the nucleic acid or on amino acid residues in the protein. (1958, p. 153; italics in the original)

This assumption about information transfer did not arise from physical considerations. They were Crick's way to give a molecular characterization of traditional neo-Darwinism: "it can be argued," he explicitly observed, "that [the protein] sequences are the most delicate expression possible of the phenotype of an organism" (p. 142). Presumably, the assumptions about simplicity and efficiency of information storage, transmission and retrieval that were implicit in the comma-free code were also based on a belief that they were obviously adaptive. However, this point was never made explicit.

Over the next few years, the implicit assumptions of Crick et al. (1957) were made explicit, and other plausible ones added, especially as mathe-

Central Dogmas of Molecular Biology 217

maticians began to get systematically involved in the attempts to decipher the code. In particular, why settle for synchronization alone? In 1958, Delbrück pointed out that, after all, genetic information ultimately resides in DNA which is double-stranded (Golomb et al. 1958). This led to an additional constraint: synchronization must hold for both strands, taking base pair complementarily into account. The "dictionary," Delbrück insisted, would be better off if this additional constraint were also satisfied—this was, as he quite correctly pointed out, a natural extension of the idea of comma-freedom. Delbrück's additional constraint came to be called "transposability." Once it was imposed, it turned out that triplet codes could no longer give 20 residue types; they gave no more than 16. Attention shifted to quadruplet codes and in 1962 Golomb reported, on the basis of computer searches, that there could be at least 57 viable coding units for transposable comma-free quadruplet codes. Twenty had disappeared as a magic number. It was now only a minimal lower constraint, but transposability was gained.

The coding problem, at this point, had not only forgotten physics, but had even abandoned attention to biological specifics in preference to formal arguments about the efficiency in storage and transmission of information. In that spirit, Golomb (1962), also introduced another scheme, that of "biorthogonal codes," based on Hadamard matrices, which had 6 nucleotides code for each amino acid residue. Its biological motivation, let alone basis, remained rather mysterious. But Golomb was sufficiently convinced of the biological potential of these new formal schemes to observe: "It will be interesting to see how much of the final solution [of the coding problem] will be proposed by the mathematicians before the experimentalists find it, and how much the experimenters will be ahead of the mathematicians" (1962, p. 100). Subsequent developments did not show much respect for the mathematicians.

9.3.3 The Experimental Denouement

What Golomb apparently was unaware of was that, by 1961, it had become clear that the code was triplet (Crick et al. 1961) showing that his speculations—and those of Delbrück—for all their analytic sophistication, had little relevance for biology. Moreover, that same year, the first codon was experimentally deciphered by Matthaei and Nirenberg (1961a,b) using a cell-free system and RNA sequences. They determined that UUU, a triplet not permitted to be meaningful by the comma-free code, coded for phenylananine. As this result was verified, and other results began to come in, it became clear that the genetic code was not remotely comma-free. It was highly degenerate, but the degeneracy had no pattern that was consistent with any of the schemes predicted by Gamow and his collaborators (see Lanni 1964 for a comprehensive contemporary review). Colinearity of the code was demonstrated in 1964 (Yanofsky et al. 1964) and, by 1966, the entire genetic code was established (Woese 1967; Ycas 1969). Synchronization has turned out to be controlled by a variety of mechanisms, none based on considerations of constraints on the flow of information. Failures in synchronization, as exemplified by the expression of frame-shifted sequences, are possible—what these mechanisms routinely do is prevent this kind of polypeptide formation, though this did not become clear until the 1980s (Atkins et al. 1992). What had gone wrong with the comma-free code was that none of the elegant properties that were imposed on the code on the basis of "information" were revered in living organisms. The attempts to decipher the code in the 1950s were recognized to be an unmitigated failure. Even the ideas that the code was fully synchronized and fully sequential eventually came to be modified through the discovery of frameshift mutations and noncoding regions of the genome (see below). The only idea to survive, besides the uniformity of codon length, was Schrödinger's original one—merely that of the existence of a genetic code.

More recent, and presumably more sophisticated, attempts to use information-based reasoning to analyze the genetic code have fared little better (see Yockey 1992). But, perhaps, in the worst insult to comma-freedom to date, it has been shown that synchronizability does not require comma-freedom (see Neveln 1990). Where Crick et al. (1957) were too restrictive was in denying the possibility of any cyclic permutation of a triplet. While this was sufficient to ensure synchronizability, the only experimental constraint that motivated this assumption of comma-freedom was the requirement of exactly 20 meaningful words. However, synchronizability can be satisfied by merely requiring, for instance, that a sequence ACGCTATGC be meaningful and not all five possibilities—ACG, CTA, TGC, CGC and TAT—also be meaningful. A code which prevents such sets of five words from being simultaneously meaningful is said to have a "stagger bound" of five. A nonoverlapping code with triplet codons and four nucleotide types can have a stagger bound up to nine. These codes

Central Dogmas of Molecular Biology 219

automatically ensure synchronizability. Even if the added restriction of at least 20 meaningful words is imposed on them, a larger number of codes are possible.

Most of these codes are degenerate. However, degeneracy does not truly present any significant biological problem for a coding scheme: an organism with a degenerate code remains viable. In contrast, synchronization is biologically important: an organism could be faced with serious difficulties if its code was not synchronized *and* it did not have mechanisms preventing incorrect expression through, for instance, inadvertent shifts of reading frame. In fact, what these points really drive home is that the assumption made in the 1950s, that degeneracy is a problem, had no biological or physical basis. It was an assumption about what should be desired of "information." It reflected a hope, and probably an expectation, that there were principles other than the ordinary ones of physics and chemistry that could be used to explain biological behavior. What the failure of these attempts shows, more than anything else, is the empirical failure of that assumption.

9.3.4 Is the Code Irrelevant?

What went wrong? In 1962 Chargaff attempted to provide an answer. The fault, he argued, lay in the very introduction of the idea of "biological information." At some point between 1937 and 1944, that concept "raised its head and began to sport a multicolored beard which has become ever more luxurious despite numerous applications of Occam's razor" (1963, p. 163). Biological information might explain the highly specific relations between nucleic acid and protein but Chargaff was skeptical that it gave any insight into the equally specific relations between cells and multicellular communities. If there was no continuous "chain of information" from the lowest level to the highest, the argued, there was no justification in claiming that "DNA is the repository of biological information" (1963, p. 165). That argument was provocative though Chargaff's reasons for rejecting any genetic code at all were well known to be poor by 1963.

However, now that the failures of the early systematic attempts to codify the code are well-known, it is worth asking again, what does the use of the concept of a code carrying information still do? At the very most, it provides a succinct look-up table on the basis of which one can predict the sequence of the polypeptide chain that would be determined by a particularly DNA chain provided at least five conditions, discovered since 1966, are fulfilled.[8] Unfortunately, if prediction is the goal, these conditions are quite debilitating:

(i) It must be known whether the code to be used is the usual one or one of the variants. Though the nonuniversality of the code is well known, this still does not, at present appear to be a particularly severe constraint by itself because the amount of known variation is not great (see Fox 1987 for a review). At present the most extensive variations have been found in mitochondrial DNA in which, for instance, across all the major kingdoms UGA codes for tryptophan rather than terminate translation (as specified by the usual code). Mitochondrial DNA is special since mitochondria probably arose as independent organisms in biological prehistory that were subsequently symbiotically incorporated into prokaryotes to form eukaryotic cells. Moreover, if prediction is all that is at stake, one could simply limit the use of the standard look-up table to nuclear DNA in eukaryotes. However, in at least four species of protozoa, UAA and UAG can code for glutamine rather than terminate translation even for nuclear DNA. For many of the species that have been studied at this level so far, UGA is also known to code for amino acid residues that do not belong to the standard set of 20. In the case of some viral DNA sequences, moreover, the UGA and UAG codons are sometimes but not always "read through," that is, ignored as termination signals. This happens within the same system, that is, in the same RNA sequence these codons sometimes result in termination, and are sometimes read through.[9] The genetic code is not nearly as universal as was thought in the 1960s. However, at present, these exceptions are rare enough and though they present a problem, it is not a severe one for the use of the genetic code for predictive purposes in most contexts. The other conditions are far more serious.

(ii) The exact point of initiation of transcription must be known. In practice this, too, is not a severe constraint by itself even though the discovery of frameshift mutations has destroyed any residual belief in a "natural synchronization" of the genetic code and has made the prospect for DNA sequence-based protein sequence prediction even harder than it would have been otherwise. Though a few examples are known, the extent to which frame shifts are present in organisms is at present largely a matter of conjecture (Atkins et al. 1991). Sometimes, frame shifts are used to begin a segment of DNA that codes for a different protein (Fox 1987).

(iii) All intron–exon boundaries, that is, boundaries between coding and noncoding regions of a segment of DNA responsible for a single polypeptide must be known because, in almost all eukaryotes, after transcription, portions of the RNA, corresponding to the introns, are spliced out. Moreover, alternative splicing (the production of different RNA segments for translation from the same original transcript) has also been found (Smith et al. 1989). There is no reason to believe at present that the distinction between introns and exons can be made on the basis of sequence information alone though that is not implausible.[10] Certainly, at present, there are simply not enough data to suggest that enough is known about these boundaries to predict sequences of amino acid residues from DNA sequences alone. It is, of course, possible to use statistical regularities to attempt such prediction for many species. However, these techniques do not only use "information" in the sense of "coding"—they make systematic use of auxiliary facts about the relation between DNA and protein. When auxiliary facts of this sort are admitted, including codon usage, it is even possible to attempt to predict DNA sequences from protein sequences but all this has little to do with the "information" carried by the code.

(iv) What has just been said about intron–exon boundaries applies with no major qualification to the boundaries between segments of DNA that code for proteins (informally, the "genes") and those that do not. These boundaries are important because a huge fraction of the DNA in the genomes of higher eukaryotes apparently is noncoding. Moreover, in eukaryotes, after transcription, bases are added as "tails" and "caps" to the mRNA.

(v) Besides splicing, several types of mRNA "editing" are also now known, and it must be known whether any of these processes are occurring before the DNA sequence can be used to read off a protein sequence (Cattaneo 1991). DNA segments producing transcripts that are subsequently edited are sometimes called "cryptic genes."[11] Moreover, even more unusual behaviors have been observed with mitochondrial RNAs. Bases can be deleted and inserted. The latter, especially, leads to a situation which can be interpreted as the formation of proteins for which there are no genes.[12]

Of course, it would be unreasonable to criticize molecular biologists for not predicting these complexities in the 1960s, long before any experimental evidence for them was discovered. Nevertheless, the point that the code—or the "information" contained in it—is of little predictive value in novel contexts seems unassailable.

Notice that these problems arise in what is still a *static* context. There is no concern yet about *dynamics*: the temporal progress of gene expression, that is, its control and regulation. But, ultimately, no matter what the relationship between gene and protein is, what matters in biology, for predictive and explanatory purposes, is the temporal sequence of events which characterizes the behavior of an organism. The theories of coding and information, even if they were universal, and even if the speculative models of the 1950s had turned out to be successful, say nothing about the dynamics. This does not mean that they are incorrect. Rather, it means that they are, in a sense, incomplete but, as long as whatever dynamical account that is developed remains consistent with these theories, they can withstand the additional concern for dynamics. However, what this limitation does underscore is that, if the actual prediction of biological behavior through an interval of time is a serious desideratum to be considered, the code by itself, even if it did its part, would stand in need of supplementation by some theory of dynamics. Can considerations about information provide such a theory? The next section will look at one such attempt: the cybernetic models of gene regulation, such as the operon model, which have been around since the 1960s.

Meanwhile, restricting attention only to the static context, the code is at present of little value for predictive purposes unless much else about he nature of the system is also specified. But set prediction aside. Perhaps the concepts of code and information permit something weaker: explanation. At one level, indeed, they do permit some explanation. Given a polypeptide sequence and the DNA sequence responsible for it, one can quite often say that the DNA sequence gives rise to that polypeptide sequence *because of* the genetic code. The trouble is that this is not a particularly revealing explanation, and cannot be pursued too far, once again because of the conditions that have to be satisfied for a successful translation from the DNA to the protein sequences. But what if such exceptions did not exist? Then the explanatory value of the code would be of some significance (and, at least, one-way prediction, from DNA to protein sequences, would be possible). Moreover, note that in such explanations, even though the concept of information is not explicitly invoked, its use is implicit because the use of the concept of a code in this way presupposes an underlying stratum of linear, symbolic information. Unfortunately, because of the absence of any pattern to the code that can be deduced from a putative "theory of information," the explanatory role of information, even in that world without exceptions, would be very limited: the code would still not be of much help in codifying a body of knowledge (which was my adequacy condition for an explanation).

9.4 Cybernetics and Gene Regulation

Recall that when Ephrussi et al. (1953) initiated the use of "information" in molecular biology, they left open the possibility that it would require a cybernetic interpretation. In the excitement following the pursuit of the possibility that information resided in DNA sequence, that other alternative was ignored. However, since the standard sequence-based model has run into trouble, it is probably reasonable, and certainly does no harm, to explore the potential of that forgotten alternative, that is, the possibility that an entirely different notion of "information," derived in some way from cybernetics, can be used to recapture the value of that concept for molecular biology.

The trouble here is that nobody seems to be sure what "cybernetics" actually encompasses (see, e.g., Pierce 1962). The term "cybernetics" was popularized with almost messianic fervor by Wiener (e.g., 1948). All that is uncontested about the discipline is that it arose out of what was called "servomechanics" during World War II. This was the study of systems such as guided missiles whose trajectories had to be corrected in flight in order to hit a target. This problem arose because the initial data about the target could only be known to a certain level of precision, and subsequent measurements had to be incorporated into a missile's path in order to correct for any error arising from the initial imprecision. Wiener, in particular, developed mathematical procedures to make such corrections in a wide variety of contexts. This type of correction of behavior was termed "regulation" or even "self-regulation" when the correction procedure was incorporated into the system from the start. The additional data that were responsible for that correction were called "feedback." Presumably what feedback provided to the system was "information"—the role of information in cybernetics never became more explicit.

Under Wiener's influence, in the early 1950s, Haldane prepared a manuscript on the application of cybernetics to genetics. It remains unpublished.[13] However, in Haldane's department at University College London,

Kalmus (1950) attempted a somewhat speculative application of cybernetics to genetics. The gene was interpreted as a message and mutations as errors in the copying of that message. Kalmus argued that the perpetuation of genes was similar to "one of the devices in electronic calculating machines, which stores sequences of signals by perpetually recreating them." However, he rejected speculations about the linear arrangement of the gene and the possibility that the message is discrete. Moreover, since the action of a particular gene was sometimes felt in a distant cell, Kalmus argued that genes acted more like "broadcasting systems" than "wired telecommunication." That paper had no discernible influence on thinking about molecular genetics in the 1950s.

9.4.1 Operons

Meanwhile, cybernetic terminology, particularly "feedback" regulation, control and inhibition crept into biology primarily through their incorporation into discussions of the dynamics of enzyme systems, apparently beginning with Umbarger (1956) and Yates and Pardee (1956). However, the real impetus for explicitly invoking cybernetics within molecular biology came from the operon model for the regulation of bacterial gene expression. The detailed study of gene regulation had begun in the late 1940s with the attempts of Monod and his collaborators to elucidate the control of enzyme formation in bacterial cells (Schaffner 1974a,b). Since about 1900 it had been known that some organisms such as yeast displayed an ability to ferment a particular type of sugar only when exposed to it for a few hours. This kind of phenomenon came to be called "enzyme induction" in the early 1950s. In the mid-1940s Monod and his collaborators had already established that the type of enzyme induction involved in the lactose metabolism of *Escherichia coli* was under genetic control. One of the enzymes involved was β-galactosidase which hydrolyses β-galactosides such as lactose. Monod et al. (1951) showed that some inducers do not interact directly or indirectly with β-galactosidase. Moreover, some substrates of β-galactosidase did not act as inducers, as Lederberg (1951) had established. Monod et al. (1952) established that the production of β-galactosidase in the cell constituted *de novo* protein synthesis, rather than the conversion of a preexisting molecule. They named the associated gene "z." By 1956, Monod and his collaborators had also established that a second enzyme, which they named "galactoside permease," was also necessary for lactose to be metabolized (Monod 1956; Rickenberg et al. 1956). As the name indicates, this enzyme was hypothesized to help the passage of lactose through the cell membrane. It was assigned a gene "y."

Meanwhile, Lederberg had found mutant strains of *E. coli* that produced β-galactosidase even in the absence of an inducer: these were called "constitutive" mutants. Monod and his collaborators had shown that the production of that enzyme could be inhibited by molecules similar to the inducer. They interpreted these findings to mean that the formation of an active enzyme required the presence of an inducer, that the inducer was naturally present in constitutive mutants, and that inhibition resulted because molecules similar to the inducer acted as a surrogate for them but led to the formation of a variant, less effective enzyme. According to this scheme, part of the specification of an active enzyme was determined by the inducer. Lederberg (1956), however, provided a strikingly different interpretation of these findings. "Dr. Monod asked whether the inducer carries information needed for the specification of the enzyme," he observed.

> One permissible view holds that the enzyme, on its critical surface, is directly molded on the inducing substrate. The alternative, which I prefer, is that all the specifications are already inherent in the genetic constitution of the cell: the inducer signals a regulatory system to accelerate the synthesis of the corresponding enzyme protein. On this notion, substrate-induced or, better, substrate-regulated enzyme formation is an evolved adaptation to relieve the organism from always having to produce a full quota of its genetic potential of enzymes regardless of their immediate utility. (p. 161)

With the idea of a genetic regulatory system, connected with associated concepts of signals and control, a new conceptual apparatus for the discussion of gene expression began to be formed.

However, progress towards a model of the lactose system was relatively slow. Monod discovered that constitutive mutants were altered only at one locus which was labeled "i." In what became a famous experiment, Pardee et al. (1959) conjugated wild-type inducible male bacteria (denoted "i^+z^+") with constitutive females (denoted "i^-z^-") to produce partially diploid (or merozygotic) i^-z^-/i^+z^+ bacteria. Two results were obtained: (i) β-galactosidase started to be produced immediately but further production stopped after about two hours in the absence of an inducer; (ii) however, the new merozygotes became inducible after that period, indicating that i^+ was dominant over i^-. The kinetics of the expression suggested that the i

and z genes interacted through a cytoplasmic molecule ultimately produced by the i gene. The interpretation of the experiment which Pardee et al. preferred was that the i^+ gene produced a "repressor" molecule that prevented the z^+ gene from being expressed. The external inducer interacted with the repressor to disable it, thus permitting expression of the z gene. The constitutive mutants (i^-) simply did not produce the correct version of the repressor.

Jacob and Monod (1959) then distinguished between genes such as i which were "regulator genes" and ordinary genes such as z which were "structural genes." What was striking about the i locus, however, was that mutations in it affected not only the z locus, but also the y, and in the same fashion in which they affected the z. Jacob et al. ([1960] 1965) showed that the repressor bound to yet another locus, termed the "operator" locus and labeled "o," which was tightly linked to y and z and controlled their expression. They called the o-y-z system an "operon": "The hypothesis of the operator implies that between the classical gene ... and the entire chromosome, there exists an intermediate genetic organization. The latter would include the *units of coordinated expression* (*operons*), comprising an operator and the group of genes for structure which it coordinates" (1965, p. 200). In common usage, however, "operon" came to signify the entire system, including the regulatory gene which need not be linked to the others.

9.4.2 "Microscopic Cybernetics"

"Feedback," "regulation," "control," "inhibition," etc.—the new concepts that had been introduced in the 1950s and were incorporated into the operon model seem to be direct applications of the conceptual apparatus of cybernetics. Nevertheless, explicit references to "cybernetics" are rare in the molecular biology literature from that period. In 1961, Chance (1961) even argued that it would be inappropriate to use "cybernetics" in biochemistry. Chance distinguished between feedback systems that were cybernetic and those that were not. The former, according to him, necessarily involved the idea of "steering" (1961, p. 289). Biochemical control, he argued, "involve[d] basically different concepts." Chance's objection does not now appear to be particularly important, or even interesting, but it remains a curious fact that none of the proponents of cybernetics ever proposed a detailed and clear definition of "cybernetic regulation."[14] In fact, in spite of Chance's strictures, "cybernetics" largely seemed to have been taken to mean little more than feedback regulation. However, even "feedback regulation" did not turn out to be a concept that was easy to characterize (Wimsatt 1971). The original attempt, inside cybernetics, by Rosenblueth et al. (1943), was unduly behavioristic, in keeping with the temper of those times. The same behaviors that they invoked were, under some formally (though not biologically) plausible assumptions, shown by Kleene (1956) not to require feedback in the sense of cyclical patterns of state determination. More recent attempts to revive a nontrivial definition of cybernetic systems attribute internal states to these systems and give them mechanisms of self-regulation and make all of these externally accessible (Shimony 1995). Such attempts trivially satisfy the strictures of behaviorism. To the extent that they show that such feedback behavior is trivially possible from the known relations between the internal states and mechanisms of a cybernetic system, they "demystify" cybernetics. However, these analyses are really about metaphysical issues of what "determines" or "causes" what; the more interesting—epistemological—question is whether cybernetic thinking affords potentially interesting explanations, either by making certain systems more intelligible, or by affording strategies for the study of new systems.

A rather remarkable discussion by Monod (1971) seemed to suggest that even the latter option is possible. For Monod, a "cybernetic system" amounted to no more than one governed by feedback regulation. In his view both allosteric enzymes and operons were examples of cybernetic systems since both were regulated through feedback.[15] For Monod, cybernetic systems were not necessarily hierarchical in organization, or even complex, because even systems as simple as allosteric enzymes could exhibit cybernetic, that is, feedback-regulated behavior. He conceived of the cell as a complex "cybernetic network [that] guarantees the functional coherence of the intracellular chemical machinery" (1971, p. 63). Even for allosteric interactions, Monod argued, cybernetic concepts became more important when regulation was effected by more than one metabolite and reactions could be branched.

Gene regulation, for Monod, was one level of organization higher than allostery since the latter regulated the metabolites whereas the former regulated the production of the allosteric enzymes themselves. Thus, what begins to emerge from this picture is a view of the organism as a hierarchically organized cybernetic system, with successive levels of parts also being cybernetic systems. Perhaps expectedly, Monod paid particular attention to the lactose operon. His description of the system was terse:

1. The regulator gene directs the synthesis, at a constant and very slow rate, of the repressor protein.
2. The repressor specifically recognizes the operator segment to which it binds, with it forming a very stable complex ...
3. In this state, synthesis of messenger ... is blocked, presumably by simple steric hindrance, the beginning of this synthesis having to occur on the level of the promoter.
4. The repressor also recognizes β-galactosides, but binds them firmly only when in a free state: hence in the presence of the β-galactosides the operator-repressor complex is dissociated, thus permitting the synthesis of messenger and consequently of protein. (Monod 1971, p. 75)

The "logic" of the system, he argued, was simple and similar to that of a computer. As he put it:

the repressor inactivates transcription; it is inactivated in turn by the inducer. From this double negation results a positive effect, an "affirmation." The logic of this negation, we may add, is not dialectical: it does not result in a new statement but in the reiteration of the original one, written within the structure of DNA in accordance with the genetic code. The logic of biological regulatory systems abides not by Hegelian laws, but, like the workings of computers, by the propositional algebra of George Boole. (Monod 1971, p. 76)

He drew three conclusions. First, the "repressor, having no activity of its own, is purely a transducer—a mediator—of chemical signals" (p. 76). Second, the role of the β-galactoside in enzyme formation was "indirect, due exclusively to the repressor's recognition properties and to the fact that two states, each exclusive of the other, are accessible to it" (p. 76). What seems to have impressed Monod is that the β-galactoside almost inadvertently initiated a chain of reactions that looped back to affect it. Third, and most important, "[t]here is no *chemically necessary* relationship between the fact that β-galactosidase hydrolyzes β-galactosides, and the fact that its biosynthesis is induced by the same compounds Physiologically useful or 'rational,' this relationship is chemically arbitrary—'gratuitous,' one might say" (pp. 76–77).

The concept of "gratuity," that is, "the independence, chemically speaking, between the function itself and the nature of the chemical signals controlling it" (p. 77), is critical to Monod's interpretation of the operon

Central Dogmas of Molecular Biology 229

(and also of allostery) as cybernetic systems. Though, no doubt, the chemical interactions determine the behavior of the operon, these chemical interactions do not explain the behavior as well as the description of the system using feedback and signals that are responsible for control and have no "chemical requirements to answer to" (p. 78). Ultimately, the reason why they exist is to be given an evolutionary explanation: such controls must have been

> selected for the extent to which they confer heightened coherence and efficiency upon the cell or organism. In a word, the very gratuitousness of these systems, giving molecular evolution a practically limitless field for exploration and experiment, enabled it to elaborate the huge network of cybernetic interconnections which makes each organism an autonomous functional unit, whose performances appear to transcend the laws of chemistry if not to ignore them altogether. (p. 78)

Monod's position, that the cybernetic account is of more explanatory value than a purely physicalist alternative, is persuasive but does it truly satisfy the adequacy condition on explanations, that is, does it help codify gene regulation in general? After all, his discussion was limited to the operon model which, in 1971, was known to hold for only a few sets of bacterial genes. The superiority of a cybernetic explanation over a purely chemical one would be demonstrated only if two conditions could be satisfied: (i) that other cases of gene regulation could also be easily described as cybernetic systems; and (ii) the claim of gratuity would continue to hold for these systems. Satisfaction of (i) is required for the cybernetic accounts to be explanations and not merely interesting redescriptions of some special cases. However, satisfaction of (i) alone would not preclude the possibility that there are equally attractive chemical accounts that even are, perhaps, of greater explanatory value. However (ii) would make such attractive chemical explanations implausible.

9.4.3 Eukaryotic Gene Regulation

I will, in fact, argue later (in section 9.6) that, far from vindicating Monod's introduction of the concept of gratuity, the operon model represents one of the most significant triumphs of ordinary chemical (or reductionist) explanations in molecular biology. For the time being, however, I wish to turn to the other question, whether the cybernetic account can even be extended beyond bacterial gene regulation so that we can attribute some explanatory role to it, whether or not it ultimately fares better than its chemical alternatives. The obvious arena to investigate is eukaryotic gene regulation. Unfortunately, the situation there is complex and poorly understood at present. Nevertheless, what is clear is that attempts to generalize the operon model to eukaryotic gene regulation have so far shown no trace of success.

The most significant attempt in that direction was that of Britten and Davidson (1969). Their model was a straightforward generalization of the operon. They postulated the existence of four classes of eukaryotic DNA: producer genes (which are the direct analogue of structural genes in bacteria), receptor genes linked to them; integrator genes (the analogue of regulatory genes in bacteria) and sensor genes linked to them. Each linked pair, receptor–producer or sensor–integrator, is contiguous on a chromosome but the two pairs may be on different chromosomes altogether. The integrator gene can produce an activator RNA transcript which binds to a receptor and activates its linked producer gene. The transcription of activator RNA is controlled by an inducing agent that may or may not be bound to a sensor. Note that altering the activator molecule from RNA to protein would not change the formal properties of the model. As with operons, both positive and negative feedback control is possible.

In more complicated versions of this model, each sensor can activate a series of integrator genes next to it; an activator molecule (from an integrator gene) can bind to more than one receptor; several activators can bind to the same receptor; and several receptors can control the same producer gene. The set of producer genes ultimately under the control of a single sensor is called a "battery." Batteries can overlap creating a complex network, and there is considerable redundancy in the interactions between integrators and receptors (mediated by the activators). Britten and Davidson suggested that repetetive DNA sequences (which Britten and Kohne 1968 had discovered) played the role of integrators and receptors, ensuring the redundancy invoked in the model.

Besides being a straightforward extension of familiar ideas, note that this model explicitly incorporated most of what was then known about eukaryotic gene expression: (i) that functionally related genes are not necessarily physically clustered in eukaryotes; (ii) that there were a large number of repeated sequences throughout the genome; (iii) that there appeared to be genomic sequences that were transcribed in the nucleus but, nevertheless, absent in the cytoplasm; and (iv) that cell differentiation, based on

Central Dogmas of Molecular Biology

differential gene expression could be induced by "simple external signals." The Britten–Davidson model was greeted with a fair degree of enthusiasm in the early 1970s (see, e.g., Lewin 1974, pp. 366–374). Its success, however, was short-lived. Regulatory mutations in eukaryotes have shown no definite pattern. No definite correlation between repetitive DNA sequences and regulatory functions has yet been proven. It is possible that further work will revive some cybernetic model, like the Britten–Davidson model, as a candidate for eukaryotic gene regulation. For the time being, however, the only reasonable conclusion is that, Monod's enthusiasm notwithstanding, cybernetics is of little value for molecular biology.

The failure of cybernetic models to capture eukaryotic gene regulation no doubt dooms the prospects for a cybernetic notion of "information" to explicate the concept of information in molecular biology. Nevertheless, it should not go unnoticed that even if some model such as the Britten–Davidson model had been successful, the notion of "information" it would have incorporated would have been radically different from that which is invoked in the three precepts that I started out with. To the extend that there is a clear notion of "information" in cybernetics, it must be what provides "feedback" for regulation. In fact, it is hard to see what "information," in this context, is other than simply feedback. But this is far from the "information" that allegedly is contained in the DNA, that gets transferred from DNA to RNA to protein and never from protein to nucleic acid. A similar negative assessment will be made for a third potential candidate for "biological information," namely, the notion of "information" that has formed part of information theory since the late 1940s.

9.5 Information Theory

Molecular biology came of age around the same time as the mathematical theory of communication, also known as information theory (Shannon 1948). The temptation to apply it to molecular biology was, therefore, probably irresistible. The first attempts to do so, however, were far from promising. According to Shannon's information theory, the amount of "information" is (roughly) measured by the logarithm of the number of choices available during a communication process. This notion of information connotes uncertainty; its numerical value is determined by an "entropy" function that is formally identical to the usual entropy function of statistical mechanics. Shannon considered a message consisting of a linear sequence of symbols obtained from a symbol set. Let p_i be the probability of occurrence of the i-th symbol. Then, the entropy (of that occurrence in a message), H, is defined by $-p_i \log_2 p_i$. Shannon (1948) justified the choice of this measure by showing that $-Kp_i \log p_i$, where K is a positive constant, is the only function that satisfied three assumptions: (i) that H be continuous in the p_i; (ii) if all the p_i are equal to $1/n$, H should be a monotonic increasing function of n, and (iii) if a choice (such as that for the choice of a symbol in the message) can be analyzed into two successive choices, H should be the weighted sum of what it is for each of these choices. K can be fixed by a choice of units. In particular, it is equal to 1 if the base of the logarithm is chosen to be 2. This definition of information as entropy was incorporated into a rigorous theory of communication systems.

Attempts to apply Shannon's information theory to molecular biology were most systematically promoted by Quastler (e.g., 1953a,b) who attempted to give information-theoretic definitions of the specificity of enzyme action and other biological interactions. The series of definitions he proposed proved to be of little use. Branson (1953) calculated the information content of a polypeptide sequences using the empirical frequencies of the various residues to calculate the probabilities of their occurrence at each position. Linschitz (1953) attempted to estimate the information content of a bacterial cell. By 1956 even proponents of information theory in molecular biology had begun to question how valuable this approach was. When Quastler (1958) summarized a round table discussion at the end of a "Symposium on Information Theory in Biology," he felt compelled to begin:

Information theory is very strong on the negative side, i.e. in demonstrating what cannot be done; on the positive side its application to the study of living things has not produced many results so far; it has not led to the discovery of new facts, nor has its application to known facts been tested in critical experiments. To date, a definitive judgment of the value of information theory in biology is not possible. (p. 399)

though he concluded, rather optimistically "that information theory is here to stay in biology" (p. 402). The absence of any insight from these early attempts precluded systematic developments along these lines, though sporadic efforts to apply information theory to molecular biology continued.[16]

9.5.1 Information and Natural Selection

Though somewhat outside what is usually considered to be molecular biology, the most intriguing of attempts to apply information theory in biology was Kimura's (1961) attempt to use it to estimate the amount of information that accumulated through natural selection. Consider an infinite population of haploid organisms. Let p be the frequency of an advantageous allele at some locus. The probability of fixation of this allele in the population by random genetic drift alone is p; by natural selection, it is 1. Thus, the uncertainty that natural selection removes is $1/p$. Therefore, Kimura argued, the information that accumulates through natural selection, $H = \log_2(1/p) = -\log_2 p = L/\ln 2$ where L is the substitutional load of the population. Kimura proposed $H = L/\ln 2$ as a definition of the information gained through natural selection in general, that is, for all organisms. The consequences of this proposal for evolutionary theory were never worked out. However, Kimura went on to estimate 10^8 bits as the amount of information that had accumulated since the Cambrian epoch. From the estimated size of chromosomes, he also estimated 10^{10} bits as the amount of information that could be stored in a typical diploid chromosome set. Redundancy in the form of repeated DNA sequences could explain this discrepancy but another possibility was that "that amount of genetic information which has been accumulated [though natural selection] is a small fraction of what can be stored in the chromosome set" (p. 135).

This was the first inkling of what Kimura (1968, 1983) would eventually advocate as the "neutral theory of molecular evolution," that is, at the molecular level, the principal mechanism of change is random genetic drift rather than natural selection. Nevertheless, when that theory came to be formulated and systematically developed, from about 1968 (see Kimura 1983), information theory played no role, though considerations of the substitutional load were critical. "Information" thus provided little more than a redescription of "load," and information theory played no explanatory role whatsoever.

Within evolutionary biology, however, in the 1960s, a systematic analysis of Kimura's calculation was given by Williams (1966) in his influential *Adaptation and Natural Selection*. Williams explicitly defined the gene to be "that which segregates and recombines with appreciable frequency" (p. 24), implicitly assumed—with no argument whatsoever—that this definition was equivalent to the gene being "any hereditary information for which there is a favorable or unfavorable selection bias equal to several or many times its rate of endogenous change [mutation]" (p. 25), and concluded, therefore, that the gene was a "cybernetic abstraction" (p. 33). Kimura's calculation was taken to be the "most notable contribution on progress as accumulation of information" (p. 35). Williams accepted Kimura's account of information accumulation but denied that information accumulation could be identified with evolution. This formed part of his general arguments against the notion of evolutionary progress. Williams's book was important because it was critical in bringing the "units of selection" problem to the forefront of biology and, eventually, philosophy. Nevertheless, Kimura's calculation seems to have received no further attention.[17]

It is impossible to predict whether Kimura's proposal may yet prove to be fruitful. For our purposes here, however, it is irrelevant: it simply does not use "information" in the way that it is invoked in the three precepts with which we started. Kimura's "genetic information" is, in the final analysis, at best a measure of the improbability of our current DNA sequences and, what is worse, if this "information" is to have any interesting applications, any two sequences (of the same length) would contain the same amount of information! That it can be simply related to the substitutional load is no doubt an interesting result, if it can be fully generalized. Nevertheless, that "information" is not the "information" of the central dogma. In fact, all three precepts that we started out with are conceptually independent of this notion of "information" and, since as far as these precepts are concerned, all DNA sequences potentially carry information, and the same amount of information if they are of the same length, it is not even clear that these precepts are consistent with Kimura's result that "information" is what gets accumulated through evolution.

9.5.2 Information Content of Sequences

Meanwhile, with the accumulation of protein and DNA sequences, information theory has begun to be used to compare sequences. Some useful techniques were developed as early as 1972 by Gatlin (1972), but the most promising results so far have come from methods developed and popularized by Schneider and his collaborators (starting with Schneider et al. 1986). For simplicity, restrict attention to DNA sequences. Suppose that several sequences, corresponding to some functional region of DNA such as a binding site for a given protein, are given. Let B be the set of nucleotide bases, A, C, G, and T. For $b \in B$, let $p(b)$ be the probability that a base is a b (as estimated from the frequencies of each of the bases in the set of sequences). Then the uncertainty of a base chosen at random is defined by $H_g = -\sum_{b \in B} p(b) \log_2 p(b)$. This is the usual Shannon measure for this situation. Now, let l be a particular position along the sequences and let $p(l,b)$ be the probability of finding base b at position l. Then the uncertainty at that position, $H_s(l) = -\sum_{b \in B} p(l,b) \log_2 p(l,b)$. The information content at that position is defined as $R_s(l) = H_g - H_s(l)$ and the total information content for the sequence is defined by $R_s = \sum_l R_s(l)$. Schneider et al. (1986), Herman and Schneider (1992), and Stephens and Schneider (1992) computed these measures for a variety of sequences from *E. coli*, phage and humans. The measure $R_s(l)$ behaved as expected: it was highest where the different sequences in each set had the same bases, and least where they varied most. Meanwhile, R_s permitted the information content of different sets to be compared.

Schneider et al. (1986) also introduced a different measure, $R_f = -\log_2 f$, where f is the frequency of a given site in a genome. R_f is therefore, roughly, the information required to locate that site in the genome. For most sets, they found that $R_s = R_f$. Drift, they argued, probably kept R_s from becoming higher than R_f, which is all that is required for recognition. Similarly, Schneider (1988) has argued that natural selection prevented R_s from becoming smaller than R_f because such sites would not be found, thus leading to a decreased fitness of the organism. These interpretations are speculative but, nevertheless, suggestive. In two cases, however, R_s was different than R_f. For a promoter site in the genome of the phage T7, R_s is almost exactly twice R_f. For the 12 *incD* repeated segment in the F plasmid, R_s is almost exactly thrice R_f (Herman and Schneider 1992). Schneider and his collaborators have interpreted these intriguing results to mean that these sites are recognized by two or three different proteins respectively. This is a prediction from their analysis: information theory has actually yielded a testable result in this context.

It is certainly too early to judge whether Schneider's program will live up to its initial promise. Nevertheless, once again, for our purposes here, even if it is successful, it is irrelevant. Schneider's "information content" is a property of sets of sequences. It is highest at those regions where the sequences are most constant, and lowest where they are most variable. The results obtained from those analyses, at best, begin to provide a method for quantifying the intuition that functionally important sequences are more likely to be conserved through evolution. If Schneider's measure continues to capture that insight, it will no doubt be a valuable innovation, especially for evolutionary theory. Nevertheless, it is not a measure of the information content of a single sequence, the "information" that is invoked in our three precepts (and which, for example, is what gets "transferred" through the genetic code). As far as capturing the notion of "information" that is customary in molecular biology, we are back where we started.

9.6 Reduction and the Physicalist Alternative

Where does all this leave us? Well, it is clear that cybernetics and whatever concept of information that might come with it are irrelevant in molecular biology. It is also clear that information theory, no matter to what use it might eventually be successfully put in molecular biology, does not involve a concept of information similar to the customary one. This only leaves "information" as sequence, as is used in the context of coding. There we saw that the concepts of "information" and "coding" no longer play any significant explanatory role, primarily due to the complexity of eukaryotic genetics, which they do not help organize in any significant way. However, had that account succeeded, and especially if the comma-free code which, as we saw, did make significant predictions, had turned out to be correct, what would have been philosophically most significant about such a development is that it would have constituted a very significant success of the nonreductionist explanation in biology.

In order to formulate this point more precisely, let me clarify exactly what I mean by "reduction." First, I am concerned here only with epistemological aspects of reduction. When I am suggesting that a particular pattern of explanation or reasoning is not reductionist, I am making no ontological claim about whether some of the underlying processes are physical or novel or whatever.[18] Second, I am leaving open the possibility that there can be several explanations of the same phenomena, some of which might be reductionist and others not. Third, as before (see section 9.1), I do not wish to get involved in disputes about reductionist explanation that are ultimately disputes about the nature of explanation.[19] Fourth, the account that I will give here is geared towards understanding reductionist explanation in molecular biology, though with some relatively trivial modifications, which are indicated in the footnotes, it can potentially be used in many other contexts. With these disclaimers, I will call an explanation—something that already satisfies whatever strictures that are put on explanations—reductionist if and only if it satisfies the following three criteria:

(i) *Physicalism*: The explanatory factors invoked in the explanation have "physical warrants," that is, they are either obtained (perhaps informally) from physical theory, or recognized as mechanisms from physical experiments alone;

(ii) *Hierarchical structure*: The complex entity whose behavior is being explained must be characterized as having a hierarchical organization in which the properties of entities at lower levels alone are used to explain its behavior;[20]

(iii) *Spatial instantiation*: The hierarchical structure of the entity must be realized in physical space, that is, entities at lower levels of the hierarchy must be spatial parts of entities at higher levels of organization.

This is, no doubt, a very stringent sense of "reduction." For instance, it has long been known that there is a fully legitimate type of "reduction" in which only criterion (i), but not the other two, is satisfied.[21] Examples include the reduction of Newtonian gravitation to general relativity, Newtonian kinematics to special relativity, geometrical optics to physical optics and, arguably, the putative reduction of psychology to neurophysiology.[22] It is also possible that criterion (ii), or criteria (ii) and (iii), are satisfied without satisfying criterion (i). All that would be required is the admission of explanatory factors that do not yet have physical warrants.

What is more interesting is that criterion (ii) can be satisfied without satisfying criterion (iii). A hierarchical organization can exist without being realized as such in physical space. Consider genetic explanations of various features of diploid organisms before the 1940s (when exactly how a gene and its effects are to be physically characterized began to be clarified). Some features were explained by one allele at one locus, others by both alleles at that locus and yet others by a set of alleles at several loci. There was a clear hierarchical structure but, in these explanations, this hierarchy was not *necessarily* a physical hierarchy. While it was known from linkage studies (that is, the study of the probabilities of different loci to be inherited together) that there was a linear order of loci on chromosomes, and it was strongly suspected that this corresponded to an underlying physical order of genes, that suspicion played no part in these explanations. Indeed, the complications of eukaryotic genetics that I have already mentioned, such as the existence of overlapping loci, show that the relation between the two orders is not one of trivial one-to-one correspondence. In fact, in 1939, hoping that ordinary mechanistic explanation would break down in genetics, Haldane (1939) predicted that the two orders would turn out to be inconsistent. Of course, that prediction has turned out to be largely false—the extent of overlap between genes that is known today does not yet significantly challenge the order established by linkage studies. By 1951, linkage studies had even led Lederberg et al. (1951) to suggest a branched genetic map in certain bacteria which, as they explicitly emphasized, was formal and did not refer to a branched chromosome.[23] It is clear that in genetics at least, the possibility that criterion (ii) could be satisfied while criterion (iii) was violated was historically a real possibility.

Moreover, during the 1930s and 1940s, the mode of action of genes—what chemicals were involved, and what mechanisms were necessary—was not known. Therefore, in these "genetic explanations," criterion (i) was also violated. Nevertheless, I think that is quite reasonable to say that features at the organismic (or, in this context, phenotypic) level were being "reduced" to genetics (that is, the genotypic level). There are thus at least two other senses of "reduction" than the one in which the satisfaction of all three of the criteria is required: one in which only criterion (i) is satisfied and one in which only criterion (ii) is satisfied. Moreover, it should be clear that (i) and (ii) can both be satisfied without satisfying (iii). Just imagine the situation when the action of genes receives a full physical characterization, but there is so much overlap and other complicating features discovered at the DNA sequence level that the genetic hierarchy does not, even at a coarse level, correspond to any physical hierarchy of nucleotide bases and groups of bases on chromosomes. Such a situation generates yet another sense of "reduction." Finally, it is possible that both criteria (ii) and (iii) are satisfied but not (i): an explanation involves a hierarchical structure, and this structure is realized in ordinary physical space, but the interactions between the parts that are invoked in the explanation do not have physical warrants.[24] I will argue below, that explanations involving information as sequence fit this pattern. In order to avoid misunderstanding, even at the expense of reiteration, let me emphasize here that such an explanation, which is not physicalist because criterion (i) is violated, nevertheless does not involve any ontological commitment to nonphysical interactions. The concern, here, is purely epistemological, with what the factors are that offer good explanations in some context.

There are thus at least five senses of "reduction" in biological explanation that can be distinguished: (i) when all three criteria are satisfied; (ii) when only criterion (i) is satisfied; (iii) when only criterion (ii) is satisfied; (iv) when only criteria (ii) and (iii) are satisfied; and (v) when only criteria (i) and (ii) are satisfied.[25] The first sense, then, is the most stringent of all of these (and, elsewhere, I have called it "strong reduction"). Nevertheless, that is the sense in which many of the most successful explanations in molecular biology have turned out to be reductionist.

However, before turning to some characteristic examples, note that the satisfaction of any or all of these criteria does not guarantee that a reductionist explanation is of much value as an instance of reduction. Such explanations routinely involve the use of approximations and limiting procedures such as assuming that a phenotypic trait is governed by an infinite number of loci. An important biological example, though one that has not received the kind of philosophical scrutiny that it deserves, is Fisher's (1918) celebrated demonstration that, if there are many loci involved, and each had an infinitely small effect, but all acted additively, a phenotypic trait (such as body weight or height) would vary continuously and follow a normal (Gaussian) distribution in a population. That famous "reduction" established the relation between biometry and ordinary Mendelian genetics. Nevertheless, such a reduction generates some unease because of the contrary-to-plausibility assumptions incorporated in the limiting procedure: an infinite number of loci, additive action, and infinitely small effects. This example suggests that there are at least some other conditions that can be used to gauge the significance of a reduction: (i) whether the approximations or limiting procedures involved are conceptually (e.g., mathematically) well defined; (ii) whether they are known to be at least approximately empirically plausible; and (iii) whether they truly are "independent" of the properties of the features to be "reduced" in the sense they have not been expressly invoked to carry out a particular reduction. It is an open problem as to whether these conditions can be made more precise in a context-independent fashion. Suffice it here merely to note that even these conditions are usually satisfied in the most of the reductionist explanations in molecular biology.[26]

If what I have said about reduction in molecular biology is true, it affords examples of the most stringent and significant instances of reductionist explanations known. Let me attempt to demonstrate the truth of this claim by just noting two striking examples. The first is the molecular explanation of allostery: the "cooperative" behavior of proteins, such as hemoglobin, which are composed of several subunits. The composite hemoglobin structure has greater affinity for oxygen or carbon dioxide, depending on the acidity (pH) of its fluid environment, than the total affinity of each of the units taken separately. The reason why this example is important is because such cooperative behavior is usually regarded as the special province of traditional antireductionists: a "whole" is said to be more than the "sum of the parts." The various contemporary models of allostery (which differ primarily in detail) each explains this phenomenon by showing how the structure of the subunits changes in the composite whole, but changes entirely due to localized ordinary physical interactions (see Perutz 1990 for a review). All of these interactions, except the slightly stronger ones involving the heme group, are entirely explained on the basis of molecular shape, underscoring how the "lock-and-key" fit generally reigns over all other physical interactions between biological macromolecules. The changed structure explains the non-linear increase in affinity. Thus, the explanation involves a hierarchical organization in ordinary space, and the usual interactions of physics and chemistry. All three criteria for reduction listed above are satisfied.

The second example is one we have already encountered: the operon. This example is important because the cell exhibits "goal-directed" behavior, also grist to the mill of antireductionists. As noted before (section 9.4), the operon model explains this phenomenon: the various genetic units are hierarchically organized in chains along DNA segments. A physical interaction between the substrate, sometimes other molecules, and one of the genetic units initiates a sequence of interactions that leads to the production of the enzyme in question. When the substrate is no longer present, the initiating step of these reactions ceases, and enzyme production stops. Moreover, molecules that have similar chemical properties to the substrate, but which do not react with the enzyme, can also induce this process.[27] To the extent that we understand the details of this interaction at present, molecular shape is of paramount importance. One again, this explanation is based on a spatially realized hierarchical structure and interactions with physical warrants. All three criteria for reduction that I gave above are trivially satisfied.

Though, in order to return soon to considerations about information, I will not pursue this point in detail here, it is also true that these explanations neither involve implausible or false approximations and limiting procedures, nor those that are designed only for biological applications. The characterization of the chemical interactions underlying allostery of operon-based gene regulation, such as van der Waals "repulsion" or hydrogen bonding, is no different from what physics or chemistry admits in a nonbiological context. The van der Waals radii of the atoms, which determine molecular shape, are not obtained from analytic solutions of Schrödinger equations. Hydrophobic bonds are only approximately explained by the statistical mechanics of water molecules. Nevertheless, the same approximations are made in ordinary inorganic chemistry and molecular physics. Perhaps what is most striking, since it is quite rare in reductions, is that the biological context, in such explanations in molecular biology, does *not* require the introduction of specially designed approximations. Reductionist explanations of this kind abound in molecular biology. In fact, at least one of the fairly well-known precepts of the field, that the "structure determines the function" of biological macromolecules, where "function" only means behavior, can be taken to be a succinct statement of the reasoning behind reductionist explanation.

This long digression about reduction has not been without a purpose. I am now in a position to defend my claim that the assumptions about coding and information, especially the rather strong ones that led to the "comma-free" code, do not satisfy all of the criteria for the strong sense of reduction that I have been using above. Thus they run counter to one of the most striking patterns of explanation in molecular biology. Consider, first, the comma-free code model of Crick et al. (1957). Criteria (ii) and (iii) for reduction are trivially satisfied. There is a hierarchy of triplets simply because there is a contiguous set of triplets. This is obviously not an interesting hierarchy in any sense but, nevertheless, it is a hierarchy. Moreover, the units, that is, the triplets, are physical objects—three bonded nucleotide bases—and their contiguity is in ordinary physical space. However, criterion (i) is violated. The restrictions on what sequences were permissible, which define the interactions between the units, were based on implicit assumptions about simplicity and efficiency of information storage and transfer, and these are clearly not explained on the basis of physical principles.[28] Explanations based on this model are not physicalist, in the epistemological sense in which that criterion was formulated above.[29] If anything, if the assumptions about information that were implicit in the comma-free code were made explicit and incorporated into a systematic account of information, there could have been a kind of reduction satisfying criteria (ii), (iii), and (i)' where (i)' is some requirement about information that replaces physicalism. However, this reduction would not be the type of physicalist reductionist explanation that, as was argued above, is characteristic of molecular biology.

What happened with the comma-free code is perhaps not very important since that code is little more than a historical curiosity any more. Nevertheless, what was just said about it can be said about our usual notion of "coding" and its associated notion of "information" as sequence. Remember that all that can be said about this code is linear, symbolic and local. Unlike the case of the comma-free code, there is no further underlying explanation of these properties from deeper assumptions about information. Consequently, even to the limited extent that it can be used for explanatory purposes, it satisfies criteria (ii) and (iii), does not satisfy (i), and there is little hope of even finding some criterion (i)' from some general account of information. Should such an account be found, then the conclusions of this paper will have to be modified. The philosophical point being made here would be lost but that is not much of a price to pay for the scientific advance that would be achieved instead.

The failure of explanations involving codes and information and the success of the usual reductionist explanations in molecular biology together suggest a rather striking possibility: abandon the notions of codes and information altogether and pursue a thoroughly physicalist reductionist account of the interactions between DNA, RNA and protein (and explain away the conceptual framework from the 1950s as an artifact of the coincidental colinearity of DNA and protein). In principle, this does not present any real difficulty. We would treat the DNA-RNA-protein system as a network of chemical reactions, and write down a system of linear differential equations to describe the process.[30]

The main difficulty with such an account is that this model would have a rather large number of variables. These variables would have to keep track of the concentrations of each different type of DNA, RNA and protein that could potentially arise in the cell, not just the ones that emerge during normal gene expression, but also those that could arise through errors. Nevertheless, in most contexts, this level of complexity is not beyond what can be quite easily numerically analyzed. The coefficients describing the formation of a particular RNA type from a given DNA type and for forming a protein type from an RNA type would incorporate the specificity of the code. Moreover, these coefficients could quite naturally incorporate all that is known to happen to an RNA segment in a particular environment including editing, as well as nonstandard translation. Moreover, such a model would be dynamic and actually allow the exact description of the concentration changes of the various components over time.

Why not abandon "codes" and "information" and pursue this possibility? There is no fully convincing reason not to do so. One possible reason why it has so far not been pursued is that only what might be called the "statics" of gene expression has usually been studied in any context. Molecular biology, especially molecular genetics, has not yet gone very far even in characterizing what genes are expressed in what part of an organism, let alone the finer details of the temporal regulation of genes within a cell except, of course, for *Escherichia coli*. In this static context, the coding account still serves some organizing function and, in spite of its increasing failures, it is being retained while few *explanations* are actually being pursued in this kind of work in genetics. Thus, the inability of the code to provide *significant* explanations is no major handicap. Another possible reason is that there is a useful sense in which "coding," "information," "translation" and related notions make the relations between DNA, RNA and protein transparent, which no dynamical account involving reaction coefficients can. The intuition, here, is that the code is "natural" in a sense that these coefficients are not. Perhaps lurking behind this usefulness there is some insight to be grabbed, which the conventional information-based account of molecular biology has grasped even if, so far, very shakily.

Nevertheless, what does seem obvious is that a dynamical account, whether it is physicalist, informational, or whatever, will eventually be necessary if even approximate accounts of gene expression, interaction and cellular behavior, let alone the development of complex organisms, are to be pursued at the molecular level.[31] There is, moreover, one development whose importance even the most conventional molecular biologists have recognized, which implicitly relies on such a dynamical model and has no concern whatsoever for coding and information even though it deals with DNA. This is the polymerase chain reaction (PCR), which permits the rapid amplification of a given (double-stranded) DNA segment. The process begins by creating complementary single strands of DNA by heating the double-stranded segments to temperatures close to boiling. Using primers, which are a few nucleotides long to start the reaction, entire double stranded segments are created from each of the single stranded segments. This results in two copies of the original double-stranded segment. After n cycles of this process, there are (on average, that is, ignoring stochastic effects) 2^n such copies and they can be used for any purpose. It has been claimed with some justice that the PCR technique has "revolutionized molecular genetics" (Watson et al. 1992, p. 79). There is no concern for coding or information in this process. What is at stake is that a particular single-stranded DNA sequence catalyzes the formation of another specific sequence, namely, the one that is complementary to it. The dynamical equations to model this process are formally those for DNA replication in general. These can be solved to calculate the rate of DNA formation. This is reductionist reasoning, through and through, and this is what molecular biologists using PCR are relying on, even if only implicitly, when they predict the time required to produce a certain amount of DNA. "Codes" and "information" are irrelevant in this context.

9.7 Conclusions

It appears, therefore, that we are faced with a quandary. The conventional account of information as sequence is of little explanatory value in the novel contexts we are beginning to encounter. The alternative physicalist account of rate equations is, at present, of use in only very few contexts. I wish to suggest that we are faced with a situation that will require a rather radical departure from the past. For a coherent account of the relations between DNA, genes and biological behavior that has significant explanatory value, we will have to avoid both these accounts. Not adopting the physicalist account is neither controversial nor difficult since it has so far never been invoked in this context except, implicitly, in the development of PCR technology. However, because of the extent to which the idea of coding has been central to how molecular biology has been conceptualized, abandoning the idea of information as sequence will have important consequences. Let me list some of these and argue that they are, in fact, desirable:

(i) If biological "information" is not DNA sequence alone then, trivially, other features of an organism can also characterize information. And this is precisely what recent developments in molecular biology indicate. In particular, the developmental fate of a cell might be largely a result of features such as methylation patterns of DNA that are not, as far as we know today, even ultimately determined only by DNA sequences. These "epigenetic" patterns can be inherited for several cell generations. Different cells in the same organism, presumably with identical DNA sequences, can be epigenetically different. Because of these differences, cell specialization and differentiation, the usual prelude to developmental changes, can take place. Epigenetic specifications are also critical in generating those differences in offspring (of sexually reproducing organisms) which depend on whether an allele is inherited from the mother or the father. Epigenetic specifications can sometimes be transmitted across organismic generations. It would be highly unintuitive not to regard these determinations as "transfers of information" if "information" is to have any plausible biological significance.

(ii) The central dogma of molecular biology, that information only flows from nucleic acid to protein, and never in the reverse direction (precept (iii) of the section 9.1), is false. However, a less grandiose claim, that protein sequences do not directly determine nucleic acid sequences in the way in which the latter determine the former, remains true as far as we currently know. No doubt this humbler claim does not have the majestic power of the central dogma but the question that I wish to raise is whether this retreat really undermines some putative insight that was enshrined in that dogma. The usual defence of the general importance of the central dogma, that is, its importance for biology in general (and not just molecular biology) has been that it is a statement, at the molecular level, of the noninheritance of acquired characteristics (precept (iv) of the section 9.1). However, this is nothing but egregious misinterpretation. Acquired characteristics are sometimes inherited, though usually not.[32] What ensures that even those acquired characteristics that involve changes in the DNA component of genomes of cells are not inherited in higher animals is the segregation of the germ-line from the soma. But plants have no germ-line, and the extent of its segregation in animals greatly varies across the phyla. Nevertheless, whatever the relationship between nucleic acid and protein may be, it is the same across the organic world: *ipso facto* the central dogma, even if it were true, could not be either an explanation or an alternative synonymous statement of the alleged noninheritance of acquired characteristics. I do not mean to deny that there is something peculiar, and extremely interesting, about how DNA resists easy change across the phyla. But this is something to be studied and understood, not something to be "explained" away on the basis of some alleged law about some incoherent notion of "information."

(iii) Many of the more influential contemporary discussions of the origin of life have concentrated on the origin of information, where information is construed simply to be nucleic acid sequences (e.g., Eigen 1992). Implicit in these discussions is the assumption that these sequences ultimately encode all that is necessary for the genesis of living forms and, therefore, a solution to the problem of the initial generation of these sequences is all that is required to solve the problem of the origin of life. The move away from sequences would put these efforts in proper perspective: to explain the possible origin of persistent segments of DNA is a far cry from explaining the origin of living cells. However, I do not wish to harp on this point since, quite justifiably, most molecular biologists think that such discussions of the origin of life are little other than idle speculation.

(iv) The position that I am advocating certainly suggests that we remove the amount of emphasis on the DNA sequence that we see today. In turn, this suggests that the sorts of arguments that were mustered to initiate the Human Genome Project, a crash program to sequence DNA blindly, that is, without first determining the functional (or behavioral) roles of the segments to be sequenced, does not have much scientific rationale. This is not a new point. It has previously been made, on the basis of other considerations, by many critics of the Human Genome Project as it was initially conceived (Sarkar and Tauber 1991; Davis 1992; Tauber and Sarkar 1992; Lederberg 1993). If my arguments here are sound, then the Human Genome Project would best be limited to its first stage, that is, the mapping of all known genetic loci to specific positions on chromosomes, and then proceeding to sequencing segments as and when they come to be known to have some functional interest.

Each of these consequences is desirable. Nevertheless, before the notion of information as sequence and the picture of molecular biology that comes with it are abandoned, some alternative is necessary. Let me end by noting two possibilities, the first would be a return to the old concept of specificity and develop it further, beyond the stereochemical theory whose limitations are, in any case, gradually becoming apparent. For instance, in the immunological context, it has already become clear that not all residues that are in contact between antibodies and antigens contribute equally to the free energy of the interaction (see Sarkar 1998). Of about twenty residues that are in contact at the interaction site, only four or less dominate the interaction. This is not a total failure of the stereochemical model. Rather, it is a modification of that part of it which asserts that only molecular shape matters.

Perhaps a similar account can be given that will explain the specific interaction between complementary base paris during DNA replication and transcription, and all this will be incorporated in a general systematic account of specificity. Coding will be retained only as a short-hand description of the usual triplet specification of amino acid residues, but it will not be assumed to have any explanatory value. Nevertheless, the special role played by DNA triplets would be incorporated in the account of specificity that would be developed. Finally, the differential equations that I mentioned above (in section 9.6) would be incorporated to provide a dynamical account of the entire DNA-RNA-protein and other interactions in the cell. The new account of specificity would remove the taint of artificiality that the reaction coefficients had. Perhaps the most interesting feature of such an account would be that it would be purely physicalist and, consequently, reductionist in the strongest sense, in sharp contrast to the informational account that is currently prevalent. So far little, if any, effort has been expended towards the elaboration of such a picture, probably because there is, as yet, no plausible candidate for such a generalized theory of specificity.

The second possibility is the elaboration of a new informational account in which "information" is construed to be broader than just DNA sequence. For instance, Shapiro (1991, 1992) has argued that the entire genome should be viewed as "a dynamic information storage system that is subject to rapid modification" (1992, p. 99). From Shapiro's point of view, there exists not only the usual genetic code relating amino acid residues to DNA base triplets, but an additional coding relation for sequences that serve regulatory and other roles. The latter code is clearly not triplet, and is not even symbolic in any conventional sense since it would be "interpreted" as a process rather than some other entity. The hierarchical organization of the entire genome determines the behavioral repertoire of a cell. "Information" is no longer determined by local sequence alone, nor is it linear, since these hierarchies can show considerable complexity. Though Shapiro is not explicit on this point, "information" in such a picture need not be constrained to DNA sequences—patterns of methylation and other heritable features of the genome could well carry information.

There is little doubt that such a picture is intriguing because it automatically retains those insights that the conventional view has, such as the existence of a peculiar triplet relationship between DNA and amino acid residues, while extending this view to incorporate recent discoveries in molecular biology. However, as Shapiro (1991) has acknowledged, no testable claim has yet emerged from this picture. Whether, eventually, any will, I cannot say. I am willing to acknowledge that "testability," as it is usually understood—that is, demanding new predictions—might well be too strong a criterion for most biological contexts. However, for a theory to be worth admission into serious discourse, it should at least allow the systematic and clear organization of known facts. Shapiro's approach has not yet even been developed to that extent. However, should an admissible theory emerge, it would, as in the case of the comma-free code, be a theory based on assumptions about information storage and utilization. It would be manifestly nonphysicalist. It would also give a new lease of life to information-oriented thinking in biology and, perhaps, even finally begin to explain what "biological information" actually happens to be.

Acknowledgments

I have benefited from extensive discussions with Angela Creager, Lily Kay, Evelyn Fox Keller, Joshua Lederberg, Richard Lewontin and William C. Wimsatt. Sections of this analysis were presented to the conference on "Methods in Philosophy and the Sciences," New York (January 16, 1993), the Boston colloquium for the Philosophy of Science (April 13, 1993) and the Department of Philosophy, McGill University (April 23, 1993). Remarks by members of the audiences, especially David Thaler and Abner Shimony, the official commentators on the first two occasions, were particularly helpful. I am grateful to all of these individuals for their comments and criticism; none of them, however, should be presumed to agree with what I say. The work reported here was partially supported by a Senior Research Fellowship from the Sidney M. Edeltstein Centre for the History and Philosophy of Science, Technology and Medicine, Hebrew University, Jerusalem, a Resident Fellowship from the Dibner Institute for the History of Science and Technology, MIT, and NIH Grant No. HG 00912-01/2. I would like to thank all these organizations for their support.

Notes

1. The distinctions between the three senses of "information" that will be used throughout this paper are apparently being *explicitly* made for the first time. As should be evident from what follows, these distinctions significantly help the philosophical analysis being attempted here. I am not suggesting that these three construals of "information" were always historically distinguished by the molecular biologists, especially in the confusing conceptual landscape of the 1950s when "information" began to infiltrate molecular biology. However, even in that rhetoric, explicit conflation of these senses is rare, and I know of no example where the different senses were conflated in any *technical* argument or inference. For an analysis of the complexities of this process which begins from a rather different methodological perspective than the one used here but, generally, reaches the same sort of conclusion, see Keller (1995). For one that reaches strikingly different conclusions by denying these distinctions, see Kay (1994).

2. Note that this is a purely epistemological construal of "physicalism" and does not invoke any of the usual ontological connotations of that doctrine that have been advocated by the later logical empiricists (e.g., Neurath or Carnap after 1931) and their followers (such as Quine). "Physicalism" will only be used in this sense in this paper.

3. Part of such a sketch, at least, has been attempted by Keller (1995). See also, Judson (1979) for a journalistic account.

4. Even before Crick's (1958) explicit identification of information as specificity, the same idea was implicitly being used quite systematically. In 1955, for instance, during a symposium on enzymes in Detroit, Mazia (1956, p. 262) argued that the role of RNA was to carry "information" from the nuclear DNA to the cytoplasm for the synthesis of proteins. At the same conference, Spiegelman (1956, p. 77) argued that

the required "informational complexity" made RNA and DNA the only two plausible candidates for being templates for protein formation and Lederberg (1956, p. 167) noted that "information" was what "specificity" was "called nowadays." See also Gamow et al. (1955) which will be discussed in the text.

5. This section is based largely on the more detailed account in Sarkar (1989).

6. There were 8 diamonds in which the top and bottom bases were the same, A, T, C or G-for each of the 4 possibilities at top and bottom, either a C-G or an A-T pair could be in the middle. Different bases at top and bottom can be chosen in $\binom{4}{2}$, that is, 6 ways. These give rise to 12 diamonds, once again because both C-G and A-T can occur in the middle. Together, there are 20.

7. If it is interpreted as a triplet code, this means the code had an overlap of two. The terminology used here is that of the 1950s: "overlapping" means an overlap of two; "partially overlapping" means an overlap of one, where only the last position of one triplet is shared with the next.

8. Any meaningful attempt at prediction in the reverse direction, that is, from polypeptide to DNA sequence is, of course, hopeless because of the degeneracy of the code, even independent of the other problems that will emerge from the discussion in the text.

9. For example, the virus $Q\beta$, which preys on *Escherichia coli* has a coat protein that is usually produced by having UGA read as a termination codon. However, 2% of the time, it is ignored, resulting in a longer coat protein whose presence turns out to be necessary for the normal behavior of the phage (Fox 1987).

10. Of course, if one had available all the DNA sequences in the world today, and all that have ever existed in organisms, and all the polypeptides these ever coded for, then it is possible to assert that, subject to an important qualification, one would be able to identify all intron–exon boundaries on basis of sequence information alone. The qualification is that novel sequences could arise through mutation, recombination, etc. So, all this information would still be of no avail for the purpose at hand. In any case, if prediction is the desideratum as it is in the discussion in the text, even without worrying about the insurmountable obstacles presented by the possibilities of mutation and recombination, the sparse information available at present about the exact relationships between known DNA and protein sequences makes attempts at prediction perilous in exactly those new contexts where they would be valuable.

11. For instance, in the intestines of mammals, the mRNA apolipoprotein undergoes a deamination of a C, converting it to a U in such a way that a stop codon is created. This behavior is tissue-specific. The same kind of deamination, and the reverse $U \to C$ amination process, have also been observed in several plant mitochondrial mRNA transcripts.

12. In the most extreme case known to date, in the human parasite *Trypanasoma brucei*, in the mRNA transcript leading to the formation of NADH dehydrogenase subunit 7 (a protein), as many as 551 U's are inserted throughout the transcript while 88 are deleted (Koslowski et al. 1990). In such a case it is hard to see why the DNA segment encoding such a transcript should be called the "gene for NADH dehydrogenase subunit 7"—by looking at the DNA sequence it would be impossible to predict beforehand that this was the protein that would eventually be produced.

13. The University of Chicago Press plans to publish this material, along with Haldane's unpublished lectures on Darwinism.

14. Consequently, it is hard to judge whether Chance's position, as well as his failure to cite Wiener as one of the originators of cybernetics, was quite as idiosyncratic as it appears to be.

15. The importance of allostery in the development of Monod's thinking has been emphasized by Angela Creager and Jean-Paul Gaudillière (personal communication; see also Creager 1994).

16. See Yockey (1992) for a recent, though unilluminating, review.

17. Though they do not refer either to Kimura or to Williams, Waddington and Lewontin (1968) presented an alternative argument for the irrelevance of this notion of information as a measure of evolutionary progress. In fact, they argued that selection would act to prevent the continued accumulation of this kind of information. This argument, too, has received no further attention, underscoring the general perception that formal information theory is irrelevant to discussions of evolution.

18. I am not, of course, denying either that epistemological success often leads to ontological commitments or that the latter generate epistemological programs. However, since, with the sole exception of some religious fanatics, there appears to be consensus that there is nothing nonphysical going on in biological systems, ontological aspects of reduction are of little more than formal interest in this context.

19. One consequence of this move is that, though the account given below generally makes no reference to "theories," it does not explicitly preclude the construal of reduction as a relation between theories. I have three reasons for no longer being willing to get involved in the dispute between "theory reduction" and "explanatory reduction" (contrary to Sarkar 1992): (i) it has become less and less clear that there is any single notion of "theory" applicable to all of science, which philosophers should try to explicate; (ii) it has also become clear that there are few, if any, universal theories. (I am even skeptical that the usual examples drawn from physics—quantum mechanics or general relativity—truly are as universal as philosophers generally believe. This will be taken up in a future paper.) Moreover, there is no harm to accept as "theories" even those claims that are acknowledged to have small domains of applicability; (iii) finally, and this is the most important of these three

reasons, the philosophical disputes about the role of theories in explanation have largely degenerated into disputes about *formal* issues and have led to little attention being paid to the *substantive*—and, I think, much more philosophically and scientifically interesting—problems with reduction. The criteria suggested below are substantive, not formal. This point is elaborated in Sarkar (1998).

20. Note that, if both criteria (i) and (ii) are satisfied, the properties referred to in (ii) must be the physical warrants invoked in (i). (We are looking at the same putative explanation). Obviously, certain kinds of explanations satisfying (i) would automatically preclude the satisfaction of (ii), for instance, if the explanation manifestly invoked properties that can only be defined by reference to higher levels of organization.

21. That there is an important distinction between reductions involving only criterion (i), and those also involving the other criteria, seems to have been first suggested by Nickles (1973) though, in his account, reduction occurs in the opposite direction than the one suggested here which follows the treatments of Wimsatt (1976) and Sarkar (1989, 1992). The relations between most of the formal accounts of reduction offered so far are reviewed in Sarkar (1989, 1992). Note that, in those previous accounts criteria (ii) and (iii) were not distinguished.

22. Note that, in the context of reductionist explanation outside the natural sciences, the major modification of this account that would be required is the replacement of criterion (i) by an assumption that the warrants for the explanatory factors that are invoked are provided by the specified theory that is supposed to play the explanatory role in that context. If criterion (ii) is also satisfied, this would mean that the warrants come from the lower level. However, criterion (i) can be satisfied without the satisfaction of criterion (ii)—In fact, the explanation might involve higher levels of organization (as noted before). In biology this is endemic in selectionist explanations. In physics, explanatorily privileged scaling laws can operate at the same level as what is being explained.

23. I am particularly indebted to David Thaler for drawing my attention to this episode. See Thaler (1996) for a more extended discussion.

24. Beyond the biological context, this last sense of reduction can be particularly important. For instance, if social behavior is to be explained on the basis of individual behavior, criteria (ii) and (iii) would be satisfied but not criterion (i) since, presumably, physical warrants would not have to be invoked to justify whatever rules are said to govern the interactions between individuals. However, in these cases, presumably, the requirement of physicalism would itself be replaced by some other kind of "fundamentalist" assumption. For instance, it might be required that the rules in question should refer only to individuals and their properties.

25. Note that since criterion (iii) is formulated in terms of a hierarchical organization, it is not possible to satisfy (iii) without satisfying (ii). This precludes the possi-

bilities of having sense of "reduction" involving the satisfaction only of criterion (iii) or only criteria (i) and (iii). There are no other logical possibilities than the ones that have so far been considered.

26. Somewhat ironically, the satisfaction of (i) is probably much more problematic in physics than in biology (see Primas 1991 and the references therein). What is at stake here is the status of the approximations routinely required in physics to establish connections between levels of organization. Leggett (1987) has called these "physical approximations." Strangely, besides Shimony (1987), no philosopher of physics seems to have acknowledged their significance.

27. It is perhaps ironic that both the processes discussed here—allostery and operon-based gene regulation—are the ones that led Monod (1971) to suggest the concept of gratuity and a role for cybernetics in molecular biology!

28. Whether or not these requirements can truly be given a selectionist interpretation is irrelevant: as noted before, selectionist arguments refer to a higher (rather than a lower) level of organization. (That they can be given a selectionist interpretation is argued in Sarkar (1989), though—I now think—not very successfully.)

29. Once again, let me emphasize that this does not mean that anything other than ordinary physical and chemical interactions are occurring in DNA in the comma-free code model. All that is at stake here is what enters into an explanation, not the ontological concern for what is "ultimately causing" some phenomena, even if that notion can be made sensible.

30. The techniques necessary for this sort of development are not particularly profound or difficult. See Sarkar (1988) for a trivial example, and Küppers (1983) for a more systematic treatment.

31. It is, of course, open to dispute whether such a pursuit will yield anything except a morass of inchoate detail. That, in turn, will show the extent of the success, and the limitations, of the type of reductionist explanations that are characteristic of molecular biology.

32. See Landman (1992) as well as Jablonka et al. (1992) for recent reviews.

References

Atkins, J. F., Weiss, R. B., Thompson, S., and Gesteland, R. F. (1991). "Towards a Genetic Dissection of the Basis of Triplet Decoding, and its Natural Subversion: Programmed Reading Frame Shifts and Hops." *Annual Review of Genetics* 25: 201–228.

Avery, O. T., Macleod, C. M., and McCarthy, M. (1944). "Studies of the Chemical Nature of the Substance Inducing Transformation of Pneumococcal Types: Induction of Transformation by a Deoxyribonucleic Acid Fraction Isolated from Pneumococcus III." *Journal of Experimental Medicine* 79: 137–157.

Branson, H. R. (1953). "A Definition of Information from the Thermodynamics of Irreversible Processes." In: Quastler, H. (ed.), *Essays on the Use of Information Theory in Biology*. Urbana: University of Illinois Press, pp. 25–40.

Brenner, S. (1957). "On the Impossibility of All Overlapping Triplet Codes in Information Transfer from Nucleic Acids to Proteins." *Proceedings of the National Academy of Sciences* 43: 687–693.

Britten, R. J., and Davidson, E. H. (1969). "Gene Regulation for Higher Cells: A Theory." *Science* 165: 349–357.

Britten, R. J., and Kohne, D. E. (1968). "Repetitive Sequences in DNA." *Science* 161: 529–540.

Cattaneo, R. (1991). "Different Types of Messenger RNA Editing." *Annual Review of Genetics* 25: 71–88.

Chance, B. (1961). "Control Characteristics of Enzyme Systems." *Cold Spring Harbor Symposia on Quantitative Biology* 26: 289–299.

Chargaff, E. (1963). *Essays on Nucleic Acids*. Amsterdam: Elsevier.

Creager, A. (1994). "Reconciling Experimental Systems and Institutions: The Invention of 'Allostery' in Paris and Berkeley, 1959–1968." Paper presented at the Fourth Mellon Workshop, Program in Science, Technology and Society, Massachusetts Institute of Technology, April 30, 1994.

Crick, F. H. C. (1958). "On Protein Synthesis." *Symposium of the Society for Experimental Biology* 12: 138–163.

Crick, F. H. C., Barnett, L., Brenner, S., and Watts-Tobin, R. J. (1961). "General Nature of the Genetic Code for Proteins." *Nature* 192: 1227–1232.

Crick, F. H. C., Griffith, J. S., and Orgel, L. E. (1957). "Codes without Commas." *Proceedings of the National Academy of Sciences (USA)* 43: 416–421.

Davis, B. D. (1992). "Sequencing the Human Genome: A Faded Goal." *Bulletin of the New York Academy of Medicine* 68: 115–125.

Eigen, M. (1992). *Steps Towards Life*. Oxford: Oxford University Press.

Ephrussi, B., Leopold, U., Watson, J. D., and Weigle, J. J. (1953). "Terminology in Bacterial Genetics." *Nature* 171: 701.

Fisher, R. A. (1918). "The Correlation Between Relatives on the Supposition of Mendelian Inheritance." *Transactions of the Royal Society of Edinburgh* 52: 399–433.

Fox, T. D. (1987). "Natural Variation in the Genetic Code." *Annual Review of Genetics* 21: 67–91.

Gamow, G. (1954a). "Possible Relation Between Deoxyribonucleic Acid and Protein Structures." *Nature* 173: 316.

Gamow, G. (1954b). "Possible Mathematical Relation Between Deoxyribonucleic Acid and Proteins." *Biologiske Meddelelser udviket af Det Kongelige Danske Videnskabernes Selskab* 22(3): 1–11.

Gamow, G., and Metropolis, N. (1954). "Numerology of Polypeptide Chains." *Science* 120: 779–780.

Gamow, G., Rich, A., and Ycas, M. (1955). "The Problem of Information Transfer from the Nucleic Acids to Proteins." *Advances in Biological and Medical Physics* 4: 23–68.

Gamow, G., and Ycas, M. (1955). "Statistical Correlation of Protein and Ribonucleic Acid Composition." *Proceedings of the National Academy of Sciences* 41: 1011–1119.

Gatlin, L. (1972). *Information Theory and the Living System.* New York: Columbia University Press.

Golomb, S. W. (1962). "Efficient Coding for the Desoxyribonucleic Acid Channel." *Proceedings of the Symposium for Applied Mathematics* 14: 87–100.

Golomb, S. W., Welch, L. R., and Delbrück, M. (1958). "Construction and Properties of Comma-Free Codes." *Biologiske Meddelelser udviket af Det Kongelige Danske Videnskabernes Selskab* 23(9): 1–34.

Haldane, J. B. S. (1939). *The Marxist Philosophy and the Sciences.* New York: Random House.

Herman, N. D., and Schneider, T. D. (1992). "High Information Conservation Implies That at Least Three Proteins Bind Independently to F Plasmid *incD* Repeats." *Journal of Bacteriology* 174: 3558–3560.

Jablonka, E., Lachmann, L., and Lamb, M. J. (1992). "Evidence, Mechanisms and Models for the Inheritance of Acquired Characteristics." *Journal of Theoretical Biology* 158: 245–268.

Jacob, F., and Monod, J. (1959). "Gènes de structure et gènes de régulation dans la biosynthèse des protéines." *Comptes Rendus des Séances de l'Academie des Sciences* 249: 1282–1284.

Jacob, F., Perrin, D., Sanchez, E., and Monod, J. [1960] (1965). "The Operon: A Group of Genes Whose Expression is Coordinated by an Operator." In Adelberg, E. A. (ed.), *Papers on Bacterial Genetics.* Boston: Little, Brown and Co., pp. 198–200.

Judson, H. F. (1979). *The Eighth Day of Creation.* New York: Simon and Schuster.

Kalmus, H. (1950). "A Cybernetical Aspect of Genetics." *Journal of Heredity* 41: 19–22.

Kay, L. E. (1994). "Who Wrote the Book of Life? Information and the Transformation of Molecular Biology, 1945–1955." In: Hagner, M. and Rheinberger, H.-J.

(eds.), *Experimental-systeme in den Biologischen-Medizinischen Wissenschaften: Objekt, Differenzen, Konjunkturen*. Berlin: Akademie Verlag.

Keller, E. F. (1995). *Refiguring Life: Metaphors of Twentieth Century Biology*. New York: Columbia University Press.

Kimura, M. (1961). "Natural Selection as the Process of Accumulating Genetic Information in Adaptive Evolution." *Genetical Research* 2: 127–140.

Kimura, M. (1968). "Evolutionary Rate at the Molecular Level." *Nature* 217: 624–626.

Kimura, M. (1983). *The Neutral Theory of Molecular Evolution*. Cambridge: Cambridge University Press.

Kleene, S. C. (1956). "Representation of Events in Nerve Nets and Finite Automata." In: Shannon, C. E. and McCarthy, J. (eds.), *Automata Studies*. Princeton: Princeton University Press, pp. 3–41.

Koslowski, D. J., Bhat, G. J., Perollaz, A. L., Feagin, J. E., and Stuart, K. (1990). "The MURF3 Gene of *T. brucei* Contains Multiple Domains of Extensive Editing and is Homologous to a Subunit of NADH Dehydrogenase." *Cell* 62: 901–911.

Küppers, B.-O. (1983). *The Molecular Theory of Evolution*. Berlin: Springer-Verlag.

Landman, O. E. (1991). "The Inheritance of Acquired Characteristics." *Annual Review of Genetics* 25: 1–20.

Landsteiner, K. (1936). *The Specificity of Serological Reactions*. Springfield, IL: C. C. Thomas.

Lanni, F. (1964). "The Biological Coding Problem." *Advances in Genetics* 12: 1–141.

Lederberg, J. (1951). "Genetic Studies with Bacteria." In Dunn, L. C. (ed.), *Genetics in the 20th Century*. New York: Macmillan, pp. 263–289.

Lederberg, J. (1956). "Comments on the Gene-Enzyme Relationship." In Gaebler, O. H. (ed.), *Enzymes: Units of Biological Structure and Function*. New York: Academic Press, pp. 161–169.

Lederberg, J. (1993). "What the Double Helix (1953) Has Meant for Basic Biomedical Science." *Journal of the American Medical Association* 269: 1981–1985.

Lederberg, J., Lederberg, E. M., Zinder, N. D., and Lively, E. R. (1951). "Recombination Analysis of Bacterial Heredity." *Cold Spring Harbor Symposia on Quantitative Biology* 16: 413–443.

Lederberg, J., and Tatum, E. L. (1946a). "Gene Expression in *Escherichia coli*." *Nature* 158: 588.

Lederberg, J., and Tatum, E. L. (1946b). "Novel Genotypes in Mixed Cultures of Biochemical Mutants of Bacteria." *Cold Spring Harbor Symposia on Quantitative Biology* 11: 113–114.

Leggett, A. J. (1987). *The Problems of Physics*. Oxford: Oxford University Press.

Lewin, B. (1974). *Gene Expression-2: Eucaryotic Chromosomes*. New York: John Wiley.

Linschitz, H. (1953). "The Information Content of a Bacterial Cell." In Quastler, H. (ed.), *Essays on the Use of Information Theory in Biology*. Urbana, University of Illinois Press, pp. 251–262.

Luria, S. E., and Delbrück, M. (1943). "Mutations of Bacteria from Virus Sensitivity to Virus Resistance." *Genetics* 28: 491–511.

Matthaei, J. H., and Nirenberg, M. W. (1961a). "Characteristics and Stabilization of DNA Sensitive Protein Synthesis in *E. coli* Extracts." *Proceedings of the National Academy of Sciences* 47: 1580–1588.

Matthaei, J. H., and Nirenberg, M. W. (1961b). "The Dependence of Cell-Free Protein Synthesis in *E. coli* upon Naturally Occurring or Synthetic Polyribonucleotides." *Proceedings of the National Academy of Sciences* 47: 1588–1594.

Mazia, D. (1956). "Nuclear Products and Nuclear Reproduction." In: Gaebler, O. H. (ed.), *Enzymes: Units of Biological Structure and Function*. New York: Academic Press, pp. 261–278.

Monod, J. (1956). "Remarks on the Mechanism of Enzyme Induction." In: Gaebler, O. H. (ed.), *Enzymes: Units of Biological Structure and Function*. New York: Academic Press, pp. 7–28.

Monod, J. (1971). *Chance and Necessity*. New York: Knopf.

Monod, J., Cohen-Bazire, G., and Cohn, M. (1951). "Sur la biosynthese de la β-galactosidase (lactase) chez *Escherichia coli* la specificite de l'induction." *Biochimicia et Biophysica Acta* 7: 585–599.

Monod, J., Pappenheimer, A., and Cohen-Bazire, G. (1952). "La cinétique de la biosynthèse de la β-galactosidase chez *E. coli* considérée comme fonction de la croissance." *Biochimica et Biophysica Acta* 9: 648–660.

Neveln, B. (1990). "Comma-Free and Synchronizable Codes." *Journal of Theoretical Biology* 144: 209–212.

Nickles, T. (1973). "Two Concepts of Inter-Theoretic Reduction." *Journal of Philosophy* 70: 181–201.

Olby, R. C. (1971). "Schrödinger's Problem: What is Life?" *Journal of the History of Biology* 4: 119–148.

Pardee, A. B., Jacob, F., and Monod, J. (1959). "The Genetic Control and Cytoplasmic Expression of 'Inducibility' in the Synthesis of β-galactosidase by *E. coli*." *Journal of Molecular Biology* 1: 165–178.

Pauling, L. (1940). "A Theory of the Structure and Process of Formation of Antibodies." *Journal of the American Chemical Society* 62: 2643–2657.

Perutz, M. (1990). *Mechanisms of Cooperativity and Allosteric Regulation in Proteins.* Cambridge: Cambridge University Press.

Pierce, J. R. (1962). *Symbols, Signals, and Noise.* New York: Harper and Brothers.

Primas, H. (1991). "Reductionism: Palaver without Precedent." In: Agazzi, E. (ed.), *The Problem of Reductionism in Science.* Dordrecht: Kluwer Academic Publishers, pp. 161–172.

Rosenblueth, A., Wiener, N., and Bigelow, J. (1943). "Behavior, Purpose and Teleology. *Philosophy of Science* 10: 18–24.

Quastler, H. (1953a). "The Measure of Specificity." In: Quastler, H. (ed.), *Essays on the Use of Information Theory in Biology.* Urbana, University of Illinois Press, pp. 41–71.

Quastler, H. (1953b). "The Specificity of Elementary Biological Functions." In: Quastler, H. (ed.), *Essays on the Use of Information Theory in Biology.* Urbana, University of Illinois Press, pp. 170–188.

Quastler, H. (1958). "The Status of Information Theory in Biology: A Round-Table Discussion." In: Yockey, H. P. (ed.), *Symposium on Information Theory in Biology.* New York: Pergamon Press, pp. 399–402.

Rickenberg, H. V., Cohen, G. N., Buttin, G., and Monod, J. (1956). "La galactoside-permease d'*Escherichia coli.*" *Annales de l'Institut Pasteur* 91: 829–857.

Sanger, F., and Thompson, O. P. (1953a). "The Amino Acid Sequence in the Glycine Chain of Insulin. 1. The Identification of Lower Peptides from Partial Hydrolysates." *Biochemical Journal* 53: 353–366.

Sanger, F., and Thompson, O. P. (1953b). "The Amino Acid Sequence in the Glycine Chain of Insulin. 2. The Investigation of Peptides from Enzymic Hydrolysates." *Biochemical Journal* 53: 366–374.

Sanger, F., and Tuppy, H. (1951a). "The Amino-Acid Sequence in the Phenylalanyl Chain of Insulin. 1. The Identification of Lower Peptides from Partial Hydrolysates." *Biochemical Journal* 49: 473–481.

Sanger, F., and Tuppy, H. (1951b). "The Amino-Acid Sequence in the Phenylalanyl Chain of Insulin. 2. The Investigation of Peptides from Enzymic Hydrolysates." *Biochemical Journal* 49: 481–490.

Sarkar, S. (1988). "Natural Selection, Hypercycles and the Origin of Life." In: Fine, A. and Leplin, J. (eds.), *PSA 1988: Proceedings of the 1988 Biennial Meeting of the Philosophy of Science Association,* vol. 2. East Lansing: Philosophy of Science Association, pp. 197–206. (Chapter 6 of this volume.)

Sarkar, S. (1989). "Reductionism and Molecular Biology: A Reappraisal." Ph.D. Dissertation. Department of Philosophy, University of Chicago.

Sarkar, S. (1992). "Models of Reduction and Categories of Reductionism." *Synthese* 91: 167–194. (Chapter 2 of this volume.)

Sarkar, S. (1998). *Genetics and Reductionism*. Cambridge: Cambridge University Press.

Sarkar, S., and Tauber, A. I. (1991). "Fallacious Claims for HGP." *Nature* 353: 691.

Schaffner, K. (1974a). "Logic of Discovery and Justification in Regulatory Genetics." *Studies in the History and Philosophy of Science* 4: 349–385.

Schaffner, K. (1974b). "The Unity of Science and Theory Construction in Molecular Biology." *Boston Studies in the Philosophy of Science* 58: 497–533.

Schneider, T. D. (1988). "Information and Entropy of Patterns in Genetic Switches." In: Erickson, G. J. and Smith, C. R. (eds.), *Maximum-Entropy and Bayesian Methods in Science and Engineering*, vol. 2. Dordrecht: Kluwer Academic Publishers, pp. 147–154.

Schneider, T. D., Stormo, G. D., Gold, L., and Ehrenfeucht, A. (1986). "Information Content of Binding Sites of Nucleotide Sequences." *Journal of Molecular Biology* 188: 415–431.

Schrödinger, E. (1944). *What Is Life?* Cambridge: Cambridge University Press.

Shannon, C. E. (1948). "A Mathematical Theory of Communication." *Bell System Technical Journal* 27: 379–423, 623–656.

Shapiro, J. A. (1991). "Genomes as Smart Systems." *Genetica* 84: 3–4.

Shapiro, J. A. (1992). "Natural Genetic Engineering in Evolution." *Genetica* 86: 99–111.

Shimony, A. (1987). "The Methodology of Synthesis: Parts and Wholes in Low-Energy Physics." In: Kargon, R. and Achinstein, P. (eds.), *Kelvin's Baltimore Lectures and Modern Theoretical Physics*. Cambridge: The MIT Press, pp. 399–423.

Shimony, A. (1995). "Cybernetics and Social Entities." *Boston Studies in the Philosophy of Science* 164: 181–196.

Smith, C. W., Patton, J. G., and Nadal-Ginard, B. (1989). "Alternative Splicing in the Control of Gene Expression." *Annual Review of Genetics* 23: 527–577.

Spiegelman, S. (1956). "On the Nature of the Enzyme-Forming System." In: Gaebler, O. H. (ed.), *Enzymes: Units of Biological Structure and Function*. New York: Academic Press, pp. 67–89.

Stephens, R. M., and Schneider, T. D. (1992). "Features of Spliceosome Evolution and Function Inferred from an Analysis of the Information at Human Splice Sites." *Journal of Molecular Biology* 228: 1124–1136.

Tauber, A. I., and Sarkar, S. (1992). "The Human Genome Project: Has Blind Reductionism Gone Too Far?" *Perspectives in Biology and Medicine* 35: 220–235.

Thaler, D. S. (1996). "Paradox as Path: Pattern as Map—Classical Genetics as a Source of Non-Reductionism in Molecular Biology." In: Sarkar, S. (ed.), *The Philosophy and History of Molecular Biology*. Dordrecht: Kluwer Academic Publishers.

Timoféeff-Ressovsky, H. A., and Timoféeff-Ressovsky, N. W. (1926). "Über das Phänotypische Manifestieren des Genotyps. II. Über Idio-Somatische Variationsgruppen bei Drosophila funebiris." *Roux Archiv für Entwicklungsmechanik der Organismen* 108: 146–170.

Umbarger, H. E. (1956). "Evidence for a Negative-Feedback Mechanism in the Biosynthesis of Isoleucine." *Science* 123: 848.

Waddington, C. H., and Lewontin, R. C. (1968). "A Note on Evolution and Changes in the Quantity of Genetic Information." In: Waddington, C. H. (ed.), *Towards a Theoretical Biology*, vol. 1. Chicago: Aldine Publishing Company, pp. 109–110.

Watson, J. D., and Crick, F. H. C. (1953a). "Molecular Structure of Nucleic Acids: A Structure for Deoxyribose Nucleic Acid." *Nature* 171: 737–738.

Watson, J. D., and Crick, F. H. C. (1953b). "Genetical Implications of the Structure of Deoxyribose Nucleic Acid." *Nature* 171: 964–967.

Wiener, N. (1948). *Cybernetics*. Cambridge, MA: MIT Press.

Williams, G. C. (1966). *Adaptation and Natural Selection: A Critique of Some Current Evolutionary Thought*. Princeton: Princeton University Press.

Wimsatt, W. C. (1971). "Some Problems with the Concept of 'Feedback.'" *Boston Studies in the Philosophy of Science* 8: 241–256.

Wimsatt, W. C. (1976). "Reductive Explanation: A Functional Account." *Boston Studies in the Philosophy of Science* 32: 671–710.

Woese, C. (1967). *The Genetic Code*. New York: Harper and Row.

Yanofsky, C., Carlton, B. C., Guest, J. R., Helinski, D. R., and Henning, U. (1964). "On the Colinearity of Gene Structure and Protein Structure." *Proceedings of the National Academy of Sciences* 51: 266–272.

Yates, R. A., and Pardee, A. B. (1956). "Control of Pyrimidine Biosynthesis in *Escherichia coli* by a Feed-Back Mechanism." *Journal of Biological Chemistry* 221: 757–770.

Ycas, M. (1969). *The Biological Code*. New York: American Elsevier.

Yockey, H. P. (1992). *Information Theory and Molecular Biology*. Cambridge: Cambridge University Press.

Yoxen, E. J. (1979). "Where Does Schroedinger's 'What is Life?' Belong in the History of Molecular Biology?" *History of Science* 17: 17–52.

10 How Genes Encode Information for Phenotypic Traits

10.1 Introduction

According to the *Oxford English Dictionary*, the term "information" (though spelled "informacion") was first introduced by Chaucer in 1386 to describe an item of training or instruction. In 1387, it was used to describe the act of "informing." As a description of knowledge communicated by some item, it goes back to 1450. However, attempts to quantify the amount of information contained in some item date back only to R. A. Fisher in 1925. In a seminal paper on the theory of statistical estimation, Fisher argued that "the intrinsic accuracy of an error curve may ... be conceived as the amount of information in a single observation belonging to such a distribution" (1925, p. 709). The role of the concept of information was to allow discrimination between consistent estimators of some parameter; the amount of "information" gained from a single observation is a measure of the efficiency of an estimator. Suppose that the parameter to be estimated is the mean height of a human population; potential estimators can be other statistical "averages" such as the median and the mode. Fisher's theory of information became part of the standard theory of statistical estimation, but it is otherwise disconnected from scientific uses of "information." The fact that the first successful quantitative theory of "information" is irrelevant in scientific contexts outside its own narrow domain underscores an important feature of the story that will be told here: "information" is used in a bewildering variety of ways in the sciences, some of which are at odds with each other. Consequently, any account of the role of informational thinking in a science must pay careful attention to exactly what sense of "information" is intended in that context.

Shortly after Fisher, and independently of him, in 1928 R. V. L. Hartley provided a quantitative analysis of the amount of information that can be transmitted over a system such as a telegraph. During a decade in which telecommunication came to be at the forefront of technological innovation, the theoretical framework it used proved to be influential.[1] Hartley recognized that, "as commonly used, information is a very elastic term," and he proceeded "to set up for it a more specific meaning" (1928, p. 356). Relying essentially on a linear symbolic system of information transmission (for instance, by a natural language), Hartley argued that, for a given message, "inasmuch as the precision of the information depends upon what other symbol sequences might have been chosen it would seem reasonable to hope to find in the number of these sequences the desired quantitative measure of information" (p. 536). Suppose that a telegraphic message is n symbols long with the symbols drawn from an alphabet of size s. Through an ingenious argument, similar to the one used by Shannon (see below), Hartley showed that the appropriate measure for the number of these sequences is $n \log s$. He identified this measure with the amount of information contained in the message.[2]

Using the same framework as Hartley, in 1948, C. E. Shannon developed an elaborate and elegant mathematical theory of communication that came to be called "information theory" and constitutes one of the more important developments of applied mathematics in the twentieth century. The theory of communication will be briefly analyzed in section 10.2 below, with an emphasis on its relevance to genetics. The assessment of relevance will be negative. When, for instance, it is said that the hemoglobin-S gene contains information for the sickle-cell trait, communication-theoretic information cannot capture such usage. (Throughout this chapter, "gene" will be used to refer to a segment of DNA with some known function.) To take another example, the fact that the information contained in a certain gene may result in polydactyly (having an extra finger) in humans also cannot be accommodated by communication-theoretic information. The main problem is that, at best, communication-theoretic information provides a measure of the amount of information in a message but does not provide an account of the content of a message—its specificity, what makes it *that* message. The theory of communication never had any such pretension. As Shannon bluntly put it

How Genes Encode Information

at the beginning of his paper: "These semantic aspects of communication are irrelevant to the engineering problem" (1948, p. 379).

Capturing *specificity* is critical to genetic information. Specificity was one of the major themes of twentieth-century biology. During the first three decades of that century, it became clear that the molecular interactions that occurred within living organisms were highly "specific" in the sense that particular molecules interacted with exactly one, or at most a very few, reagents. Enzymes acted specifically on their substrates. Mammals produced antibodies that were highly specific to antigens. In genetics, the ultimate exemplar of specificity was the "one gene–one enzyme" hypothesis of the 1940s, which served as one of the most important theoretical principles of early molecular biology. By the end of the 1930s, a highly successful theory of specificity, one that remains central to molecular biology today, had emerged. Due primarily to L. Pauling (see, e.g., Pauling 1940), though with many antecedents, this theory claimed: (i) that the behavior of biological macromolecules was determined by their shape or "conformation"; and (ii) what mediated biological interactions was a precise "lock-and-key" fit between shapes of molecules. Thus the substrate of an enzyme had to fit into its active site. Antibodies recognized the shape of their antigens. In the 1940s, when the three-dimensional structure of not even a single protein had been experimentally determined, the conformational theory of specificity was still speculative. The demonstration of its approximate truth in the late 1950s and 1960s was one of early molecular biology's most significant triumphs.

Starting in the mid-1950s, assumptions about information provided an alternative to the conformational theory of specificity, at least in the relation between DNA and proteins (Lederberg 1956). This relation is the most important because proteins are the principal biological interactors at the molecular level: enzymes, antibodies, molecules such as hemoglobin, molecular channel components, cell membrane receptors, and many (though not most) of the structural molecules of organisms are proteins. Information, as F. H. C. Crick defined it in 1958, was the *"precise* determination of sequence, either of bases in the nucleic acid or of amino acid residues in the protein" (Crick 1958, p. 153; emphasis in the original). Genetic information lay in the DNA sequence. The relationship between that sequence and the sequence of amino acid residues in a protein was seen to be mediated by the genetic "code," an idea that, though originally broached by E. Schrödinger in 1943, also dates from the 1950s. The code explains the specificity of the one gene–one enzyme relationship elegantly: different DNA sequences encode different proteins, as can be determined by looking up the genetic code table. Whatever the appropriate explication of information for genetics is, it has to come to terms with specificity and the existence of this coding relationship. Communication-theoretic information neither can, nor was intended to, serve that purpose.

Surprisingly, a comprehensive account of a theory of information appropriate for genetics does not exist. In the 1950s there were occasional attempts by philosophers—for instance, by R. Carnap and Y. Bar-Hillel (1952)—to explicate a concept of "semantic" information distinct from communication-theoretic information. However, these attempts were almost always designed to capture the semantic content of linguistic structures and are of no help in the analysis of genetic information. Starting in the mid-1990s, there has been considerable skepticism, at least among philosophers, about the role of "information" in genetics. For some, genetic information is no more than a metaphor masquerading as a theoretical concept (Sarkar 1996a,b; Griffiths 2001). According to these criticisms, even the most charitable attitude toward the use of "information" in genetics can only provide a defense of its use in the 1960s, in the context of prokaryotic genetics (i.e., the genetics of organisms without compartmentalized nuclei in their cells). Once the "unexpected complexity of eukaryotic genetics" (Watson, Tooze, and Kurtz 1983, chapter 7) has to be accommodated, the loose use of "information" inherited from prokaryotic genetics is at least misleading (Sarkar 1996a). Either informational talk should be abandoned altogether or an attempt must be made to provide a formal explication of "information" that shows that it can be used consistently in this context and, moreover, is useful.

Section 10.3 gives a sketch of one such attempted explication. A category of "semiotic" information is introduced to explicate such notions as coding. Semiotic information incorporates *specificity* and depends on the possibility of *arbitrary* choices in the assignment of symbols to what they symbolize as, for instance, exemplified in the genetic code. Semiotic information is not a semantic concept. There is no reason to suppose that any concept of *biological* information must be "semantic" in the sense that philosophers use that term. Biological interactions, at this level of organization, are about the rate and accuracy of macromolecular interactions. They are not about meaning, intentionality, and the like; any demand that such notions be explicated in an account of biological information is no more than a signifier for a philosophical agenda inherited from manifestly nonbiological contexts, in particular from the philosophy of language and mind. It only raises spurious problems for the philosophy of biology.

Section 10.3 also applies this framework to genetics at the macromolecular level of DNA and proteins. It concludes that there is a sense in which it is appropriate and instructive to use an informational framework for genetics at this level. However, proteins are often far removed from the traits that are usually studied in organismic biology; for instance, the shape, size, and behavior of organisms. Section 10.4 explores the extent to which informational accounts carry over to the level of such traits. Much depends on how "trait" is construed, and there is considerable leeway about its definition. Given a relatively inclusive construal of "trait," section 10.4 concludes that, to the extent that a molecular etiology can at all be attributed to a trait, a carefully circumscribed informational account remains appropriate.

Finally, section 10.5 cautions against any overly ambitious interpretation of the claims defended earlier in this chapter. They do not support even a mild form of genetic reductionism (that genes alone provide the etiology of traits), let alone determinism. They do not support the view that DNA alone must be the repository of biological information. Perhaps, most important, they do not support the view that the etiology of traits can be fully understood in informational terms from a predominantly genetic basis.

10.2 The Theory of Communication

Shannon conceived of a communication system as consisting of six components:

(i) an information source that produces a raw "message" to be transmitted;

(ii) a transmitter that transforms or "encodes" this message into a form appropriate for transmission through the channel;

(iii) the channel through which the encoded message or "signal" is transmitted to the receiver;

(iv) the receiver that translates or "decodes" the received signal back into what is supposed to be the original message;

(v) the destination or intended recipient of the message; and

(vi) sources of the noise that acts on the channel and potentially distorts the signal or encoded message. Obviously this is an undesirable component, but one that is unavoidable in practice.

The most important aspect of this characterization is that it is abstracted away from any particular protocol for coding as well as any specific medium of transmission. From the point of view of the theory of communication, information is conceived of as the choice of one message from a set of possible messages with a definite probability associated with the choice: the lower this probability, the higher is the information associated with the choice, because a greater uncertainty is removed by that choice. The central problems that the theory of communication sets out to solve include the efficiency (or relative accuracy) with which information can be transmitted through a channel in the presence of noise, and how the rate and efficiency of transmission are related to the rate at which messages can be encoded at the transmitter and to the capacity of the channel.

To solve these problems requires a quantitative measure of information. Suppose that a message consists of a sequence of symbols chosen from a basic symbol set (often called an "alphabet"). For instance, it can be a DNA sequence, with the basic symbol set being $\{A,C,G,T\}$. In a sequence of length n, let p_i be the probability of occurrence of the i-th symbol in that sequence. Then the information content of the message is $H = -\sum_{i=1}^{n} p_i \log p_i$. (In what follows, this formula will be called the *Shannon measure of information*.) Consider the sequence "ACCTCGATTC." Then, at the first position, $H_1 = -p_1 \log p_1$, where $p_1 = \frac{2}{10} = 0.2$ because "A" occurs twice in the sequence of ten symbols (which is all that is known about the relative frequencies of the occurrence of the symbols). $H = \sum_{i=1}^{10} H_i$, computed in this way, is the information content of the entire sequence.

Shannon justified this choice for the measure of information by showing that $-K \sum_{i=1}^{n} p_i \log p_i$, where K is a constant, is the only function that satisfies the following three reasonable conditions: (i) the information function is continuous in all the p_i; (ii) if all the p_i are identical and equal to $1/n$, then the function is a monotonically increasing function of n; and

(iii) if the choice involved in producing the message can be decomposed into several successive choices, then the information associated with the full choice is a weighted sum of each of the successive choices, with the weights equal to the probability of each of them. K is fixed by a choice of units. If the logarithm used is of base 2, then K is equal to 1.

Formally, the Shannon measure of information is identical to the formula for entropy in statistical physics. This is not surprising since entropy in statistical physics is a measure of disorder in a system and Shannon's information measures the amount of uncertainty that is removed. Over the years, some have held that this identity reveals some deep feature of the universe; among them are those who hold that physical entropy provides a handle on biological evolution (e.g., Brooks and Wiley 1988). Most practitioners of information theory have wisely eschewed such grandiose ambitions (e.g., Pierce 1962), noting that the identity may be no greater significance than, for instance, the fact that the same bell curve (the normal distribution) captures the frequency of measurements in a wide variety of circumstances, from the distribution of molecular velocities in a gas to the distribution of heights in a human population. However, the entropic features of the Shannon measure have been used effectively by T. D. Schneider (see, e.g., Schneider 1999) and others to identify functional regions of DNA sequences: basically, because of natural selection, these regions vary less than others, and so a relatively invariant sequence with a high probability and, therefore, a low information content is expected to be found in these regions.

Shannon's work came, as did the advent of molecular biology, at the dawn of the computer era when few concepts were as fashionable as that of information. The 1950s saw many attempts to apply information theory to proteins and DNA; these were an unmitigated failure (Sarkar 1996a). Within evolutionary genetics, M. Kimura (1961) produced one intriguing result: the Shannon measure of information can be used to calculate the amount of "information" that is accumulated in the genomes of organisms through natural selection. A systematic analysis of Kimura's calculation was given by G. C. Williams in 1966. Williams first defined the gene as "that which segregates and recombines with appreciable frequency" (Williams 1966, p. 24) and then assumed, offering no argument, that this definition is equivalent to the gene being "any hereditary information for which there is a favorable or unfavorable selection bias equal to several or many times its rate of endogenous change [mutation]" (p. 25). From this, Williams concluded that the gene is a "cybernetic abstraction" (p. 33). Williams's book lent credence to the view that the gene, already viewed informationally by molecular biologists, can also be so viewed in evolutionary contexts.

Nevertheless, interpreting genetics using communication-theoretic information presents insurmountable difficulties even though the most popular objection raised against it is faulty:

(i) The popular objection just mentioned distinguishes between "semantic" and "causal" information (Sterelny and Griffiths 1999; Maynard Smith 2000; Sterelny 2000; Griffiths 2001). The former is supposed to require recourse to concepts such as intentionality. As mentioned in the last section, and as will be further demonstrated in the deflationary account of information given in the next section, such resources are not necessary for biological information. (It is also not clear why a notion of information based on intentionality should be regarded as "semantic" if that term is supposed to mean what logicians usually take it to mean.) Flow of the latter kind of information is supposed to lead to covariance (in the sense of statistical correlation) between a transmitter and receiver. Communication-theoretic information is supposed to be one type of such "causal" information. (It is also misleading to call this type of interaction "causal," since it presumes nothing more than mere correlation.) With this distinction in mind, the objection to the use of "causal" information in genetics goes as follows: in genetics, the transmitter is supposed to consist of the genes (or the cellular machinery directly interpreting the DNA—this choice does not make a difference), the receiver is the trait, and the cytoplasmic and other environmental factors that mediate the interactions between the genes and the trait constitute the channel. Now, if environmental conditions are held constant, the facts of genetics are such that there will indeed be a correlation between genes and traits. However, one can just as well hold the genetic factors constant, treat the environment as the transmitter, and find a correlation between environmental factors and traits. The kernel of truth in this argument is that, depending on what is held constant, there will be a correlation between genetic or environmental factors and traits. In fact, in most circumstances, even if neither is held entirely constant, there will be correlations between both genetic and environmental factors and traits. All that follows is that, if correlation suffices as a condition for information transfer, both genes and environments carry information for traits. Thus, if genes are to be informationally privileged, more than correlation is required. However, mathematical communication or information theory is irrelevant to this argument. The trappings of Shannon's model of a communication system are extraneously added to a relatively straightforward point about genetic and environmental correlations, and do no cognitive work. This argument does not establish any argument against the use of communication-theoretic information to explicate genetic information.

(ii) The main reason communication-theoretic information does not help in explicating genetic information is that it does not address the critical point that genetic information explains biological specificity. This is most clearly seen by looking carefully at what the various components of Shannon's model of a communication system do. Three of these are relevant here: the transmitter, the channel, and the receiver. There is a symmetry between the transmitter and the receiver; the former encodes a message to be sent, the latter decodes a received message. Specific relations are at play in Shannon's model only during encoding and decoding. In most cases, though not all, encoding and decoding consists of something like translation using a dictionary. The transmitter obtains from the source a message in some syntactic form. The encoding process uses this as an input and produces as output some sequence of entities (another syntactic form) that is physically amenable for propagation through the channel. This process may consist of using the value of some syntactic object in the input (for instance, the intensity of light of a particular frequency) to select an appropriate output; however, most often, it consists of using a symbolic look-up table. In either case, there is a specific relationship between input and output. (In information theory, this is the well-studied "coding" problem.) A similar story holds for decoding. The critical point is that the Shannon measure of information plays no role in this process. That measure applies only to what happens along the channel. If a highly improbable syntactic object (the signal) is the input and, because of noise, a less improbable entity is the output, information has been lost. Specificity has nothing to do with this.

(iii) Even leaving aside specificity, note that high communication-theoretic information content is the mark of a message of low probability. Yet, as a result of natural selection, DNA sequences that most significantly affect the functioning of an organism, and are thus most functionally informative, become increasingly frequent in a population. Thus, communication-theoretic information is negatively correlated with functional information. This conclusion raises a peculiar possibility. Communication-theoretic information may be a guide to functional information if, as a measure of functional information, its inverse is used. If functional information is taken to be a measure of genetic information (which is reasonable since, during evolution, functional information is primarily transmitted through DNA), communication-theoretic information provides access to genetic information through this inversion. Nevertheless, this attempt to rescue communication-theoretic information for genetics fails because those genes that are not selected for are no longer carriers of information. These include not only genes that may be selected against but nevertheless persist in a population (for instance, because of heterozygote advantage or tight linkage—that is, being very close on a chromosome to a gene that is selected for) but also the many genes that are simply selectively neutral. Understanding the role of information in genetics will require a different approach altogether.

10.3 Semiotic Information

The use of the concept of information pervades contemporary genetics with the gene as the locus of information. DNA is supposed to carry information for proteins, possibly for traits. The so-called central dogma of molecular biology states: "once 'information' has passed into protein *it cannot get out again*. In more detail, the transfer of information from nucleic acid to nucleic acid, or from nucleic acid to protein may be possible, but transfer from protein to protein, or from protein to nucleic acid is impossible" (Crick 1958; emphasis in the original). J. Maynard Smith (2000) has claimed that DNA is the sole purveyor of biological information in organisms. Routinely, talk of information is intertwined with linguistic metaphors from both natural and artificial languages: there is a genetic *code* because a triplet of DNA (or RNA) nucleotides codes for each amino acid residue in proteins (polypeptide chains); there are alternative *reading frames*—DNA is *transcribed* into RNA, RNA is *translated* into protein, RNA is *edited*; and so on. The use of such talk is so pervasive that it almost seems impossible that, short of pathological convolution, the experimental results of genetics can even be communicated without these resources.

What is largely forgotten is that there is good reason to believe that such talk of information in genetics may not be necessary: "information" was introduced in genetics only in 1953, just before the DNA double-helix model (Ephrussi et al. 1953). "Information" was supposed to introduce some frugality in a field that had become unusually profligate in terminological innovation (for instance, "transformation" and "transduction") to describe genetic interactions in bacteria. The main reason the informational framework became central to the new molecular biology of the 1950s and 1960s was the characterization of the relationship between DNA and proteins as determined by a universal genetic code, "universal" in the sense that it is supposed to be the same for all species. (There is, however, some variation, though the extent of it is not known—see chapter 9 and Fox 1987.) A triplet of nucleotide bases specifies a single amino acid residue. Since there are four nucleotide bases (A, C, G, and T) in DNA, there are 64 possible triplets. If all triplets code for an amino-acid residue, since there are only 20 standard amino-acid residues in proteins, the code must be degenerate (or partially redundant): more than one triplet must code for the same amino-acid residue. Three factors make the informational interpretation of this relationship illuminating: (a) the relationship can be viewed as a symbolic one, with each DNA triplet being a symbol for an amino acid residue; (b) the relationship is combinatorial, with different combinations of nucleotides potentially specifying different residues; and, most important, (c) the relationship is arbitrary in an important sense. Functional considerations may explain some features of the code—why, for instance, an arbitrary mutation tends to convert a hydrophilic amino-acid residue to another such residue—but they do not explain why the code is specifically what it is. The physical mechanisms of translation do not help either. The genetic code is *arbitrary*. Along with specificity, this arbitrariness is what makes an informational account of genetics useful.

To explain how this account works will now require an explication of semiotic information (Sarkar 2000). This explication will proceed in two stages. After the structure of an information system is defined, conditions will first be laid down to establish specificity; second, more conditions will be imposed to capture the type of arbitrariness also required of semiotic information.

A formal information system consists of a relation, ι (the information relation), between two sets, A (for instance, a set of DNA sequences) and B

(for instance, a set of polypeptide sequences) and will be symbolized as $\langle A, B, \iota \rangle$. The relation ι holds between A and B because it holds between their elements. The following simplifying assumptions will be made: (i) ι partitions A into a set of equivalence classes, with all member of an equivalence class being informationally equivalent to each other (in the case of DNA and protein, an equivalence class consists of all the triplets that specify the same amino acid residue); and (ii) A and ι exhaust B—that is, for every element of B, there is some element in A that is related to it by ι. One consequence of this assumption is that ι partitions B into a set of equivalence classes, with each equivalence class of B corresponding to one of A.

With this background, for an information system to allow for specificity, the most important condition on $\langle A, B, \iota \rangle$ is:

(I1) *Differential specificity*: Suppose that a and a' belong to different equivalence classes of A. Then, if $\iota(a,b)$ and $\iota(a',b')$ hold, then b and b' must be different elements of B.

Condition (I1) suffices to capture the most basic concept of a specific informational relation holding between A and B.[3] An additional condition will sometimes be imposed to characterize a stronger concept of specificity:

(I2) *Reverse differential specificity*: if $\iota(a,b)$ and $\iota(a',b')$ hold, and b and b' are different elements of B, then a and a' belong to different equivalence classes in A.

If every equivalence class in A specifies exactly one element of B through ι, condition (I2) is automatically satisfied provided that (I1) is satisfied. That A and B covary (that is, there are statistical correlations between the occurrences of elements of A and B) is a trivial consequence of either condition. The justification for these conditions is that they capture what is customarily meant by information in any context: for instance, they capture the sense in which the present positions and momenta of the planets carry information about their future positions and momenta.

By definition, if only condition (I1) holds, then A will carry *specific information* for B; if both conditions (I1) and (I2) hold, A *alone* carries specific information for B. In the case of prokaryotic genetics, both conditions hold: DNA alone carries information for proteins. Some care has to be taken at this stage, and the discussion must move beyond formal specification to the empirical interpretation of ι in this (genetic) context. The claim that DNA alone carries information for proteins cannot be interpreted as saying that the presence of a particular DNA sequence will result in the production of the corresponding protein no matter what the cellular environment does. Such a claim is manifestly false. In certain cellular contexts, the presence of that DNA sequence will not result in the production of any protein at all. Rather, the claim must be construed counterfactually: if the presence of this DNA sequence were to lead to the production of a protein, then ι describes which protein it leads to. In general, the production of the relevant protein requires an enabling history of environmental conditions. This history is not unique. Rather, there is a set M of "standard" histories that result in the formation of protein from DNA. A complete formal account of biological information must specify the structure of this set. This is beyond what is possible given the current state of empirical knowledge: M cannot be fully specified even for the most studied bacterium, *Escherichia coli*. However, because the relevant informational relation is being construed counterfactually in the way indicated, the inability to specify M does not prevent the use of that relation. (It does mean, though, that the information content of the DNA, by itself, does not suffice as an etiology for that protein.) The structure of the relation between M and the protein set will be a lot more complicated than $\langle A, B, \iota \rangle$, where A is the DNA set and B the protein set; this is already one sense in which genes are privileged.

So far, only specificity has been analyzed; it remains to provide an account of arbitrariness. This requires the satisfaction of two conditions that are conceptually identical to those imposed by Shannon on communication systems in order to achieve a suitable level of abstraction. It shows an important commonality between communication-theoretic and semiotic information:

(A1) *Medium independence*: $\langle A, B, \iota \rangle$ can be syntactically represented in any way, with no preferred representation, so long as there is an isomorphism between the sets corresponding to A, B, and ι in the different representations.[4]

An equation representing the gravitational interaction between the earth and the sun is a syntactic representation of that relation. The actual earth and sun, along with the gravitational interaction, also make up a syntactic representation of the relation. The latter representation is preferred because the equation is a representation of the earth, sun, and their gravitational interaction in a way in which the latter is not a representation of the former. Medium-independence for information denies the existence of any such asymmetry. From the informational point of view, a physical string of DNA is no more preferred as a representation of the informational content of a gene than is a string of As, Cs, Gs, and Ts on this sheet of paper. (The most useful analogy to bear in mind is that of digital computation, where the various representations, whether it be in one code or another or as electrical signals in a circuit, are all epistemologically on a par with each other.) This is also the sense of medium-independence required in Shannon's account of communication.

The second condition is:

(A2) *Template assignment freedom*: let A_i, $i = 1, \ldots, n$, partition A into n equivalence classes, and let $\langle A_1, A_2, \ldots, A_n \rangle$ be a the sequence of these classes that are related by ι to the sequence $\langle B_1, B_2, \ldots, B_n \rangle$ of classes of B. Then the underlying mechanisms of gene expression allow for any permutation $\langle A_{\sigma(1)}, A_{\sigma(2)}, \ldots, A_{\sigma(n)} \rangle$ (where $\sigma(1) \, \sigma(2) \ldots \sigma(n)$ is a permutation of $1 2 \ldots n$) to be mapped by ι to $\langle B_1, B_2, \ldots, B_n \rangle$.

This condition looks more complicated than it is; as in the case of Shannon's communication systems, it shows that any particular protocol of coding is arbitrary. (However, as emphasized in the last section, coding and decoding are not part of information transfer in the theory of communication.) In the case of DNA (the template), all that this condition means is that any set of triplets coding for a particular amino acid residue can be reassigned to some other residue. However, there is a subtle and important problem here: such a reassignment seems to violate a type of physical determinism that no one questions in biological organisms, namely, that the chemical system leading from a piece of DNA to a particular protein is deterministic. (Indeed, conditions [I1] and [I2] of specificity require determinism of this sort.) Condition (A2) must be interpreted in the context of genetics as saying that all these different template assignments were evolutionarily possible.[5]

The customary view, that the genetic code is an evolutionarily frozen accident, supports such an interpretation. However, actually demonstrating that different template assignments or genetic codes were evolutionarily possible, beyond arguing for the plausibility of the frozen accident model, is not easy. Suppose that there is a selectionist story to be told about the evolution of the genetic code at the level of the energetic properties of the chemical reaction systems responsible for gene expression (which is one of many possibilities). The first assumption that has to be made is that the chemical processes that lead to gene expression do not preclude a different genetic code (set of template assignments). The second assumption is that a particular chemical gene expression scheme, with a specified code, and the various transcription and translation mechanisms, is energetically preferred to immediate variants. Thus each possible code is at a local optimum in some space of chemical reactions. Many local optima and, therefore, many different codes are possible. That some mechanism of this sort must have been operative during the evolution of the genetic code is underscored by the fact that different species and different organelles within an individual of the same species, have genetic codes that are not identical with the canonical one. Moreover, the standard frozen code need not be the global optimum; it happens to be the optimum that was accessed first in evolutionary history. That is the sense in which it is an accident. It deserves emphasis that this argument, although plausible, remains speculative in the absence of a detailed empirically adequate reconstruction of the evolution of the genetic code and, in that sense, is not fully satisfactory.

To make this argument sound requires a demonstration—in gory chemical detail—that each such gene expression scheme is indeed energetically preferable to near variants. Modeling such a process will not be easy. The classical approach to modeling metabolic systems mathematically consists of following the temporal trajectories of the values of the state variables, leaving systemic parameters such as the reaction rates, as well as the stoichiometry, constant.[6] From the evolutionary perspective, however, what is critical are the evolutionary changes in stoichiometry. These must be the targets of the models that are necessary to substantiate the argument of the previous paragraph.[7] If only the initial and final stages of a reaction pathway are considered, stoichiometric evolution will result in changes of rate "constants" over evolutionary time. If, faced with several alternatives, one rate constant becomes much higher than others, then there is evolution toward specificity. Presumably, this is how the specificity of the relation between nucleic acids and proteins evolved.

In the present context, what will have to be modeled to show the arbitrariness in genetic code assignments are all the neighboring stoichiometries (also leading from nucleic acids to proteins) of the stoichiometry of the reaction pathways of the standard genetic code and its known variants (henceforth called the "usual stoichiometries"). The energetic costs of each of these will have to be assessed to determine whether the usual stoichiometries are local optima of energetic cost. To demonstrate that some stoichiometry is a local optimum merely requires exploration of all of its one-step neighbors. However, to get a better sense of the structure of the space requires a more extended exploration. The predictions of this account are: (i) each of the usual stoichiometries is at a local optimum; and (ii) there are many such optima. If (i) is correct, there is a trivial sense in which (ii) is correct, since it is empirically known that there are several stoichiometries: the one using the standard code and all the ones using the variant codes, for instance, in mitochondria. However, much more than this is meant here: there should also be unoccupied such optima to be consistent with the idea that the standard code is truly a frozen accident. Whether or not this particular speculative account of the origin of the genetic code is correct, it has the advantage of being testable.

By definition, if conditions (I1), (A1), and (A2) hold, then ι is a coding relation, and A encodes B. Thus, DNA encodes proteins; for prokaryotes, DNA alone encodes proteins. The formal characterization given above lays down adequacy conditions for any ι that embodies semiotic information. It does not specify what ι is; that is, which as are related to which bs. The latter is an empirical question: coding and similar relations in molecular biology are empirical, not conceptual, relations. This is a philosophically important point. That DNA encodes proteins is an empirical claim. Whether conditions (I1), (I2), (A1), or (A2) hold is an empirical question. Under this interpretation, these conditions must be interpreted as empirical generalizations with the usual *ceteris paribus* clauses that exclude environmental histories not belonging to M. The relevant evidence is of the kind that is typically considered for chemical reactions: for the coding relation, subject to statistical uncertainties, the specified chemical relationships must be shown to hold unconditionally. First suppose that, according to ι, a string of DNA, s, codes for a polypeptide, π. Now suppose that, as an experimental result, it is found that some π' different from π is produced in the presence of s. There must have been an *error*, for instance, in transcription or translation: a *mistake* has been made. All this means is that an anomalous result, beyond what is permitted due to standard statistical uncertainties, is obtained. It suggests the operation of intervening factors that violate the *ceteris paribus* clauses; that is, the reactions took place in an environmental history that does not belong to *M*. Thus, the question of what constitutes a mistake is settled by recourse to experiment. There is nothing mysterious about genetic information, nothing that requires recourse to conceptual resources that are extraneous to the ordinary biology of macromolecules.

As was noted, the informational account just given allows DNA alone to encode proteins in prokaryotic genetics. Genetically, prokaryotes are exceedingly simple. Every bit of DNA in a prokaryotic genome either codes for a protein or participates in regulating the transcription of DNA. For the coding regions, it is straightforward to translate the DNA sequence into the corresponding protein. Consequently, the informational interpretation seems particularly perspicuous in this context. This is the picture that breaks down in the eukaryotic context. Eukaryotic genetics presents formidable complexities including, but not limited to, the following (for details, see chapter 9):

(i) the nonuniversality of the standard genetic code—some organisms use a slightly different code, and the mitochondrial code of eukaryotic cells is also slightly different;

(ii) frameshift mutations, which are sometimes used to produce a variant amino-acid chain from a DNA sequence;

(iii) large segments of DNA between functional genes that apparently have no function at all and are sometimes called "junk DNA";

(iv) similarly, segments of DNA within genes that are not translated into protein, these being called introns, while the coding regions are called exons —after transcription, the portions of RNA that correspond to the introns are "spliced" out and not translated into protein at the ribosomes; but

(v) there is alternative splicing by which the same RNA transcript produced at the DNA (often called "pre-mRNA") gets spliced in a variety of ways to produce several proteins—in humans, it is believed that as many as a third of the genes lead to alternative splicing; and

(vi) there are yet other kinds of RNA "editing" by which bases are added, removed, or replaced in mRNA, sometimes to such an extent that it becomes hard to say that the corresponding gene encodes the protein that is produced.

In the present context, points (i), (ii), (iv), (v), and (vi) are the most important. They all show that a single sequence of DNA may give rise to a variety of amino-acid sequences even within a standard history from *M*. Thus, in the relation between eukaryotic DNA and proteins, condition (I2) fails. Nevertheless, because condition (I1) remains satisfied, eukaryotic DNA still carries specific information for proteins. However, this formal success should not be taken as an endorsement of the utility of the informational interpretation of genetics. As stated, condition (I1) does not put any constraint on the internal structure of the sets *A* (in this case, the set of DNA sequences) or *B* (in this case, the set of protein sequences). If both sets are highly heterogeneous, there may be little that the informational interpretation contributes. In the case of the DNA and protein sets, heterogeneity in the former arises only because of the degeneracy of the genetic code. This heterogeneity has not marred the utility of the code. For the protein set, heterogeneity arises because a given DNA sequence can lead to different proteins with varied functional roles. However, leaving aside the case of extensive RNA editing, it seems to be the case that these proteins are related enough for the heterogeneity not to destroy the utility of the informational interpretation.

The main upshot is this: even in the context of a standard history from *M*, although genes encode proteins and carry specific information for them, this information is not even sufficient to specify a particular protein. Knowledge of other factors, in particular so-called environmental factors, is necessary for such a specification. Whether these other factors should be interpreted as carrying semiotic information depends on whether they satisfy at least conditions (I1), (A1), and (A2). So far, there is no evidence that any of them satisfy (A1) and (A2). This is yet another sense in which genes are privileged.

10.4 Traits and Molecular Biology

Proteins are closely linked to genes: compared to organismic traits such as shape, size, and weight, the chain of reactions leading to proteins from genes is relatively short and simple. It is one thing to say that genes encode proteins, and another to say that they encode information for traits. It depends on how traits are characterized. The first point to note is that "trait" is not a technical concept within biology, with clear criteria distinguishing those organismic features that are traits from those that are not (Sarkar 1998). In biological practice, any organismic feature that succumbs to systematic study potentially qualifies as a trait. Restricting attention initially to structural traits, many of them can be characterized in molecular terms (using the molecules out of which the structures are constructed). These occur at varied levels of biological organization, from the pigmentation of animal skins to receptors on cell membranes. Proteins are intimately involved in all these structures. For many behavioral traits, a single molecule, again usually a protein, suffices as a distinguishing mark for that trait. Examples range from sickle-cell hemoglobin for the sickle-cell trait to huntingtin for Huntington's disease. Thus, for these traits, by encoding proteins, genes trivially encode information for them at the molecular level.

However, at the present state of biological knowledge it is simply not true that, except in the case of a few model organisms, most traits can be characterized with sufficient precision at the molecular level for the last argument to go through. This is where the situation becomes more interesting. The critical question is whether the etiology of these traits will permit explanations at the molecular level—in particular, explanations in which individual proteins are explanatorily relevant. (Note that these explanations may also refer to other types of molecules; all that is required is that proteins have some etiological role. The involvement of these other molecules allows the possibility that the same protein [and gene] may be involved in the genesis of different traits in different environmental histories—this is called *phenotypic plasticity*.) If proteins are so relevant, then there will be covariance between proteins and traits although, in general, there is no reason to suppose that any of the conditions for semiotic information is fulfilled. Proteins will be part of the explanation of the etiology of traits without carrying information for traits. Allowing genes to carry information does not endorse a view that organisms should be regarded as information-processing machines. Informational analysis largely disappears at levels higher than that of proteins.

Nevertheless, because of the covariance mentioned in the last paragraph, by encoding proteins, genes encode information for traits (while not encoding the traits themselves). The critical question is the frequency with which proteins are explanatorily relevant for traits in this way. The answer is "Probably almost always." One of the peculiarities of molecular biology is that the properties of individual molecules, usually protein molecules such as enzymes and antibodies, have tremendous explanatory significance, for instance, in the structural explanations of specificity mentioned in section 10.1. Explanations involving systemic features, such as the topology of reaction networks, are still forthcoming. Unless such explanations, which do not refer to the individual properties of specific molecules, become the norm of organismic biology, its future largely lies at the level of proteins.

This is a reductionist vision of organismic biology, not reduction to DNA or genes alone, but to the entire molecular constitution and processes of living organisms (Sarkar 1998). Such a reductionism makes many uncomfortable, including some developmental biologists imbued in the long tradition of holism in that field, but there is as yet no good candidate phenomenon that contradicts the basic assumptions of the reductionist program. It is possible that biology at a different level—for instance, in the context of psychological or ecological phenomena—will present complexities that such a reductionism cannot handle. However, at the level of individual organisms and their traits, there is as yet no plausible challenge to reductionism.

10.5 Conclusions

Many of the recent philosophical attacks on the legitimacy of the use of an informational framework for genetics have been motivated by a justified disquiet about facile attributions of genetic etiologies for a wide variety of complex human traits, including behavioral and psychological traits, in both the scientific literature and in the popular media. The account of semiotic information given here does not in any way support such ambitions. Although, to the extent that is known today, genes are privileged over all other cellular factors as carriers of semiotic information involved in the etiology of traits, they may not even be the sole purveyors of such information. Moreover, genetic information is but one factor in these etiologies. Traits arise because of the details of the developmental history of organisms, in which genetic information is one resource among others. Thus, even when genes encode sufficient information to specify a unique protein (in prokaryotes), the information in the gene does not provide a sufficient etiology for a trait; at the very least, some history from M must be invoked.

Thus, an informational interpretation of genetics does not support any attribution of excessive etiological roles for genes. The two questions are entirely independent: whether genes should be viewed as information-carrying entities, and the relative influence of genes versus nongenetic factors in the etiology of traits. This is why the latter question could be investigated, often successfully, during the first half of the twentieth century, when "information" was yet to find its way into genetics. To fear genetic information because of a fear of genetic determinism or reductionism is irrational.

Acknowledgments

Thanks are due to Justin Garson and Jessica Pfeifer for helpful discussions.

Notes

1. Garson (2002) shows how this technological context led to the introduction of informational concepts in neurobiology during roughly the same period.

2. Even earlier, H. Nyquist (1924) had recognized that the logarithm function is the appropriate mathematical function to be used in this context. Nyquist's "intelligence" corresponds very closely with the modern use of "information."

3. Because ι may hold between a single a and several b, this relation is not transitive. A may carry information for B, B for C, but A may not carry information for C. As will be seen later, this failure of transitivity results in a distinction between encoding traits and encoding information for traits.

4. Here, ι is being interpreted extensionally for expository simplicity.

5. However, condition (A2) is stronger than what is strictly required for arbitrariness, and may be stronger than what is biologically warranted. For arbitrariness, all that is required is that there is a large number of possible alternative assignments. Biologically, it may be the case that some assignments are impossible because of developmental and historical constraints.

6. This strategy goes back to work on enzyme kinetics around the turn of the twentieth century. Haldane (1930) provides an early synthesis. In the context of general cellular metabolism, the pioneering work was that of Garfinkel and Hess (1964).

7. Efforts in this direction are relatively novel. See, for instance, Ebenhöh and Heinrich (2001) for an optimality model of the evolution of the stoichiometry of ATP and NADH producing systems.

References

Brooks, D. R., and Wiley, E. O. 1988. *Evolution as Entropy: Towards a Unified Theory of Biology*, second ed. Chicago: The University of Chicago Press.

Carnap, R., and Bar-Hillel, Y. 1952. "An Outline of a Theory of Semantic Information." Technical Report no. 247. Research Laboratory of Electronics, Massachusetts Institute of Technology.

Crick, F. H. C. 1958. "On Protein Synthesis." *Symposia of the Society for Experimental Biology* 12: 138–163.

Ebenhöh, O., and Heinrich, R. 2001. "Evolutionary Optimization of Metabolic Pathways. Theoretical Reconstruction of the Stoichiometry of ATP and NADH Producing Systems." *Bulletin of Mathematical Biology* 63: 21–55.

Ephrussi, B., Leopold, U., Watson, J. D., and Weigle, J. J. 1953. "Terminology in Bacterial Genetics." *Nature* 171: 701.

Fisher, R. A. 1925. "Theory of Statistical Estimation." *Proceedings of the Cambridge Philosophical Society* 22: 700–725.

Fox, T. D. 1987. "Natural Variation in the Genetic Code." *Annual Review of Genetics* 21: 67–93.

Garfinkel, D., and Hess, B. 1964. "Metabolic Control Mechanism VII. A Detailed Computer Model of the Glycolytic Pathway in Ascites Cells." *Journal of Biological Chemistry* 239: 971–983.

Garson, J. 2002. "The Introduction of Information in Neurobiology." M. A. Thesis, Department of Philosophy, University of Texas at Austin.

Griffiths, P. 2001. "Genetic Information: A Metaphor in Search of a Theory." *Philosophy of Science* 67: 26–44.

Haldane, J. B. S. 1930. *Enzymes*. London: Longmans, Green.

Hartley, R. V. L. 1928. "Transmission of Information." *Bell Systems Technical Journal* 7: 535–563.

Kimura, M. 1961. "Natural Selection as a Process of Accumulating Genetic Information in Adaptive Evolution." *Genetical Research* 2: 127–140.

Lederberg, J. 1956. "Comments on the Gene–Enzyme Relationship." In O. H. Gaebler (ed.), *Enzymes: Units of Biological Structure and Function*. New York: Academic Press, 161–169.

Maynard Smith, J. 2000. "The Concept of Information in Biology." *Philosophy of Science* 67: 177–194.

Nyquist, H. 1924. "Certain Factors Affecting Telegraph Speed." *Bell Systems Technical Journal* 3: 324–346.

Pauling, L. 1940. "A Theory of the Structure and Process of Formation of Antibodies." *Journal of the American Chemical Society* 62(2): 643–657.

Pierce, J. R. 1962. *Symbols, Signals, and Noise*. New York: Harper.

Sarkar, S. 1996a. "Biological Information: A Skeptical Look at Some Central Dogmas of Molecular Biology." In S. Sarkar (ed.), *The Philosophy and History of Molecular Biology: New Perspectives*. Dordrecht: Kluwer, 187–231. (Chapter 9 of this volume.)

———. 1996b. "Decoding 'Coding'—Information and DNA." *BioScience* 46: 857–864. (Chapter 8 of this volume.)

———. 1998. *Genetics and Reductionism*. New York: Cambridge University Press.

———. 2000. "Information in Genetics and Developmental Biology: Comments on Maynard-Smith." *Philosophy of Science* 67: 208–213.

Schneider, T. D. 1999. "Measuring Molecular Information." *Journal of Theoretical Biology* 201: 87–92.

Shannon, C. E. 1948. "A Mathematical Theory of Communication." *Bell Systems Technical Journal* 27: 379–423, 623–656.

Sterelny, K. 2000. "The 'Genetic Program' Program: A Commentary on Maynard Smith on Information in Biology." *Philosophy of Science* 67: 195–202.

——— and Griffiths, P. E. 1999. *Sex and Death: An Introduction to Philosophy of Biology*. Chicago: The University of Chicago Press.

Watson, J. D., Tooze, J., and Kurtz, D. T. 1983. *Recombinant DNA: A Short Course*. New York: W. H. Freeman.

Williams, G. C. 1966. *Adaptation and Natural Selection*. Princeton, N.J.: Princeton University Press.

Part IV Evolution

11 Neo-Darwinism and the Problem of Directed Mutations

11.1 Introduction

In the first edition of the *Origin* (1859), natural selection acting upon blind variation was Darwin's favored mechanism for evolution. By the "blindness" of variation, Darwin seems to have meant little more than that its causes were unknown. When he said that the variations were "due to chance," he emphasized that this, "of course, is a wholly incorrect expression, but it serves to acknowledge plainly our ignorance of the cause of each particular variation" (1959, p. 131). But there is implicit in Darwin's discussions a critical assumption that goes beyond a mere assertion of ignorance about the causes of variation. Blind variation, for Darwin, is just as likely to lead to the birth of less fit individuals as those that are more fit. Though Darwin did not put it in this way, the blindness of variation requires that there is no correlation between the genesis of a variation and the fitness of the individual exhibiting it.

Though Darwin, in 1859, emphasized blind variation and natural selection, he allowed some effects of use and disuse, habit, and environmental induction to be inherited and thus generate variation in populations. These Lamarckian mechanisms of change, however, only had subsidiary importance. In the following decade Darwin's almost exclusive reliance on blind variations came under attack. Like most others of his time, Darwin believed in "blending inheritance," that is, that offspring from two parents differing in some character would possess it to an intermediate degree. Though Mendel published his famous paper on the particulate nature of inheritance in 1865, it went generally unnoticed for over thirty years. Meanwhile, Jenkin (1867), published a devastating criticism of the possibility of evolution through blind variation. If variants in a population arise entirely by chance, a fitter variant will only arise very rarely. But, then, the quality that made this variant fitter will be swamped out in a few generations through blending. Evolution will, therefore, be a very slow process (as Darwin had assumed). Its pace will be limited by the very rare simultaneous appearance of similar blind variants and the possibility of such individuals mating together. Jenkin pointed out that the physicists' thermodynamic estimate of the age of the earth was only about 98 million years, far too small for Darwinian evolution to have occurred.

Darwin found this objection serious. His response was to emphasize the Lamarckian mechanisms at the expense of natural selection acting on blind variations. If heritable variation can arise in response to the environment, through direct induction, use and disuse, or habit, the simultaneous generation of many similar variants is far more likely than through blind variation. Moreover, the same variants are likely to occur in the same area and the chance of their mating is also enhanced. Evolution can thus speed up and escape the limits apparently put by thermodynamics. By the sixth edition of the *Origin* (1872), Darwin's emphasis on natural selection and blind variation had been replaced by a much more varied account of evolutionary change.

In the generation after Darwin, some biologists, notably Romanes, tried to maintain a "pluralistic" interpretation of Darwinism, which permitted the inheritance of environmentally induced changes. In sharp contrast, Weismann (*e.g.*, 1892) was adamant that no acquired characteristic could be inherited. Weismann's account of inheritance was based primarily on his theory of development. He generalized from observations on animals such as hydroids to postulate that, in general, the germ-line segregates from the soma very early in development; that, in some sense, the germ-line is immortal; and that a molecular barrier prevents changes in the somatic cells to propagate into the germ-line. Since the somatic cells are the only ones that interact with the environment, this theory precludes the inheritance of acquired characteristics. Romanes (1896) coined the term "neo-Darwinism" to describe Weismann's views and distinguish them from the pluralism of Darwin's later writings.

In the first half of the twentieth century, neo-Darwinism emerged as the dominant interpretation of evolution. Mendelian inheritance was rediscovered around 1900 and the genesis of novel variation was discovered to be through the mutation of genes. The new population genetics of the

1920s, which attempted to found evolutionary theory on Mendelian genetics, implicitly assumed the non-inheritance of acquired characteristics and the blindness of mutagenesis. Moreover, claims of the inheritance of acquired characteristics fell into disrepute thanks not only due to the accumulation of a vast amount of contrary data, but also because of some notorious cases of fraud and malfeasance on the part of those making these claims. For instance, Kammerer's (1924) claims of inducing a heritable formation of nuptial pads in midwife toads by raising successive generations of toads in water turned out to be a case of blatant fraud (see Hull 1984). Even worse was the well-known Lysenko affair in the erstwhile Soviet Union during which Mendelian genetics was banned and geneticists persecuted in the name of a new "proletarian" science that required the inheritance of acquired characteristics (see Joravsky 1970 and Soyfer 1989).

Recent experiments, however, have begun to challenge this neo-Darwinian consensus. What has been put to question, again, are both the noninheritance of acquired characteristics and the blindness of mutagenesis. The purpose of this paper is to discuss the latter development, and related conceptual and interpretive issues. The inheritance of acquired characteristics has already been extensively discussed in a recent review (Landman 1991). To set the stage for that discussion, the following six aspects of the conceptual structure of the debates about neo-Darwinism should be explicitly noted:

(i) Critics of neo-Darwinism fully admit that much of evolutionary change proceeds through blind mutagenesis and the non-inheritance of acquired characteristics. The focus of the dispute is whether all inheritance can be interpreted in this fashion.

(ii) The two claims, that there is no inheritance of acquired characteristics, and that mutagenesis is blind, are assumptions that are independent of the theory of natural selection. That theory only requires phenotypic variation in a population, differential fitness of variant phenotypes and the heritability of phenotypes (see, *e.g.*, Endler 1986). The source of variation and the mode of inheritance are left open. Neo-Darwinism thus goes beyond the theory of natural selection in these additional assumptions and these are the ones that have now come up for question.

(iii) Moreover, the non-inheritance of acquired characteristics and the blindness of mutagenesis are conceptually different claims. By itself, the possibility of the inheritance of acquired characteristics permits them to be inherited irrespective of whether they are adaptive or not. Moreover, if there is preferential adaptive mutagenesis, that is "directed" mutagenesis, it is not clear that this should be called an "acquired characteristic" in such simple organisms such as bacteria or yeast. If it should be so called, then the inheritance of acquired characteristics is a broader category than directed mutagenesis.

(iv) Part of the trouble comes from an ambiguity in the notion of an "acquired characteristic." If radiation induces a mutation, is it not "acquired"? The task of clarifying this notion is beyond the scope of this paper (see Landman 1991 for an important attempt in that direction).

(v) There is a similar problem with the notion of a "directed mutation." The definition that will be used here is that a mutation is directed it if occurs (or occurs more often) in an environment where it enhances the fitness of the organism than in an environment in which it does not do so, all other factors being equal (Sarkar 1990). In other words, for a mutation to be directed, the rate of its occurrence must increase in an environment in which it is beneficial. Note that there is no assumption about possible mechanisms in this definition.

(vi) What has been called "blind" variation or mutagenesis here is also very often called "random." There is no harm in doing this as long as it is remembered that "random," in this context, does not mean anything more than the absence of correlation between a mutation rate and the fitness of the induced phenotype. Randomness, here, does not imply the equiprobability of all mutations. In fact, it has long been known that certain parts of the genome are more prone to mutation than others, mutations at one site can be controlled by those at another, that certain types of substitutions are preferred to others, *etc.* (see Drake 1991 for a recent review).

11.2 Directed Mutagenesis

It is customary to associate the inheritance of acquired characteristics with the views of Lamarck. From a historical point of view, this is improper. Though Lamarck certainly believed in the inheritance of acquired characteristics, so did virtually all other biologists of the first half of the nineteenth century (Zirkle 1946). What is peculiar to Lamarck is the view that

Neo-Darwinism and Directed Mutations

the environment can be responsible, directly or indirectly, for the genesis of variations that are adaptive to it. Sarkar (1991b) has used "strong neo-Lamarckism" to refer to the position that *all* variations arise in this way, reserving "weak neo-Lamarckism" to refer to the position that *some* variation has such an origin. The former position is clearly untenable in the light of all the experimental evidence that has accumulated in this century. But the latter position has been dramatically revived by some recent experiments with bacteria.

The orthodox position, that bacterial mutagenesis is blind, was ushered in by a path-breaking experiment of Luria and Delbrück (1943) who explored the transformation of bacteria from virus sensitivity to virus resistance. They devised the "fluctuation test" to distinguish between the possibilities that this transformation occurs independently of any exposure to the virus and that it is induced by interaction with the virus. The former possibility corresponds to neo-Darwinism; the latter to (strong) neo-Lamarckism because the transformation is obviously adaptive. In the fluctuation test a large number of test-tubes containing nutrients are each seeded with a virus-sensitive bacterium. All of the cells used are assumed to be genetically identical. The cells are allowed to reproduce in the test-tubes forming clones. The content of each of the test-tubes is then separately inoculated onto a plate containing nutrients and virus. Sensitive cells die while resistant ones flourish and form visible plaques. The number of such plaques on a plate is, therefore, equal to the number of resistant cells in the source test-tube at the time of plating.

If the transformation is induced by the presence of virus, then it is due to a small, but finite, probability that this happens. Since this is assumed to be a random process, the number of mutants across the plates will follow the Poisson distribution with the Poisson parameter equal to the product of the probability of mutation of a cell and the number of cells present in a test-tube at the time of plating. If the transformation is independent of the presence of the virus, the number of mutants observed on the plates depends on the time at which the mutation occurred in the source test-tube. If it occurred early in clonal growth, successive replications would lead to a large number of mutants. If it occurred late, there would be a lot fewer mutants observed after plating. Thus, a large variance in the expected number of mutants is generated. Luria and Delbrück could not solve this distribution. However, they obtained an estimate for its variance which is much greater than that expected from the Poisson distribution (whose variance equals its mean). Haldane, in 1946, provided a partial solution of the distribution which was never published (see Sarkar 1991a). Finally, Lea and Coulson (1949) gave an approximate procedure for calculating what they baptized as the "Luria-Delbrück distribution." A more exact, and particularly simple, recursion relation has since been given by Sarkar *et al.* (1992).

Luria and Delbrück applied the fluctuation test to the interaction of phage T1 to a strain of *Escherichia coli* normally sensitive to it. The variance-to-mean ratio observed was much higher than 1. They concluded, on the basis of this observation, that the transformation under question was entirely due to blind mutagenesis. Over the next decade this observation, and the associated interpretation, was extended to many other cases of bacterial mutagenesis (see Sarkar 1991b for details of the history). Independent confirmation of this result also came from redistributive plating (Newcombe 1949), chemostat (Novick and Szilard 1950), replica plating (Lederberg and Lederberg 1952) and sib-selection (Cavalli-Sforza and Lederberg 1956) experiments. For many bacteriologists, wary of the statistical reasoning of Luria and Delbrück, these independent experiments, especially replica plating, were the decisive ones. Only Ryan's (1952a,b) results struck a potentially discordant note by showing that the observed distributions sometimes deviated significantly from the Luria–Delbrück as calculated on the basis of Lea and Coulson's (1949) procedures.

In retrospect, it is easy to see at least three flaws in the design and interpretation of the experiments by which the almost universal consensus in favor of universally blind mutagenesis was reached:

(i) Luria and Delbrück (1943), and those that followed them, seem to have been willing to consider only two possibilities, that *all* mutations were blind (random) or that *all* were directed (adaptive). That *some* mutations could be directed while others, perhaps most, were blind does not seem to have been entertained as a serious possibility (Sarkar 1990, 1991b). In fluctuation test experiments a large enough variance was taken to be enough to indicate that the Luria–Delbrück distribution was being observed. Most of the other types of experiments simply did not have the resolving power to decide whether there was some directed mutagenesis (Sarkar 1991b).

(ii) When deviations from the Luria–Delbrück distribution were observed, these could be easily accounted for by invoking any of a myriad of complicating factors such as differential fitnesses (growth rates) of mutant and nonmutant strains in the test-tubes or phenotypic lag (delayed phenotypic expression) (Armitage 1952). These factors could potentially be ruled out on experimental grounds but confidence in the orthodox position seems to have been sufficient enough for these additional experiments not to have been systematically pursued.

(iii) Luria and Delbrück's (1943) original experiment, and many of the others, involved an environment that was lethal to nonmutant strains. Consequently, if an induced directed mutation can only arise over a period of time, these would not be detected in such experiments. Delbrück (1946), at least, seems to have realized this possibility. In fact, the discordant results of Ryan (1952a,b) were from a situation in which the environment was not lethal to nonmutants: one of the mutations he studied was that from lactose-nonutilizing (Lac^-) to lactose-utilizing (Lac^+) in a strain of *Escherichia coli*.

Ryan's work provided the basis for the fluctuation test experiments of Cairns *et al.* (1988) which, more than any other report, has been responsible for shattering the orthodox consensus. Studying the same transformation ($Lac^- \rightarrow Lac^+$), they observed that:

(i) There were significant deviations from the Luria–Delbrück distribution.

(ii) The deviations seemed to follow a Poisson distribution. However, this has never been satisfactorily quantitatively demonstrated and seems to amount to little more than an observation that the variance of the deviations is small compared to what would be expected from the Luria–Delbrück distribution.

(iii) The capacity to generate these mutations seemed to be under genetic control. Some strains of bacteria exhibited this property, others did not.

(iv) Many more post-plating mutants continued to appear on the plates than could be explained by phenotypic lag.

Because a considerable fraction of the mutants were still arising from blind mutations, the effects observed by Cairns *et al.* would have escaped detection in the nonquantitative experiments that had established the orthodox consensus. Moreover, as noted above, the quantitative experiments of Ryan were consistent with the new observations. In unfortunately provocative language, Cairns *et al.* argued that these observations show bacteria "can choose which mutations they should produce." What really is at stake, however, is the much more mundane question of whether the mutation is "directed" in the sense indicated above.

As expected, Cairns *et al.* (1988) drew extended criticism from traditionalists committed to blind mutagenesis (see Sarkar 1990, 1991b for reviews). It was pointed out that many different possible mechanisms including differential growth rates of mutants and nonmutants in the preselective media, phenotypic lag or even plating inefficiency would decrease the variance of the expected distribution and, in that sense, result in a deviation from the Luria–Delbrück in the direction of the Poisson. Sarkar (1990) noted that though these possibilities could potentially be experimentally ruled out, virtually any complicating factor would shift a Luria–Delbrück distribution to one with a lower variance simply because of the extremely high variance of that distribution. He argued, therefore, that the only sure way to resolve the controversy is to find the mechanisms responsible for this type of mutagenesis. That has not been done yet though several intriguing putative mechanisms have been suggested. Meanwhile, most (though not all) further experiments, especially those of Hall (1990, 1991) seem to corroborate and extend the conclusions of Cairns *et al.* (see Foster 1991, 1992 for reviews). What is most remarkable about Hall's (1991) results is that even composite double mutation events appear to be directed. At present, it is at least uncontroversial that there is more to the modulation of bacterial mutagenesis than had previously been suspected (Drake 1991). Unless one has boundless confidence in the orthodoxy, moreover, it is hard not to suspect the existence of directed mutations.

Two broad classes of mechanisms have been suggested for the genesis of directed mutations. The first of these required mutations be "instructed" in the sense that the mechanisms directly and preferentially produce specific adaptive sequences (*e.g.*, by Cairns *et al.* 1988). What is important about such mechanisms is that they would involve the transfer of information from protein to DNA and thereby violate the central dogma of molecular biology which specifically precludes such a transfer. Though they remain a possibility, there is as yet no positive evidence for such mechanisms. The probability of success of such a mechanism in instructing mutagenesis would, under most imaginable schemes, be small but nonnegligible. Therefore, the resultant distribution of such mutants would be Poisson, as observed.

The second class of possible mechanisms is more interesting. These rely on variant strands of DNA being routinely—and randomly—produced in a starved nondividing cell.[1] When a functional strand arises, the cell begins to digest food, divides and fixes that variant as a mutation. There have been many varieties of these "trial-and-error" models. For example, Hall (1990) suggested that starved cells may occasionally enter a "hypermutable" state in which variants would be routinely produced. The probability of such an entry would increase with the age of the culture. In the hypermutable state, a product that was mutagenic only to such cells might accumulate giving rise to the increased rate of genesis of variants. An adaptive variant would be fixed as a mutation through cell division. All other hypermutable cells would die.[2] It turns out that the distribution of mutants even in this case is Poisson (for a proof, see Sarkar 1993).

Thus, statistical tests will not be able to distinguish whether a directed mutation is instructed or not. This further underscores and extends the methodological point made by Sarkar (1990) that, in order to resolve the disputes over directed mutagenesis, the exact mechanisms of mutagenesis must be investigated. Hall (1991) failed to detect variants of the sort that this model requires. However, it is controversial whether his discriminatory strategy had enough sensitivity (Foster 1992) and these "trial-and-error" models remain a plausible potential mechanism for directed mutagenesis.

There are four interpretive points about these developments that are particularly important:

(i) Note that even weak neo-Lamarckism contradicts neo-Darwinism if, as is usual, the latter doctrine is construed to require that *all* mutagenesis is blind. However, the "trial-and-error" models of directed mutagenesis, though neo-Lamarckian, can peacefully coexist with neo-Darwinism to the extent that it still maintains a critical aspect of neo-Darwinism, namely, that "blind chance" remains the ultimate source of variation (Sarkar 1991b). What is perhaps even more interesting about these models is that they illustrate the important biological principle that what appears to be blind at one level is directed at another. The same principle is the one that accounts for the directedness of organic evolution, so apparent in evolutionary history, by invoking blind mutagenesis followed by natural selection. It is also at the core of some of the recent accounts of development (Buss 1987).

(ii) The failure of statistical tests to discriminate between alternative models is a failure of nonreductive modes of research in biology, that is, types of research which do not proceed by investigating smaller and smaller parts of organism, ultimately at the molecular level, and invoking nothing more than standard physical and chemical interactions (Sarkar 1990, 1991b). Progress in understanding the modulation of mutagenesis depends on an investigation of its molecular mechanisms. This interpretive issue is of some importance because the nature and relevance of reductionism in molecular biology remains philosophically controversial (Sarkar 1989, 1992).

(iii) It should be obvious that a mechanism for directed mutagenesis, as a response to environmental stress, would be of benefit to any organism and could itself be selected for (Davis 1989). In this respect, the striking analogy between some of the possible mechanisms for directed mutagenesis and the SOS response (Radman 1974; Walker 1985) is particularly suggestive though it remains to be analyzed in detail.

(iv) The type of directed mutagenesis that has been considered here has so far been observed at least at six loci of *Escherichia coli*. It has also been observed at a locus responsible for the synthesis of histidine in yeast (*Saccharomyces cerevisiae*) (Hall 1992). Moreover, what might well be similar phenomena have been observed in other cases though these have not yet been scrutinized in detail. For instance, yeast whose mitochondrial DNA has been affected by azide is no longer sensitive to that agent (Nagai et al. 1961). In flax (*Linum usitatissimum*), heritable and apparently adaptive genomic changes can be induced by environmental agents (Cullis 1987). The mechanism for the generation of these changes is not known. All cases such as these deserve more scrutiny.

11.3 Conclusions

Beyond an assessment of the experimental situation in the study of potentially directed mutagenesis, a more general lesson can be drawn from these developments. This is simply that orthodox neo-Darwinism of the sort associated with Weismann is not nearly as universally tenable as was generally supposed a generation ago. Both the recent attention to cases of the inheritance of acquired characteristics (Landman 1991) and the existence of directed mutations point to a much more complicated picture of the origin of variations than neo-Darwinism has supposed.

This should come as no surprise. Weismann's doctrinaire rejection of the possibility of the inheritance of acquired characteristics was based on the false assumption of the very early segregation of germ-line from soma in all organisms. Not only do plants and fungi not have segregated germ-lines but, as Buss (1983) and others have pointed out, an early segregation of germ-line from soma does not routinely hold for most phyla. Therefore, there is no theoretical reason that precludes the inheritance of acquired characteristics.

Further, in organisms where such inheritance is possible, it is easy enough to see that a mechanism that could adaptively control the generation of mutations, would be of selective value especially if, as is generally believed, most induced mutations are detrimental. If mechanisms of directed mutagenesis are discovered in bacteria, what will be most interesting is the search for potential analogs in other organisms where such mutations could be inherited. Only after that possibility is fully explored will it be possible to assess the importance of such phenomena in evolution in general. No matter how important, and prevalent, directed mutagenesis turns out to be, its existance shows the necessity of an expansion of the conceptual framework of evolutionary biology beyond the confines of orthodox neo-Darwinism. Whether this expansion constitutes an implicit rejection of neo-Darwinism is an interpretive issue that can only be decided after the mechanisms of directed mutagenesis have been worked out. If such mutagenesis is also instructed, not much gets retained of the neo-Darwinian conceptual framework. If, however, directed mutagenesis occurs though one of the "trial-and-error" mechanisms, then, as suggested above, there can be peaceful coexistence of the expanded framework with neo-Darwinism.

Finally, it should also come as no surprise that these new conceptual possibilities have arisen because of the interaction of traditional evolutionary biology with molecular biology. In general, molecular biology permits a more subtle and detailed analysis of biological phenomena than was possible before its advent. For evolutionary biology, in the first half of this century, a mutation was a single type of event. However, once the molecular understanding of the gene was achieved, it immediately became apparent that many different kinds of changes at the molecular level were being collected together as a "mutation" by classical genetics. Mutation had turned out to be a much more complex process than what classical genetics had ordained. That the modulation and control of mutations are also revealing new complexities was only to be expected.

Acknowledgments

This is Contribution No. BTBG-92-9 from the Theoretical Biology Group, Boston University, 745 Commonwealth Ave., Boston, MA 02215. Thanks are due to James F. Crow, Richard E. Lenski, and Scott Williams for comments on an earlier draft of this paper.

Notes

1. The intuition behind this class of mechanisms is reminiscent of the often-invoked idea that some organisms, such as rotifers, which normally reproduce asexually, often switch to sexual reproduction under adverse circumstances such as starvation. Sexual reproduction allows new varieties, produced through recombination, to be tried out against such an environment. (Note that this argument is manifestly group-selectionist whereas the argument for the selection of a mechanism that generates variants in bacterial cells is not.)

2. Other "trial-and-error" models mainly differ from Hall's (Stahl 1988; Boe 1990) in not requiring the death of unsuccessful hypermutable cells.

References

Armitage, P. 1952. The Statistical Theory of Bacterial Populations Subject to Mutation. *Journal of the Royal Statistical Society* 14: 2–40.

Boe, L. 1990. Mechanism for Induction of Adaptive Mutations in *Escherichia coli*. *Molecular Microbiology* 4: 597–601.

Buss, L. W. 1983. Evolution, Development and the Units of Selection. *Proceedings of the National Academy of Sciences (USA)* 80: 1387–1391.

Buss, L. W. 1987. *The Evolution of Individuality*. Princeton: Princeton University Press.

Cairns, J., Overbaugh, J., and Miller, S. 1988. The Origin of Mutants. *Nature* 335: 142–145.

Cavalli-Sforza, L. L. and Lederberg, J. 1956. Isolation of Preadaptive Mutants by Sib-Selection. *Genetics* 41: 367–381.

Cullis, C. A. 1987. The Generation of Somatic and Heritable Variation in Response to Stress. *American Naturalist* 130: S62–S73.

Darwin, C. 1859. *On the Origin of Species*. London: John Murray.

Davis, B. 1989. Transcriptional Bias: A Non-Lamarckian Mechanism for Substrate-Induced Mutations. *Proceedings of the National Academy of Sciences (USA)* 86: 5005–5009.

Delbrück, M. 1946. [Discussion to article by A Lwoff]. *Cold Spring Harbor Symposia in Quantitative Biology* 11: 154.

Drake, J. 1991. Spontaneous Mutation. *Annual Review of Genetics* 25: 125–146.

Endler, J. A. 1986. *Natural Selection in the Wild*. Princeton: Princeton University Press.

Foster, P. 1991. Directed Mutation in *Escherichia coli*: Theory and Mechanisms. In Tauber, A. I., ed., *Organism and the Origins of Self*. Dordrecht: Kluwer, pp. 213–234.

Foster, P. 1992. Directed Mutation: Between Unicorns and Goats. *Journal of Bacteriology* 174(6): 1711–1716.

Hall, B. G. 1990. Spontaneous Point Mutations That Occur More Often When Advantageous Than When Neutral. *Genetics* 126: 5–16.

Hall, B. G. 1991. Adaptive Evolution That Requires Multiple Spontaneous Mutations: Mutations Involving Base Substitutions. *Proceedings of the National Academy of Sciences (USA)* 88: 5882–5886.

Hall, B. G. 1992. Selection-Induced Mutations Occur in Yeast. *Proceedings of the National Academy of Sciences (USA)* 89: 4300–4303.

Hull, D. 1984. Lamarck among the Anglos. In Lamarck, J. B., *Zoological Philosophy*. Chicago: University of Chicago Press, pp. xl–lxvi.

Jenkin, F. 1867. The Origin of Species. *North British Review* 42: 149–171.

Joravsky, D. 1970. *The Lysenko Affair*. Cambridge, MA: Harvard University Press.

Lamarck, J. B. 1984. *Zoological Philosophy*. Chicago: University of Chicago Press.

Landman, O. E. 1991. The Inheritance of Acquired Characteristics. *Annual Review of Genetics* 25: 1–20.

Lea, D. E. and Coulson, C. A. 1949. The Distribution of the Number of Mutants in Bacterial Populations. *Journal of Genetics* 49: 264–285.

Lederberg, J. and Lederberg, E. 1952. Replica Plating and Indirect Selection of Bacterial Mutants. *Journal of Bacteriology* 63: 399–406.

Lenski, R. E. 1989. Are Some Mutations Directed? *Trends in Ecology and Evolution* 4: 148–151.

Luria, S. E. and Delbrück, M. 1943. Mutations of Bacteria from Virus Sensitivity to Virus Resistance. *Genetics* 28: 491–511.

Nagai, S., Yanagashima, N., and Nagai, H. 1961. Advances in the Study of Respiration-Deficit Mutation in Yeast and Other Microorganisms. *Microbiological Review* 25: 404–426.

Newcombe, H. B. 1949. Origin of Bacterial Variants. *Nature* 164: 150–151.

Novick, A. and Szilard, L. 1950. Experiments with the Chemostat on Spontaneous Mutation in Bacteria. *Proceedings of the National Academy of Sciences (USA)* 36: 708–719.

Radman, M. 1974. Phenomenology of an Inducible Mutagenic Repair Pathway in *Escherichia coli*: SOS Repair Hypothesis. In Prakash, L., Sherman, R., Miller, M., Lawrence, C., and Tabor, H. W., ed., *Molecular and Environmental Aspects of Mutagenesis*. Springfield: Charles C. Thomas, pp. 128–142.

Romanes, G. J. 1896. *Life and Letters*. London: Longmans, Green.

Ryan, F. J. 1952a. Adaptation to Use Lactose in *Escherichia coli*. Journal of General Microbiology 7: 69–88.

Ryan, R. J. 1952b. Distributions of Numbers of Mutant Bacteria in Replicate Cultures. *Nature* 169: 882–883.

Sarkar, S. 1989. Reductionism and Molecular Biology: A Reappraisal. Ph.D. Dissertation, Department of Philosophy, University of Chicago.

Sarkar, S. 1990. On the Possibility of Directed Mutagenesis in Bacteria: Statistical Analyses and Reductionist Strategies. In Fine, A., Forbes, M., and Wessels, L., ed., *PSA 1990: Proceedings of the 1990 Biennial Meeting of the Philosophy of Science Association*. 1. East Lansing: Philosophy of Science Association, pp. 111–124.

Sarkar, S. 1991a. Haldane's Solution of the Luria Delbrück Distribution. *Genetics* 127: 257–261.

Sarkar, S. 1991b. Lamarck *contre* Darwin, Reduction *versus* Statistics: Conceptual Issues in the Controversy over Directed Mutagenesis in Bacteria. In Tauber, A. I., ed., *Organism and the Origins of Self*. Dordrecht: Kluwer, pp. 235–271. (Chapter 12 of this volume.)

Sarkar, S. 1992. Models of Reduction and Categories of Reductionism. *Synthese* 91: 167–194. (Chapter 2 of this volume.)

Sarkar, S. 1993. Beyond Neo-Darwinism: the Challenge of Directed Mutations. *Philosophical Studies from the University of Tampere* 50: 69–84.

Sarkar, S., Sandri, G. V., and Ma, W. T. 1992. On Fluctuation Analysis: A New, Simple and Efficient Method for Computing the Expected Number of Mutants. *Genetica* 85: 173–179.

Soyfer, V. N. 1989. New light on the Lysenko era. *Science* 339: 415–420.

Stahl, F. W. 1988. A Unicorn in the Garden. *Nature* 335: 112–113.

Walker, G. C. 1985. Inducible DNA Repair Systems. *Annual Review of Biochemistry* 54: 425–457.

Weismann, A. 1892. *Das Keimplasma: Eine Theorie der Vererbung.* Jena: Gustav Fischer.

Zirkle, C. 1946. The Early History of the Idea of the Inheritance of Acquired Characteristics and Pangenesis. *Transactions of the American Philosophical Society* 35: 91–151.

12   Lamarck *contre* Darwin, Reduction versus Statistics: Conceptual Issues in the Controversy over Directed Mutagenesis in Bacteria

12.1   Introduction

The purpose of this chapter is to examine some conceptual aspects of the controversy over the possibility of directed mutagenesis in bacteria that has erupted since the publication of some provocative results by Cairns, Overbaugh, and Miller.[1] The conceptual issues that are important here occur at least at two levels, the first of which is, in a sense, *metaphysical* and the second, *epistemological*. *First*, the possibility of directed mutagenesis challenges the core of the current orthodox framework of evolutionary theory. Thus the sense in which mutations can indeed be "directed" is of considerable foundational importance to evolutionary theory. To the extent that such foundational issues are "metaphysical," in the sense that they concern the most general and universal underlying features of the world explored by science, these conceptual issues properly belong to metaphysics. *Second*, much of the evidence on which the current controversy thrives is statistical evidence about the number of mutant bacteria. Experimental methods which rely on such evidence are nonreductive in the sense that they attempt to understand what occurs at a lower level—for instance, that within a bacterial cell—by making observations at a higher level—in the example, that of the cell. The interesting question is whether such nonreductive strategies can resolve the controversy being considered here. This issue is epistemological because it concerns the powers and limitations of different scientific methods. It is particularly interesting in the present context because the role played by reductionism in molecular biology remains a matter of considerable controversy.[2]

Section 12.2 of this chapter attempts to explicate the nature of "neo-Darwinism" and "neo-Lamarckism" and to put them in their proper conceptual contexts. In particular it attempts to distinguish various senses of "neo-Lamarckism" which will ultimately be relevant to the characterization of various potential positions that might be adopted in this controversy. Though it briefly touches on the long and variegated history of the persistent disputes between neo-Darwinians and neo-Lamarckians, it is by no means even a representative summary of this history, let alone a complete one. Section 12.3 gives a short account of the early history of the actual problem of concern here, namely, the nature of the origin of bacterial mutants. The account is intended as a critical one in the sense that both the experiments performed and their interpretations are carefully appraised. Section 12.4 is a very short account of the new experimental work on the topic which forms part of the new controversy over such mutants.

Section 12.5 takes up, in detail, the question of the interpretation of all the experimental results, that is, those described or mentioned in the preceding two sections. Given the various interpretive possibilities encountered, it argues that mere statistical data are unlikely to be able to resolve the controversy: the discovery of the appropriate mechanisms of mutagenesis will be necessary for that purpose. It then analyzes various suggested, but as yet unestablished, mechanisms that might account for the empirical data. The conceptual framework developed in section 12.2 is utilized to clarify the various issues that arise. It is also suggested that there is a sense in which neo-Darwinism and neo-Lamarckism might peacefully coexist. Section 12.6 turns to the issue of reduction. It uses a model of reduction, based on explaining wholes in terms of their parts by invoking only physically characterized interactions, to argue that the type of research that was found to be most likely to resolve this controversy, that is, the search for mechanisms, is fundamentally reductionist.[3] In section 12.7 the various conclusions drawn during the course of this paper are reiterated. A final important, but often ignored, moral is drawn: certain metaphysical and epistemological questions, such as those mentioned above, can be answered, and perhaps only be answered, experimentally.

12.2 Neo-Darwinism and Neo-Lamarckism

The dominant intellectual current within contemporary evolutionary biology is neo-Darwinism, an interpretation of evolution in which the emphasis is on natural selection acting on random variation. What constitutes evolution by natural selection is easy enough to explicate. As formulated by Richard Lewontin in 1970, a population of individuals evolves by natural selection if three criteria are satisfied.[4] *First*, there must be phenotypic variation: "Different individuals in a population have different morphologies, physiologies, and behaviors".[5] Were all individuals identical with respect to all properties that interact with their environment, there would either be complete stasis or change due to some process rather than natural selection. *Second*, these different phenotypes, or sets of properties of the individual that interact with the environment, must have, associated with them, different fitnesses, that is, "different rates of survival and reproduction."[6] Here these rates and, consequently, fitness itself is to be construed probabilistically, that is, the fitness of an individual is to be construed as the expected frequency of its offspring in the next generation. *Third*, fitnesses must be heritable. The fitnesses of offsprings of an individual must depend on the fitness of the individual itself in that environment.[7]

The notion of randomness that is appropriate for evolutionary biology, however, is quite murky.[8] The considerations adduced here will use as weak a notion of randomness as possible. It will only be assumed that the type of variation that arises through a mutation of a gene is random if and only if the probability of its occurrence in an environment has no correlation with the fitness of the phenotype induced by it in that environment. Thus, though certain chemical mutagens or types of radiation are known to induce entire classes of mutations with a fairly predictable frequency, these mutations would still be regarded as random because the sort of correlation just mentioned does not exist. Once the randomness of a mutation is construed in this fashion, all mutations that are *undirected* by the environment are obviously random. Such a weak notion of randomness is being used here for two reasons. *First*, in any analysis, it is generally better to assume as little as possible so as not to make the success of the analysis depend at all on dispensable features. *Second*, no stronger notion of randomness in evolutionary biology seems to be empirically warranted. Certainly the possibility of enhancing mutation rates in general through the action of radiation and chemical mutagens shows that not all mutation events are equiprobable at the physical level.

Though random variation and natural selection are being emphasized here as the dominant aspects of the conceptual framework of neo-Darwinism there are also other features of neo-Darwinism that are often regarded as being quite important. In fact, from some points of view, it is these other features that are central to the neo-Darwinian orthodoxy. Some see neo-Darwinism as requiring that *all* evolutionary change be caused by natural selection.[9] For such points of view, the existence of other mechanisms of evolutionary change such as allometry, or differential rates of growth of different parts of an organism, or pleiotropy or multiple, partly nonselective, effects of genes, or even neutral mutations at the molecular level, constitute a challenge to neo-Darwinism. Others view neo-Darwinism as requiring that evolution proceeds by continuous slight variations. Indeed, in the debate between the biometricians (who claimed to be following Darwin strictly) and the Mendelians at the turn of this century, it was this feature of neo-Darwinism that was considered central.[10] These other features are not emphasized here because they are easily incorporated into the neo-Darwinian framework without any perceptible rupture of its conceptual framework. It is largely uncontroversial today that mechanisms other than natural selection are operative in evolutionary change though opinion is divided on how important these other features may be.[11] Further, the reduction of biometrical insights to Mendelian genetics by Haldane, Fisher and Wright in the 1920s, which largely created the current version of neo-Darwinian orthodoxy, has already shown one way in which debates over the size of variations might be resolved.[12] Yet, the neo-Darwinian framework has largely survived.

The two claims that are being construed here to be central to neo-Darwinism, that is, that natural selection is operative, and that mutations are random, are independent. Each could be violated without the other. Mutations might occur at random but all mutants might have identical fitnesses. Natural selection would thus not be operative. Similarly, natural selection could be operative in a population but the mutations that occur in it might not be random. This is the position that corresponds to that of neo-Lamarckism but, in order to specify neo-Lamarckism exactly enough to evaluate its claims as an alternative to neo-Darwinism, three other distinctions are necessary.

First, neo-Darwinism is being construed here to require that *all* mutations are random. This usage is standard. Neo-Lamarckism could potentially be analogously construed as requiring that *all* mutations are not random (or even directed which, as will be discussed below, is a stronger claim).

Such a construal of neo-Lamarckism, which might be called "strong neo-Lamarckism," would be uninteresting since it is quite trivially false: that at least *some* mutations are random is uncontroversial. Neo-Lamarckism will, therefore, be construed here as only requiring that some mutations are not random. If a distinction between the two ways of construing "neo-Lamarckism" is needed, this weaker notion might be called "weak neo-Lamarckism." Weak neo-Lamarckism then just amounts to a denial of the universalization over *all* mutations that is part of the standard construal of neo-Darwinism. Indeed, historically, neo-Lamarckian positions have usually been construed in this fashion.[13] Only during the height of the justly infamous Lysenko affair did most of Lysenko's defenders suggest that all mutations had to be specifically induced by the environment and would not, therefore, be random.[14] In the discussion below, unless otherwise indicated, "neo-Lamarckism" will be taken to be "weak neo-Lamarckism."

Second, neo-Lamarckism, as it is usually construed, and as it will be used here, actually requires something stronger than the assertion that mutations are not all random. Nonrandom mutations could either enhance or decrease the fitness of the associated phenotypes in that environment. Neo-Lamarckism requires the former. It is precisely in this sense that neo-Lamarckism requires that some mutations are *directed*. The sense of direction that is involved here is that these mutations only occur, or at least occur more frequently, in those environments in which the associated phenotype has an enhanced fitness. Thus there is, basically, a positive correlation between the mutation rate and the fitness of the associated phenotype.

Third, it is necessary to distinguish carefully between two ways in which a directed mutation might be defined. A strong definition would be one that requires that a mutation be considered "directed" *if and only if* it occurs (or occurs more frequently) in an environment where its associated phenotype has an enhanced fitness.[15] A weaker definition would be one that dropped the "only if" clause: a mutation is then considered "directed" if it occurs (or occurs more frequently) in the fitness-enhancing or "selective" environment as characterized above.[16] In the context of particular mutations, the former corresponds to strong neo-Lamarckism and the latter to weak neo-Lamarckism, since the former excludes any random (or selectively neutral) occurrence of these mutations. Only the latter will be used here because of the considerations given during the discussion of the first point above and because it suffices to capture the disagreements that characterize the controversy over directed mutagenesis in bacteria.

"Neo-Lamarckism," when construed in this fashion, makes no reference to what has historically been a fairly standard conception of the position during debates between neo-Darwinians and neo-Lamarckians, namely, that of the "inheritance of acquired characteristics." There are two reasons for this. *First*, in the current controversy regarding the possibility of directed mutagenesis in bacteria, which is the immediate concern of this paper, there is usually no reference to the inheritance of acquired characteristics.[17] Construing the issues involved in terms of "directed mutations" thus makes it easy to remain faithful to the debates. *Second*, the "inheritance of acquired characteristics" is a concept too vague to serve any useful purpose here.[18] Interpreted broadly enough, the "inheritance of acquired characteristics" can include so much that it becomes trivially true. In some plants, such as flax, environmentally induced changes in the genome are routinely inherited.[19] These clearly should be considered as the inheritance of acquired characteristics. In some sense, even radiation induced damage to chromosomes is an "acquired" characteristic that is obviously inherited. What is missing from such examples, however, is that they leave open the question of the correlation of these variations with the fitnesses of the associated phenotypes. It is exactly the last issue that is of interest in the controversy being examined here. Thus, to construe it in terms of the inheritance of acquired characteristics is unilluminating whereas the characterizations of neo-Lamarckism given above capture the spirit of the disputes.[20]

The terms "neo-Darwinism" and "neo-Lamarckism" obviously refer to Darwin and Lamarck and are intended to capture some part of their views. However, the terms as explicated above (and those explications are consistent with how the terms are usually used) have very significant differences with the views that Darwin and Lamarck actually held. Lamarck suggested in 1809 that "a more frequent and continuous use of any organ gradually strengthens, develops and enlarges that organ"; disuse has the opposite effect; and that such changes are inherited.[21] An animal thus has the property of being able to induce variations that, in some sense, enhance its ability to survive in a particular environment. There is thus some resonance between Lamarck's views and the explication of "neo-Lamarckism" given above but it cannot be suggested that this explication constitutes a reconstruction of Lamarck's actual views.

Exactly fifty years after Lamarck, Darwin published the *Origin*, in which he argued that variations usually arose at random.[22] However, Darwin did not exclude the possibility of some variations arising through the "use and disuse" mechanism of Lamarck or through direct environmental induction.[23] Moreover, under the pressure of various objections to his theory on the grounds that there could not have been enough time for evolution to have taken place purely through natural selection acting upon random variations, he began to emphasize such mechanisms over random variation in later editions of the *Origin*.[24] The conception of neo-Darwinism that excludes all possibilities of variation except random mutations is largely due to Weismann.[25] The term "neo-Darwinism" seems to have been coined to describe Weismann's views by one of his principal opponents, Romanes, who tried to maintain Darwin's more "pluralist" approach to the sources of variation.[26] Thus, once again, while there is some consonance between "neo-Darwinism," as used above, and Darwin's own views, the explication is not to be taken as one of Darwin's actual views. In fact, while these historical associations are interesting, it should be emphasized that what truly matters in the adoption of the given analyses of "neo-Darwinism" and "neo-Lamarckism" is their utility for conceptual clarification and not whether they accurately reflect historical developments.[27]

12.3 The "Classic" Experiments

During the late 1920s and 1930s, considerable biological attention shifted to microorganisms such as viruses, thanks to the research programs initiated by Delbrück and others.[28] Out of these developments came such crucial events as the formation of the Phage Group and Lederberg's discovery of bacterial recombination which would both play a critical role in the emergence of molecular biology. At this point it was already known that certain viruses, called "bacteriophages" or "phages" preyed upon bacteria that were "sensitive" to them. It was also known that when such viruses and bacteria were grown together, occasionally, strains of bacteria that were "resistant" or immune to the viruses arose. Apparently, some bacteria had somehow been transformed from virus sensitivity to virus resistance and this transformation was heritable. The cause of this transformation, however, remained unknown and subject to considerable controversy.[29] D'Herelle, one of the discoverers of phage, was among those who argued that the presence of the virus induced this transformation while Gratia and Burnet, among others, argued that the transformation in the bacteria arose before exposure to the virus.[30]

The former position conforms to neo-Lamarckism, the latter to neo-Darwinism as these terms have been explicated in the last section. If it is necessary that the virus be present in order to induce the observed transformation in a bacterium, then there is a correlation between the occurrence of this transformation and the environment of the bacterium. Further, this transformation, by permitting the bacteria which would otherwise fall prey to the virus to survive and reproduce, obviously enhanced the fitness of the bacteria. Thus the transformation is clearly neo-Lamarckian. If, however, the transformation occurred spontaneously in the sense that the presence of the virus was irrelevant to it, then the occurrence of such transformations would have no correlation with the presence of virus. Further, assuming that these transformations did not affect the fitness of the bacteria in some unknown way other than permitting virus resistance, there would also be no correlation with the fitness of any associated phenotype in the environment. Thus they would be random in the sense explicated in section 12.2. Therefore, this position is genuinely neo-Darwinian.

The canonical attempt to resolve this dispute was through the use of the so-called "fluctuation test" devised by S. Luria and M. Delbrück.[31] The basic idea was not new though Luria and Delbrück seem to have arrived at it independently. As early as 1934, Yang and Bruce White had observed that there were wide fluctuations in the occurrence of "rough" variants of *V. cholerae*, that is, variants which formed morphologically rough colonies, and had concluded that the occurrence of a large number of variant colonies signified the spontaneous, that is, random occurrence of the variation and its subsequent clonal expansion during growth.[32] They did not develop the idea mathematically and their work seems to have been largely ignored.[33] Luria and Delbrück are responsible for developing the same insight and exploiting it to its full in order to establish the neo-Darwinian position noted above.

If the transformation of bacteria from virus sensitivity to virus resistance occurs as a result of variations induced by the presence of the virus, then there is a finite probability that such an event occurs in each bacterium after exposure to the virus. Subsequently, this variation is inherited. Luria and Delbrück called this possible outcome the hypothesis of "acquired hereditary immunity".[34] If the transformation of bacteria from virus sensitivity to virus resistance occurs as the result of a spontaneous variation, then variants arise at random during the growth of the bacteria. The number of variants arising from a particular initial event resulting in a variant grows exponentially with time because variants geometrically give rise to more variants in the next generation by normal cell division. Luria and Delbrück called this possibility the hypothesis of "mutation."[35] They note with care, however, that they were not claiming that such mutations were the same, or even very strongly similar, to what were commonly called "mutations" in higher organisms.[36]

They then devised the so-called "fluctuation test" to distinguish between these two hypotheses (see fig. 12.1). A number of "sister" cultures of bacteria are grown to a specified density in liquid media in different test-tubes, ideally from a single ancestral cell that is known to be sensitive to the virus, though, in practice, from a few such cells that are assumed to be identical in their genotype. The number of resistant bacteria in each culture after exposure to the phage is then determined by mixing each culture (or a fixed fraction of it) with an excess of phage on a single test plate. Sensitive bacteria die while the resistant ones form colonies that are visually recognizable on the plate. The number of such colonies on a particular plate is, therefore, equal to the number of resistant bacteria in the corresponding culture (or the fraction of it that was plated).

Under the hypothesis of acquired hereditary immunity, the resistant mutant bacteria arose because of the finite though small probability that a bacterium would be induced to resistance by exposure to the phage. The distribution of the number of mutants would, therefore, follow the well-known Poisson distribution with a mean equal to the product of that probability and the number of bacteria in the culture (see fig. 12.2). Since the distribution is Poisson, the mean would be equal to the variance. Under the hypothesis of mutation, however, the number of mutants in a culture would vary depending on how early in the history of that culture the

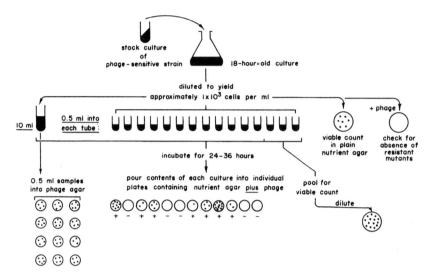

Figure 12.1

The Fluctuation Test. A single bacterial cell from a stock culture is taken and incubated. Sister cultures are then grown in parallel, each from a single cell, in different test-tubes. These are then plated on the selective medium, (phage-sensitive) agar. Mutants showing phage resistance are the only cells that can form colonies. Thus the number of colonies on a plate equals the number of resistant cells in the corresponding test-tube before plating. © W. B. Saunders. From Braun, W. *Bacterial Genetics*, W. B. Saunders and Company.

mutation had occurred. The mutations that occurred early would be represented by an enormous number of mutants due to exponential growth. These are usually called "jackpot" cultures. Because of the presence of such "jackpot" cultures the "Luria–Delbrück" distribution has a much greater variance-to-mean ratio than the Poisson (for which this ratio is 1). Thus the fluctuations in the number of mutants would be much greater in this case—hence the name "fluctuation test."

Assuming that the mutant and non-mutant strains of bacteria grow at the same rate before exposure to the phage, which occurs if they had the same fitness, Luria and Delbrück managed to derive a formula for this variance which showed that it is much greater than 1. However, they could neither give an explicit expression for the distribution nor specify a procedure for calculating the probability that a given culture would have a certain number of mutants. Lea and Coulson first solved the latter prob-

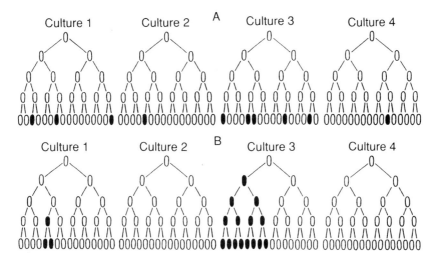

Figure 12.2

The Poisson (A) and Luria–Delbrück (B) distributions. The ovals represent bacteria growing clonally. Note how the Poisson distribution arises when mutations occur only after exposure to the selective environment. The variance is equal to the mean. The Luria–Delbrück distribution has one "jackpot" culture in which a mutation occurred in the first generation and the number of mutants grew with the clone. © W. H. Freeman and Company. From Stent, G. S. *Molecular Genetics*, W. H. Freeman and Company.

lem by finding an iterative procedure for calculating these probabilities.[37] Recently, Stewart, Gordon, and Levin have given a systematic treatment of this distribution under a wider set of assumptions than initially conceived by Luria and Delbrück, such as differential reproductive rates of mutant and nonmutant strains and inefficient plating techniques.[38] Their procedure permits the computation of the actual distribution expected in each of these cases. No explicit analytic formula for the distribution has yet been found.

Luria and Delbrück applied the fluctuation test to the action of the phage T1 on a strain of the bacterium, *E. coli*, that was normally sensitive to it. The distribution for the number of mutants obtained in "sister" cultures had a far higher variance to mean ratio than 1. They concluded, on the basis of this statistical test, in favor of the hypothesis of mutation. Between 1943 and 1949 the Luria–Delbrück experiment was repeated (in the course of other work) by Newcombe, and also by Demerec and Fano, for the same system.[39] In fact, the last two authors systematically applied fluctuation analysis to the case of resistance to phages T3 to T7 by *E. coli*. Witkin similarly studied the interaction of *E. coli* with radiation while Ryan studied its histidine independence.[40] Demerec applied fluctuation analysis to the resistance of *Staphylococcus aureus* to penicillin; Oakberg and Luria to its resistance to sulphathiazole; and Demerec, later, to its resistance to streptomycin.[41] Ryan, Schneider and Ballentine applied the method to the study of uracil independence of *Clostridium septicum*; Alexander and Reidy to streptomycin resistance of *Hemophilus influenza*; and Curcho to the tryptophan independence of *Eberthella typhosa*.[42]

In general, the evidence was taken to show that the observed distributions corresponded to the Luria–Delbrück distribution within the limits of experimental error. Sometimes, however, the deviations were obvious as in the case of Newcombe, who then invoked phenotypic lag, though only Ryan, in the cases of the interaction of *E. coli* with radiation and its mutations to histidine independence, argued that the deviations were significant.[43] However, in the former of Ryan's examples, Witkin and Newcombe were explicitly satisfied with the usual interpretation that the deviations were not significant.[44] In the case of mutations of *Clostridium* to uracil independence, Ryan, Schneider and Ballentine observed that the mutation occurred in spite of the presence of uracil in the environment, and thus was not directed, but that the data did not permit an assumption that the mutations were random in the sense that the distribution of mutants followed the Luria–Delbrück distribution.[45]

In fact, it was Ryan who struck a further slightly discordant note in the growing consensus, among those using fluctuation analysis, in favor of the neo-Darwinian hypothesis that bacterial mutations occurred at random.[46] He studied mutations of two strains of bacteria from Lac$^-$, a phenotype which does not have the ability to digest lactose to Lac$^+$, a phenotype which does.[47] In both strains, there was some divergence from the expected Luria–Delbrück distribution but Ryan considered only one of these significant enough to report.[48] However, he did not interpret this divergence to suggest that any nonrandom process was occurring. He argued that the assumptions of Luria and Delbrück were too simplistic and seems to have been suggesting that factors like differential fitness of mutants and nonmutants in the original medium or phenotypic lag (the latter of which had already been shown by Armitage to lead to some divergence from the Luria–Delbrück distribution) might be operative.[49]

However, not all were convinced by fluctuation analysis. Partly this must have been because of the abstract mathematical nature of the argument. The most strident critics, however, were those who, like Hinshelwood, held that there were no "genes" in bacteria analogous to genes in higher organisms even as late as 1957.[50] According to Hinshelwood, bacterial "genetics" could be accounted for in terms of complex patterns of chemical reaction networks, and the response of these to environmental stimuli was responsible for transformations like the one from virus sensitivity to virus resistance. The large variations observed in the fluctuation analysis experiments were to be explained on grounds of uncontrolled environmental differences between the various fluctuation test tubes, a claim initially made on experimental grounds by Eriksen, and subsequently repeated by Hinshelwood and his collaborators.[51]

Cavalli and Mitchison attempted to provide an independent statistical argument for the spontaneous origin of the mutants by studying secondary characteristics of the mutants, that is, characteristics other than the one directly selected for in the fluctuation tests.[52] Such characteristics would presumably differ in descendants of mutants from different mutations, that is in different test tubes of the fluctuation experiment if the hypothesis of mutation was correct.[53] Further all mutants within a test tube, having then arisen from the same mutation, would be very likely to have such characteristics to the same degree. Indeed, they found a strong statistical correlation for such characteristics for mutants within a single test tube and much greater variation between test tubes. This strongly suggested the clonal nature of the mutants as would be expected if they arose from single spontaneous mutations in each test tube. Cavalli also argued that the absence of a correlation between the number of mutants detected by the fluctuation test and the average resistance of the mutants (the secondary characteristic) further suggested that the environmental factors were irrelevant since it seemed likely that the same environmental factors would be relevant to both.[54]

Meanwhile Newcombe carried out a different experiment on *E. coli* bacteria normally sensitive to the phage T1.[55] Bacteria sensitive to phage were incubated for a few hours on several plates in order to allow them to grow. After this, on half the plates, the cells were redistributed, more or less uniformly, using a little salt solution. The other half of the plates were left undisturbed. Phage was then sprayed on to all the plates and the plates were incubated again. Under the hypothesis of acquired hereditary immunity, the respreading should have no effect on the appearance and number of resistant colonies on the plates since the mutants would only arise due to interaction with the phage after exposure. Under the hypothesis of (spontaneous) mutation, however, the respreading would make a difference. Mutants would have arisen in clones which would have remained as part of the same colony before respreading. Respreading would spatially redistribute clones whose members would then form separate colonies. Thus the plates in which respreading occurred would be expected to have more colonies than the plates in which it did not occur. The experimental results provided striking confirmation of this expectation.

Newcombe recommended his method for the demonstration that mutations arose at random over the fluctuation test because the experimental method seemed to be more "direct."[56] Undoubtedly, this directness helped establish that result. However, the most decisive experiments were two equally direct experiments involving "indirect selection," where the mutants were obtained without any prior exposure to the selective media of any of their ancestors. These experiments, involving the use of "replica plating" and "sib-selection" in liquid media were performed by Lederberg and his collaborators.[57] This work involved some brilliant experimental designs and will be described in some detail, here, because of their role in establishing the orthodox neo Darwinian position.

Replica plating permits the direct transfer of a pattern of bacterial growth on one plate to several others. A piece of velveteen is pressed against the plate containing the original bacteria. The pattern of growth on the plate is thus transferred on to the velveteen. The imprinted velveteen can then be used to inoculate, serially, several plates. Thus several replicas of the original pattern are obtained on these plates. It would be the case that if these new plates contained phage, bacterial colonies at corresponding sites would show phage resistance on them. Were resistance being induced by the phage, resistant colonies would be randomly distributed in each of these new plates. It turned out that resistance occurred at corresponding sites, which established that it had arisen in the colonies at these sites prior to exposure to the phage. This result already provided quite conclusive evidence that the mutation from phage sensitivity to phage resistance was occurring spontaneously. However, Lederberg and Lederberg went on to extend it even further to select mutants that had never been exposed to phage, as will shortly be described.

Lederberg and Lederberg carried out this test not only for phage resistance of bacteria but also for resistance to streptomycin. In the case of phage resistance, they also managed to use replica selection to isolate resistant variants without their, or their ancestors, ever having been exposed to phage. Replicas from the original plate were imprinted on one plate containing phage and another without it. A colony from the latter plate whose site corresponded to a resistant one in the former plate was selected. Cells from this colony were then allowed to grow and plated again. Replicas were then plated onto plates with and without phage. Iteration of this procedure permitted purification of the resistant line. Thus a phage-resistant strain of bacteria was "indirectly" selected without its ever having been exposed to phage.

Results obtained from replica plating do not permit quantitative analysis. However, Cavalli-Sforza and Lederberg initiated a scheme that permits quantitative investigation of the problem by devising a method of indirect selection, usually called "sib selection," using liquid media. If one starts out with a culture in a liquid medium where a resistant cell is so rare that only 1 in 10 test tubes would contain a resistant cell, this sample can be detected by exposure of a fraction of its contents to the selective medium. Discarding the other 9 test-tubes leads to a 10-fold enhancement of the concentration of resistants in the population. Dilution and repeated iteration of this procedure permits, in principle, the isolation of a pure line of resistants that had no contact with the selective media.[58] Not only that, the extent of the gradual enrichment of resistants could be precisely calculated from the hypothesis that all mutations were spontaneous. Cavalli-Sforza and Lederberg carried out this experiment for resistance to streptomycin and chloramphenicol. For streptomycin resistance, they found full agreement between theory and data (within the limits of the statistical power of their experiment). For chloramphenicol resistance, a large amount of statistical error did not permit detailed comparison of expected and observed enrichments. However, the data was consistent with all mutants arising spontaneously.

There are three possible criticisms of these experiments. All three are particularly relevant to the resurgence of the controversy over the possibility of directed mutagenesis in bacteria during the last three years and will, therefore, be discussed in some detail. The *first* criticism, which has been emphasized by Stahl, is that the selective media used were lethal to nonmutants.[59] Thus, these experiments do not truly explore the possibility that environmentally directed mutations might occur, especially if they occur at a slow enough rate. Exceptions to this general practice were the experiments of Ryan, but those experiments were precisely the ones that showed some significant deviation from the Luria–Delbrück distribution.[60] Thus the experimental data of the 1950s might not have been quite as conclusive as it then seemed.[61]

Second, all these experiments seem to have been designed to test whether *all* mutations were random or whether all mutations were directed. The possibility that some mutations could be directed while others, perhaps most, were random seems not to have been taken seriously. In other words, the experiments do not seem to concern themselves with what was proposed to be (weak) neo-Lamarckism in section 12.2. Luria and Delbrück explicitly assumed that one of two extreme hypotheses hold: all mutations are spontaneous (random) or all are directed, as if each of these is the negation of the other.[62] The fluctuation test as devised, and used by them, only distinguished these two cases. Indeed, until the work of Lea and Coulson in 1949, there was no procedure to calculate the Luria–Delbrück distribution and any observation that the variance of an experimentally obtained distribution was significantly greater than its mean was taken to be indicative that the Luria–Delbrück distribution was obtained.[63] Further, the work of Armitage in 1952 made it even more difficult to be more precise about when the Luria–Delbrück distribution holds because deviations from it could already be attributed to factors other than the occurrence of directed mutations.[64] This point will be extremely important below. Even the quantitative sib-selection experiments of Cavalli-Sforza and Lederberg is flawed in the sense that it, too, would fail to distinguish between any except the extreme hypotheses of all spontaneous or all directed mutations.[65] Since the proportion of nonmutants decreases in these experiments with every iteration of the dilution procedure, the statistical limits of the experiment would not permit the identification of additional mutants arising from the nonmutants after exposure to the selective medium, whether or not the selective medium was lethal to the nonmutants. Finally, the replica plating experiments and Newcombe's experiment are simply not quantitative in their nature and can be of no help in deciding this issue.[66]

Third, and least important, there is a further difficulty with the design of these experiments, though it can be easily avoided with some secondary choices. *Directly*, all that they test is whether certain mutations occur *before* or *after* exposure to certain media. The choice whether the medium is selective is made independently of the experimental design. The experiments discussed chose media that are trivially selective. However, what is truly of significance in order to distinguish the neo-Lamarckian and neo-Darwinian hypotheses is the response of the bacteria precisely to selective media. In practice, this difficulty was avoided by ensuring that the medium used was trivially selective for the mutants. However, this does not permit consideration of subtle fitness differences that might be involved. This will also be of some importance below.

12.4 Recent Experiments

The first new experimental evidence that suggests the occurrence of directed mutations[67] came from Shapiro.[68] Shapiro studied a strain of *E. coli* in which the regulatory part of the arabinose operon was separated from the terminal portion of the lac operon by a short segment of the transposable phage Mu which contained signals for the termination of transcription of DNA to RNA. When the intervening phage segment is deleted, cells of this strain can digest and grow on lactose provided that arabinose is also present to induce transcription. This phenotype is symbolized Lac (Ara)$^+$; the phenotype with an inability to do this is symbolized Lac (Ara)$^-$. Shapiro observed that when this strain of bacteria, still containing the intervening phage sequence between the parts of the ara and lac operons, was plated in media containing minimal arabinose and lactose, Lac (Ara)$^+$ mutants, which can grow and form colonies on this medium, occur only after a delay of about three days. However, if these plates are seeded with strains of preexisting Lac (Ara)$^+$ cells, they grow and form colonies immediately. It appears from this experimental result that the formation of the Lac (Ara)$^+$ mutants, presumably by the deletion of the intermediate phage sequence, did not occur before the exposure of the original strain to arabinose and lactose, and occurred because of some interaction with the environment afterwards. Moreover, since the Lac (Ara)$^+$ mutants are selected for in this environment because of their ability to digest lactose, this is a case when the mutation is one that enhances fitness. This result is thus potentially a case of neo-Lamarckian mutagenesis.

Cairns, Overbaugh and Miller have presented two sets of experiments designed to show that some directed mutations occur in bacteria.[69] *First*, they investigated the problem of the origin of mutants in bacteria using the ability to digest lactose as the phenotypic marker somewhat along the lines initiated by Ryan.[70] They studied two strain of *E. coli*. The first had a mutation in the structural gene of the lac operon that prevented the production of the enzyme that digests lactose. The second contained an additional mutation, a deletion in the so-called *"uvrB-bi"* region of the bacterial chromosome. The distribution of mutants for these that "reverted" to a form that could digest lactose (the so-called "revertants") in a minimal lactose medium apparently turned out to have less variance than that was expected from the Luria–Delbrück distribution. Further, the deviation was much more significant for the second than the first. Cairns, Overbaugh, and Miller interpreted this result to suggest that both random mutations before exposure to lactose and directed mutations after such exposure were occurring. This interpretation was based on the observation that the distribution obtained was consistent with a Poisson component "added on" to a Luria–Delbrück distribution as obtained from other strains. Further, the distribution of the mutants appearing after exposure to lactose, according to them, followed a Poisson distribution, though Cairns, Overbaugh, and Miller did not actually perform any statistical analysis. Further, that the second strain had a much greater deviation indicated an influence of environmental factors.

In a *second* set of experiments, Cairns, Overbaugh, and Miller replicated the observations of Shapiro just discussed and extended them further. First they showed that strains of *E. coli* that were Lac (Ara)$^-$, not because of the inserted phage sequence, but because of a point mutation in either the lac or the ara operon, showed no delay in producing the revertant Lac (Ara)$^+$ colonies. They interpret this to argue that there must be some special property of the intervening phage sequence (at that position) that permits its deletion (and thus reversion to Lac (Ara)$^+$) in the presence of lactose and arabinose. They also report a series of experiments in liquid cultures that showed the same effect.

Meanwhile Hall has experimented with a strain of *E. coli* that required two mutations in the *β*-glucosidase (*bgl*) operon in order to have the ability to digest salicin.[71] One of these (here referred to as the first mutation) is the excision of an intervening sequence of nucleotides from the structural gene that codes for the protein that actually participates directly in the digestion of salicin. The other (here referred to as the second mutation) is one in the regulatory gene of the operon. Note that neither mutation by itself confers the ability to digest salicin. Hall found that the first (excision) mutation rate in a strain with the second mutation was very low in the absence of salicin. Yet, Hall found that when this strain of bacteria with neither mutation is grown in agar rich in salicin, the first (excision) mutation rate is greatly enhanced. This observation is particularly odd because the first (excision) mutation, by itself, does not confer any ability to digest salicin.[72] One possibility is that the intermediate genotype, that is, the one with the first (excision) but not the second mutation, enhances fitness in some way other than the ability to digest salicin. Hall tested this possibility by growing mixtures of cells with or without the first (excision) mutation, but not the second mutation, in an environment containing salicin. No difference in fitness associated with the intermediate genotype was observed. Though Hall does not explicitly interpret these results as favoring directed mutation, he notes the "striking similarity" between his observations and those of Cairns, Overbaugh, and Miller discussed above.

Levin, Gordon and Stewart have repeated and generalized the experiments of Cairns, Overbaugh, and Miller, though for a different strain of *E. coli* and different culture media.[73] They report deviations from the Luria–Delbrück distribution similar to those observed by Cairns and his collaborators. However, they generally interpret their results to suggest that the mutations had already occurred in the fluctuation test tubes prior to plating and the colonies had appeared late after plating due to "physiological processes" associated with their interaction with high densities of a stationary parental population of nonmutants. Meanwhile, Mittler and Lenski have repeated the experiments of Shapiro and Cairns, Overbaugh, and Miller on the Lac (Ara)$^+$ mutation on *E. coli*.[74] They verified the reported observations but ran a control that casts serious doubts on the interpretation favored by Cairns, Overbaugh and Miller. While the mutant colonies in question were present in media with lactose and arabinose, they found a comparable number of Lac (Ara)$^+$ colonies produced even in plates where both were absent. These colonies were detected by spraying plates with lactose and arabinose after different intervals of time during which these sugars had been absent. Similar results were obtained with liquid cultures. They interpret these findings to show that the appearance of Lac (Ara)$^+$ mutants does not require the presence of lactose and arabinose but seems do depend on the physiological state of the cells. However, it is not clear yet whether these results can be reproduced.[75]

12.5 Interpretation

The results briefly touched upon in the last sections show that the experimental situation is still quite uncertain. However, this uncertainty is the result not so much of disagreements about the observations themselves as about their interpretation. These interpretive problems arise, to a very large extent, because of the three limitations, discussed at the end of section 12.3, in the design of the experiments that established the orthodox neo-Darwinian position regarding bacterial mutagenesis and which remain equally problematic for most of the experiments mentioned in the last section. The first of those limitations is not particularly important at this stage since the current experiments all study the response of nonmutant bacteria in media which are selective for the mutants but not lethal to the nonmutants. The third problem was that those experiments only directly test whether certain mutations occur before or after exposure to certain media and not whether there is any correlation between these mutations and the fitness of the resultant phenotype in that medium. The requirement that a medium be selective is not, in any direct sense, part of the experimental designs. Thus these experiments do not help in resolving questions of the sort raised by Mittler and Lenski.[76] In practice, as in the experiments described in section 12.3, the medium is chosen to be trivially selective. The second limitation is the most interesting and needs to be discussed in greater detail. In discussing it, attention will, however, be confined to fluctuation analysis since that is the only method that is relevant to the new experiments.

The source of the trouble is that virtually any factor operative in cultures that influences the rate of colony formation shifts a Luria–Delbrück distribution to one with less variance which can be interpreted as being a shift in the direction of the Poisson. This makes it extremely difficult, if not impossible, to use the fluctuation test even indirectly to argue that some directed mutations are occurring by showing such deviations from the Luria–Delbrück distribution as Cairns, Overbaugh, and Miller have attempted.[77] In other words, this unfortunate property of the Luria–Delbrück distribution is precisely what limits the potential use of the fluctuation test to decide whether *some* directed mutations might be occurring. The effect of various factors on the Luria–Delbrück distribution is best seen by exploiting the general procedure for computing these distributions given by Stewart, Gordon, and Levin.[78] They have computed the distributions under a variety of circumstances. These include differential fitness of mutant and nonmutants, a mutation rate dependent on the nutritional state of the cell, and the possibility that plating efficiency is less than 1. In all cases, there is a shift from the Luria–Delbrück distribution to one with less variance, which can be interpreted as a shift in the direction of the Poisson and, therefore, as a mixture of the two distributions. Thus, the type of observation reported by Cairns, Overbaugh, and Miller can be accounted for without recourse to the hypothesis of directed mutation. Some simple qualitative arguments demonstrate these effects. These will now be given for: (i) differential fitness of mutants and nonmutants; (ii) phenotypic lag; (iii) plating efficiency less than 1; and also (iv) for multiple mutations and selection.[79]

The effect of differential fitness of mutants and nonmutants:[80] If the fitness of the mutant is lower than that of the nonmutant in the original medium, that is, not the medium in which the mutant is being selected for, then the distribution of mutants as observed during a fluctuation analysis experiment will be shifted from a Luria–Delbrück one in the direction of a Poisson distribution. This is easy enough to motivate in an extreme case. Consider the situation when the fitness of the mutant is so low that it is not even viable for one generation. Then the distribution obtained, even if mutants only arose at random, would be a Poisson one, because the only mutants that would be observed through colony formation after plating on the selective medium would be the ones that arose during the last generation. In less extreme cases, this tendency towards the Poisson is just less extreme.[81] This can be shown using the iterative procedure developed by Stewart, Gordon, and Levin as has been mentioned above.[82] The same effect is also seen in a simulation reported by Lenski, Slatkin, and Ayala where they showed that the distribution obtained from a model with spontaneous and some directed mutation could be mimicked by one with differential fitnesses.[83] However, the differential fitness model needs to assume a much higher spontaneous mutation rate than the directed mutation model. This, too, can be seen intuitively. Since the mutants have a lower fitness, they grow more slowly than the nonmutants in the original medium. Thus to achieve the same concentrations as in the directed mutation model (where mutants and nonmutants have the same fitness), when mutants have lower fitness, they have to arise more frequently. Further, other mutants must also arise to compensate for those apparently arising after exposure to the selective environment in the directed mutation model.

The effect of phenotypic lag: Phenotypic lag is observed when a mutant phenotype is not expressed for some generations after a mutation occurs. This is best illustrated by the following example: suppose that a mutation makes some functionally necessary protein abnormal; suppose also that the functionally necessary protein is present in quantities much greater than strictly necessary in the cytoplasm when the mutation occurs. Then, for some generations of cell division, the resultant cells would continue to have functional proteins in their cytoplasm in sufficient quantities to preclude loss of function because of the division of the cytoplasm of the original cell. The phenotypic change—loss of the function associated with the protein—would only be observed late. Phenotypic lag obviously explains the late appearance of mutants. The reason why it seems to shift the Luria–Delbrück distribution in the direction of the Poisson, however, is a little complicated. Such a shift occurs only when the selective medium favors the mutants and the mutants are relatively rare, as is true in the cases being investigated experimentally. Mutants grow faster than nonmutants, and this process disproportionately affects the relative concentration of mutants to nonmutants in those cultures where mutants are rare and has very little effect on jackpot cultures. Thus, the variance of the distribution decreases.

The effect of plating efficiency being less than 1: It is easy to see why this would also decrease the variance of a distribution. Assume that the loss due to nonideal plating efficiency is the same for all cultures. Assume also that this loss is proportional to the total number of cells, mutant or nonmutant. Then the proportional loss of mutants in a jackpot culture is the same as in a culture with few or no mutants. However, the number of mutant cells lost in a jackpot culture is much greater than in cultures with few or no mutants. This decreases the variance of the distribution, thus shifting a Luria–Delbrück distribution in the direction of the Poisson.

The effect of multiple mutations and selection: Lenski, Slatkin, and Ayala have also proposed, and simulated, a different model which gives rise to a similar distribution of eventual mutants as the one obtained from the directed mutation model.[84] In this model, g_0, the initial nonmutant which cannot grow on the selective media first undergoes a mutation to g_1, which can only grow slowly on those media, and then to g_2 which grows rapidly on those media and whose distribution is the one that is measured. Colonies of g_2 appear late because of the slow growth of the intermediate genotype, g_1. Thus, though the late appearance of the colonies would have been inconsistent with the measured growth rate of g_2, the discrepancy can be explained by invoking the slow rate of growth of g_1. The source of this delay is, of course, the slow growth of the second genotype. The result of such a process would be that the initial results from the fluctuation test, which uses colony counts, would show less variance than is actually warranted. However, Levin, Gordon, and Stewart have failed to detect such multiple mutation events (though they did not systematically search for them).[85]

The variant interpretive possibilities of the same fluctuation analysis data that has just been summarized highlight the point that fluctuation analysis alone cannot resolve the dispute over whether some directed mutations occur in bacteria, that is, whether there is some neo-Lamarckian mutagenesis. Any observation of a deviation from the Luria–Delbrück distribution can be explained by invoking subsidiary interactions such as those of differential fitnesses, phenotypic lag or inefficient plating while remaining strictly within the neo-Darwinian framework. Of course, hypothesizing such subsidiary interactions consists of making additional empirical claims and these, in turn, require experimental support. However, the kind of experiment these require must, in most cases, go beyond mere statistical analysis of the number of mutants and nonmutants.[86] Those favoring directed mutations, meanwhile, can experimentally rule out the existence of the subsidiary interactions posited by the neo-Darwinians. If literally all such explanations are ruled out, and it is shown that a certain mutation occurs (or occurs more frequently) if the original strain is exposed to a media in which the phenotype associated with the mutation has enhanced fitness, then this evidence can be taken to suggest the plausibility of the neo-Lamarckian position.

However, given the variety of neo-Darwinian alternatives that have already been posited, such a strategy seems unlikely to generate easy acceptance of directed mutations. What the neo-Lamarckian needs to do, in order to convince the skeptic, is to find *mechanisms* for directed mutations. This will involve both theoretical speculation—imagining physically plausible mechanisms for directed mutagenesis—and experimental corroboration. Cairns, Overbaugh, and Miller have already begun the former process and have suggested three possible mechanisms.[87] Stahl and Davis have suggested others.[88] Some of these mechanisms have already been experimentally ruled out or, at least, been made quite implausible.[89] They are still discussed here in order to show how the possibility of directed mutagenesis can quite easily and plausibly be modelled at the molecular level.[90] All of these merit some further discussion because they illustrate that there might be nothing genuinely mysterious, or even fundamentally new, about some neo-Lamarckian processes.[91] In the discussion of these mechanisms the distinction of "variant" strands of DNA or RNA and "mutations" is maintained scrupulously since it will be important later. "Variant" strands of DNA are simply alternative strands (from that which is initially part of the genome) that arise through a variety of mechanisms such as failure of the repair of errors or the action of the environment when a single-stranded segment of DNA is exposed to it. Similarly, variant strands of RNA can also arise by a variety of mechanisms including transcription errors. Such variants might or might not get expressed. Only when such a variant gets incorporated into the genome when the cell divides has a "mutation" taken place.

Cairns, Overbaugh and Miller discuss three scenarios.[92] The first of these will be called a "specific reverse transcription" mechanism. When somewhat generalized from the detailed description of this mechanism given by them, it operates along the following lines: the cell produces a highly variable set of messenger RNA molecules (possibly through processes like inaccurate transcription). Of these, one codes for the functionally best protein. This messenger RNA molecule would then be reversely transcribed into DNA. Thus a variant DNA molecule would arise by an obviously directed mutation; as the cell then divided, this variant DNA molecule would give rise to a mutation in the sense that it would become part of the genome.

This mechanism requies three things: (i) a special organelle for reverse transcription such as the gag-pol complex of retroviruses which contains reverse transcriptase; (ii) a method by which messenger RNA molecules are monitored for performance of proteins coded for, that is, some kind of a feedback loop back from protein to nucleic acid; and (iii) some interaction from this feedback process to the organelle in which reverse transcription would take place.[93]

The second scenario presented by them will, in analogy to the first, be called a "nonspecific reverse transcription" mechanism: in this mechanism, a highly variable set of messenger RNA molecules, after being produced, would be routinely reverse transcribed into DNA. However, reverse transcription would only be initiated when a cell began growing. This would amount to a positive feedback loop that would ensure that the variant sequence that permitted the cell to grow out of a situation of stress would be the one that was reverse transcribed. This model also requires two things: (i) the existence of some reverse transcriptase in the cell; and (ii) a regulatory mechanism that initiates reverse transcription when cell growth begins. Though this mechanism would be less efficient than the last one, it is a little more plausible for two reasons: (i) it assumes less specificity before reverse transcription begins; and (ii) the regulatory mechanism that correlates reverse transcription and cell growth does not seem too improbable. The third scenario will be called a "variant DNA copies" mechanism: here the DNA itself is assumed to give rise to mutant copies and positive feedback is assumed from the best mutant. No other detail was specified.

Stahl presents another possible scenario which might be best referred to as the "correction enzyme cannibalization" mechanism.[94] In normal cells, routine errors during replication are normally corrected by mismatch repair enzymes. In starving cells these enzymes might well be cannibalized for energy. In that case variant strands of DNA can easily occur during the repair synthesis of DNA in a nondividing cell. Should one fortuitously code for a protein that digests some substrate, the cell replicates the DNA as, perhaps, it divides. Thus the variant strand in question is stabilized as a mutation. The positive feedback from protein to DNA is thus mediated by cell division. Davis has presented yet another scenario which will be called the "transcriptional bias" mechanism.[95] According to this mechanism, single stranded DNA, exposed during transcription routinely gives rise to variant strands. Should one of the variants be such that an inducer, when some little energy is available, can initiate transcription, it does so. Thus lactose would initiate transcription when a DNA strand corresponding to Lac$^+$ arises. This direct induction of transcription would constitute the positive feedback from the environment to the DNA. Now, as the cell with this variant has the ability to digest lactose, energy becomes available cell growth takes place and the mutation is stabilized. Thus indirect feedback from the protein to the DNA occurs.

There are three important points to note about these mechanisms. *First*, all the mechanisms easily satisfy the requirements of (weak) neo-Lamarckism.[96] In each of these cases the mutation rate for a certain mutation is enhanced if the environment, that is, the medium, is selective, that is, it is one in which the associated phenotype has a higher fitness. The associated phenotype is the presence of a protein capable of digesting a particular substance and thereby providing the energy necessary for growth which obviously increases the fitness of the cell. It might, however, be argued that, for example, in the "correction enzyme cannibalization" and the "transcriptional bias" mechanisms, that the underlying variations are occurring spontaneously (randomly) and are only being selected for by the environment. However, even here, selection occurs within a generation, and is a selection of *variants*, not *mutations*, unlike the usual neo-Darwinian processes.

However, because of this underlying selection of variants, these mechanisms illustrate very well the *second* important point that needs to be noticed, namely, that there might not be any serious conflict between neo-Darwinism and neo-Lamarckism. In this sense, at least, there might be *peaceful coexistence* between neo-Darwinism and neo-Lamarckism in such circumstances. The mutations are arising in a directed fashion. However, in the "correction enzyme cannibalization" and the "transcriptional bias" mechanisms, the underlying genesis of variants is, *formally*, neo-Darwinian in the sense that there is random variation and natural selection, although that selection is extremely rapid. However, it is not completely clear whether the same can be said about the mechanisms hypothesized by Cairns, Overbaugh and Miller.[97] The "specific" and "nonspecific reverse transcription" mechanisms do postulate the existence of new organelles and processes. How they fit with neo-Darwinism can only be fruitfully investigated if such entities are discovered or the mechanisms precisely specified. The conclusion arrived here might seem surprising given that, in section 12.2, neo-Lamarckism was explicated as the denial of neo-Darwinism. However, there is no confusion here because the formally neo-Darwinian process is occurring at the level of the genesis of variants whereas the neo-Lamarckian effect is only seen at the level of the occurrence of mutagenesis. Thus, all that is being suggested here is that, in some circumstances, directed mutagenesis can take place because basically random changes that can be incorporated in the genome occur and, they are subsequently so incorporated because the environment is fiercely selective for them in the sense that only one variant is viable.[98] *Third*, all these mechanisms rely on random processes to some extent. However, this still does not preclude the occurrence of directed mutagenesis simply because the randomness is at the level of the genesis of variants. However, it is conceivable that there could be models of directed mutagenesis that do not rely at all on random genesis of variants. Should such models be formulated, and then subsequently found to capture what occurs in nature, they would present greater difficulties for the potentially peaceful coexistence of neo-Darwinism and neo-Lamarckism that is being suggested here.

12.6 Reduction

It was argued in the last section that it is unlikely that fluctuation analysis alone would be able to resolve the controversy about the possibility of some directed mutagenesis in bacteria. Further, it was argued that the resolution of that controversy would have to await the discovery and specification of the mechanisms that account for the experimental data. It will now be argued that the search for such mechanisms constitutes *reductionist* research. This point is important for two reasons. *First* the role of reduction in molecular biology has been extremely controversial.[99] In applying physical (including chemical) techniques to biology, both at the theoretical and the experimental levels, molecular biologists are obviously contributing to a physical understanding of biological processes. This makes it quite plausible to suggest that, in some sense, molecular biologists are reducing biology to physics and chemistry. However, such a suggestion, however plausible, is problematic. There are two main reasons for this: (i) some of the techniques used such as statistical analyses, for example, fluctuation analysis or sib selection, as will be argued below, are definitely nonreductionist; and (ii) there is remarkable disagreement among philosophers as to what constitutes reduction and it is certain that at least some of the models of reduction that have been proposed by philosophers fail to capture the nature of research or explanation that is involved in molecular biology.[100] *Second*, Cairns, Overbaugh, and Miller have explicitly connected the possibility of the occurrence of directed mutations in bacteria with a purported failure of a reductionist account of contemporary research in molecular biology. "The early triumphs of molecular biology strongly supported the reductionists," they argue. However, "when we come to consider what mechanism might be the basis for the forms of mutation described in this paper we find that molecular biology has, in the interim, deserted the reductionist. Now, almost anything seems possible. In certain systems, information freely flows back from RNA into DNA; genomic instability can be switched on under conditions of stress, and switched off when stress is over; and instances exist where cells are able to generate extreme variability in localized regions of the genome. The only major category of informational transfer that has not been described is between proteins and the messenger RNA (mRNA) molecules that made them. If a cell discovered how to make that connection, it might be able to exercise some choice over which mutations to accept and which to reject."[101] They go on to describe the scenarios described in the last section.

In discussing questions related to that of reductionism, two issues need to be precisely formulated. *First*, since there are many questions in whose answers reductionsm is at stake, the actual question that is being considered needs to be precisely specified. For example, the question whether the research strategy of molecular biology is uniformly reductionist is distinct from the question whether the explanations afforded by it are reductionist. The two might receive different answers as will be seen below. The discussion here will focus on the questions whether fluctuation analysis (or other statistical analyses at the same "level," that is, of individual bacteria), and the search for mechanisms, are reductionist as experimental methods in the sense whether they seek reductionist explanation. Only a few remarks will be made about the question whether the structure of explanations in molecular biology is reductionist and the two questions will be systematically kept distinct. These questions are each *epistemological*. *Ontological* questions regarding whether some entity is "formed" of, or "consists entirely" of other entities or types of entities will be completely ignored because they do not seem to be particularly interesting in this context.

Second, since there are a large number of models of reduction, it needs to be stated clearly which model is being used. The model of reduction adopted here is due to Sarkar,[102] based to a large extent on the earlier work of Kauffman and partly on Wimsatt.[103] Reductionist explanation is construed here as (i) an explanation based on the constituent parts of the entity under investigation; and (ii) an explanation that only posits those interactions between these parts that have a "physical warrant" in the sense of being derivable from physical (or chemical) theory or being corroborated by experiments involving only inanimate matter. It is assumed in this model that the explanation afforded satisfies all criteria required for its adequacy as a scientific explanation. What is of interest, here, is whether it satisfies the two additional requirements noted above to be reductionist explanations. The basic intuition, which is incorporated in the first of these requirements, is that the explanation is of a whole in terms of its parts. This intuition is not the same as that of an "upper-level" theory or phenomenon being explained in terms of more general "lower-level" theories which lay behind many of the classic accounts of reduction.[104] The intuition incorporated in the second requirement is that the interactions between the parts invoked in the explanation are physical in the sense that they are known on the basis of physical theory or experimentation.

Explanation of observed distributions of mutant cells by invoking mechanisms of mutagenesis *within* a cell obviously satisfies the first of these two criteria since the explanation of the behavior of a cell is being given at the level of the nucleic acids and the proteins inside it. However, to the extent that the explanation invokes the notion of "information," in the sense that it continuously refers to "codes," "translation," and so forth, it initially appears to violate the second criterion because the properties of biological information are obviously neither derivable from physical theory nor learned from the type of experiment that is envisioned in the statement of that criterion. However, "information," as used here is not a technical concept and, indeed, has never received sufficient explication in biology to become such a concept: it is, in fact, quite dispensable in the considerations that are at stake here. Once the mechanism for a directed (or, for that matter, a random) mutation is specified, as in the case of the examples discussed in the last section, the concept of information is redundant and the explanation can be given in terms of molecular interactions and reaction rates. It is customary to adhere to the usual terminology only for the sake of clarity. Thus, the explanation of the occurrence of mutations, directed or random, by the invocation of mechanisms of mutagenesis satisfies both requirements of reductionist explanation. Thus, the type of research that is involved in searching for these mechanisms is also fundamentally reductionist. Assuming that such explanations in terms of mechanisms are eventually obtained, both the epistemological questions mentioned above receive the same answer in favor of reductionism.

In contrast, fluctuation analysis or other such statistical analyses are not reductionist in their approach. By fluctuation analysis, conclusions about subcellular processes are drawn from statistical examination at the cellular level which, in the case of bacteria, is the organismic level. In effect, Luria and Delbrück were doing for bacteria what Mendel did for peas.[105] Properties of the whole were neither being explained nor being investigated by examining properties of parts. In fact, fluctuation analysis turns the investigation the other way around. Thus the first criterion in the model of reduction specified above is violated. Such research strategies are, therefore, nonreductionist. Moreover, it is not even clear that the results obtained from fluctuation analysis (or other such statistical analyses) would constitute explanation, even if these results were unequivocal with respect to their interpretation. In a sense, all that would have been obtained is a description of what must be happening at the subcellular level (which is, of course, extremely important) but not an explanation. Consequently, the question whether the results of such analyses permit reductionist explanation cannot be fruitfully asked.

The failure of fluctuation analysis and the necessity for the search for mechanisms to resolve the controversy of directed mutagenesis in bacteria underscores the importance of reductionist research in molecular biology. In fact, contrary to the claims of Cairns, Overbaugh, and Miller, this controversy further illustrates the increasing success of reductionist research in this field.[106] This need not have been so: many of the founders of molecular biology were not redutionists. Delbrück and Stent, motivated by the views of Bohr, hoped that the laws of physics might fail at a certain level that would be discovered through biological research.[107] "Complementary" biological laws would be discovered. The research strategy of Delbrück and his followers reflected these nonreductionist hopes, at least to some extent, as the design of the fluctuation test makes apparent.[108] The explanations they hoped to find would require laws beyond the domain of physics and chemistry, that is, the laws governing the behavior of inanimate matter. Such explanations, too would not have been reductionist.

It is precisely this hope that has not been realized. Explanations offered by molecular biology in the 1950s and 1960s were routinely reductionist.[109] Thus, the issue of reductionism in the questions regarding research strategy and the structure of explantions was resolved differently in this case. This is precisely why the questions needed to be kept separate in the first place. Since the 1960s, moreover, molecular biology has progressively become more reductionist in its research strategies. It has generally been successful at explanation and these explanations have continued to be routinely reductionist. Few would doubt that, in the process, a deeper understanding of some fundamental biological phenomena has been obtained.

12.7 Conclusion

This chapter has attempted to establish two sets of claims. *First*, it has attempted to clarify the differences between neo-Darwinism and neo-Lamarckism by proposing a certain (weak) explication of "neo-Lamarckism" that would permit serious consideration of neo-Lamarckian hypotheses in the sense that such hypotheses would not be trivially false or uninteresting. It has then used this notion of neo-Lamarckism to give criteria which bacterial mutagenesis would have to satisfy in order to be neo-Lamarckian. It has also examined possible mechanisms of mutagenesis to see whether they satisfy these criteria. It has also suggested that, in some cases, neo-Darwinist and neo-Lamarckist hypotheses do not stand in serious conflict. Nothing substantive depends on whether the term "neo-Lamarckist" is used to describe the relevant positions. Those who object to the use of the term for ideological reasons, or for the connotations of the Lysenko affair that it still often carries, might well prefer to used a myriad of terms like "directed mutations." What is important is that the criteria for distinguishing neo-Darwinist and conflicting positions are clearly explicated: that is the substantive issue.

Second, it has been argued that the type of research that is likely to resolve the controversy under discussion is reductionist in the precise sense of reduction discussed in section 12.5. It has also been argued that the type of explanation that would be obtained, if the research strategy suggested succeeds, would also be reductionist. These arguments should not, however, be construed to suggest either that reductionist research strategies are useful, or even viable, in all contexts or that reductionist explanations can always be obtained. What research strategy is appropriate in which context is something that can only be decided by a detailed consideration of many factors, including the nature of the questions asked, the complexities of the entities being investigated, and the theoretical and experimental resources available. Both types of strategy have had long and illustrious careers in biology: in molecular biology, however, the reductionist currently has the upper hand and it is probably reasonable to suggest that this situation will continue for the foreseeable future. Whether all phenomena can eventually be afforded reductionist explanation, whether in biology or elsewhere, is also an open question. Evolutionary explanations routinely violate reductionist criteria by appealing to composite systems that affect fitness.[110] It would be rash to say that all such explanations will eventually receive reductionist support. It would be equally rash to assume they never will. Only further work can decide this question.

The issues in the first set of claims are metaphysical since they involve the most fundamental aspects of the framework of evolutionary theory, as was pointed out in section 12.2. To the extent that evolutionary theory, as it is formulated at present, is correct, these aspects of evolutionary theory reflect fundamental features of the natural world. Put in another way, the fundamental question simply is whether there is, even to a very limited extent, any direction in evolution. Perhaps it is precisely because the question can be formulated in this way, loaded with ideological and even theological connotations, that the behavior of a strain of bacteria, often under stress, has generated so much controversy. However, at this level, the question is of scant scientific interest. This is exactly why such broad formulations of the question have here been avoided. The issues in the second set of claims are epistemological in a straightforward sense: they are issues concerning the nature of scientific explanations and strategies of research and, therefore, strategies for obtaining further knowledge.

Both sets of issues will be resolved only by experience. It is obvious that whether any neo-Lamarckian processes exist is an empirical question to be settled by further experimentation. However, whether reductionist research is necessary or whether statistical analyses such as fluctuation analysis can resolve the controversy being discussed here is also an empirical question.

The future history of this problem will determine the answer. Perhaps a truly important moral, and one which is not sufficiently recognized, that can be drawn from all these considerations is that metaphysical and epistemological questions of this sort can be, and perhaps only be, answered from experience. Perhaps this is truly the sense in which the interaction between philosophy and the world of experience, that is, the world described by science is dialectical: each progresses in response to the development of the other.

Acknowledgments

Thanks are due to John Cairns, Robert S. Cohen, James F. Crow, Patricia Foster, Stephen Jay Gould, David Hull, Joshua Lederberg, Richard Lenski, Heidi Lindquist, Bruce R. Levin, Richard Lewontin, Abner Shimony, John Maynard Smith, Alfred I. Tauber and William Wimsatt for helpful discussions and comments on earlier versions of parts of this analysis. Comments by members of the audience at a colloquium at Boston University, Spring 1990, where this analysis was presented, were also very helpful.

Notes

1. Cairns, J., J. Overbaugh, and S. Miller. The origin of mutants. *Nature* 335: 142–145, 1988.

2. Ibid.; Sarkar, S., Reductionism and molecular biology: a reappraisal. Ph. D. Dissertation. Department of Philosophy, University of Chicago, 1989.

3. Sarkar, Reductionism and molecular biology; Kauffman, S. A., Articulation of parts explanation in biology and the rational search for them. *Boston Studies in the Philosophy of Science* 8: 257–272, 1972; Sarkar, S., Models of reduction and categories of reductionism. *Synthese* 91: 167–194. (Chapter 2 of this volume.)

4. Lewontin, R. C., The units of selection. *Annual Review of Ecology and Systematics* 1: 1–18, 1970.

5. Ibid., p. 1.

6. Ibid.

7. Note that this explication of evolution by natural selection makes no mention of what the individuals of the population are. They could be molecules, organelles within a cell, cells, individuals (which in multi-cellular organisms would not be

identical with cells), kins, groups, populations or species. What the appropriate units of selection are is a matter of considerable biological and philosophical controversy. In the case of bacteria, which is all that is directly of concern in this article, the levels of cells and individuals are identical thus partly avoiding this controversy. For details of this controversy, see Brandon, R. N. and R. N. Burian (eds.), *Genes, Organisms, Populations: Controversies Over the Units of Selection*. Cambridge, MA: MIT Press, 1984.

8. See Wimsatt, W. C., Randomness and perceived randomness in evolutionary biology, *Synthese* 43: 287–329, 1980, for an attempt to analyze the difficulties with the notion of randomness appropriate for evolutionary biology.

9. Kimura (Kimura, M. *The Neutral Theory of Molecular Evolution*. Cambridge: Cambridge University Press, 1983), in his more polemical moments seems to suggest such a viewpoint.

10. See Provine, W., *The Origins of Theoretical Population Genetics*. Chicago: University of Chicago Press, 1971, for details of this fascinating history.

11. See Maynard Smith, J., *Evolutionary Genetics*, Oxford: Oxford University Press, 1989, for an emphasis on natural selection and Lewontin's review (Lewontin, R. C., A natural selection, *Nature* 339: 107, 1989) for a critique.

12. Haldane, J. B. S., *The Causes of Evolution*. London: Harper & Brothers, 1932; Huxley, J., *Evolution: The Modern Synthesis*. London: George Allen & Unwin, 1942.

13. Not one of the figures discussed by Hull (Hull, D. Lamarck among the Anglos. In J. B. Lamarck, *Zoological Philosophy*. Chicago: University of Chicago Press, 1984, pp. xl–lxvi) in his colorful, though short, account of the history of neo-Lamarkism seems to have suggested that all mutations were directed.

14. For details of Lysenkoist claims, see Hudson, P. S., and R. H. Richens, *The New Genetics in the Soviet Union*. Cambridge: Imperial Bureau of Plant Breeding and Genetics, 1946. Among those at least partly sympathetic to Lysenko's claims, only Haldane (Haldane, J. B. S., Lysenko and genetics. *Science and Society* 4: 433–437, 1940) seems to have been fully clear that all that was needed to explain the kind of effects Lysenko claimed to be observing was that some mutations could be induced in such a fashion.

15. This strong notion of directedness has been championed by Lenski, R. E., Are some mutations directed? *Trends in Ecology and Evolution* 4: 148–150, 1989.

16. This weaker definition has been used by Sarkar, S. On the possibility of directed mutations in bacteria: statistical analyses and reductionist strategies. In *PSA 1990*. A. Fine, editor, 1990. Philosophy of Science Association, East Lansing. In press.

17. Many of the participants in the controversy would also reject the use of "neo-Lamarckism" on various grounds including vagueness. However, vagueness, at least, has been removed if the distinctions and definitions elaborated in the text are clear

enough. For responses to some other objections to the use of "neo-Lamarckism," see Sarkar, On the possibility of directed mutations.

18. However, for an important treatment of the possible inheritance of acquired characteristics where the concept is not used vaguely, see Jablonka, E., and M. J. Lamb, The inheritance of acquired epigenetic variations, *Journal of Theoretical Biology* 139: 69–83, 1989.

19. Cullis, C. A., The generation of somatic and heritable variation in response to stress. *American Naturalist* 130: S62–S73, 1987.

20. The reasons given for not using the notion of the inheritance of acquired characteristics are also those for not using another related distinction due to Mayr, namely, that between "hard" and "soft" inheritance (Mayr, E., *The Growth of Biological Thought*. Cambridge, MA: Harvard University Press, 1982, pp. 687–689). These choices, however, are made for the purpose of conceptual clarification. No suggestion is being made here that these other construals of the differences between neo-Darwinism and neo-Lamarckism can be avoided when the history of these disputes, especially during the first half of this century, is considered.

21. Lamarck, J. B., *Zoological Philosophy*. Chicago: University of Chicago Press, 1984, p. 113.

22. Darwin, C., *On the Origin of Species*. London: John Murray, 1859.

23. See, for example, Darwin, C., *On the Origin of the Species*, pp. 134–139. Mayr lists nine other examples (Mayr, E. Introduction. In C. Darwin, *On the Origin of Species*. Cambridge, MA: Harvard University Press, 1964, p. xxvi).

24. See Eiseley, L., *Darwin's Century*. New York: Doubleday Anchor, 1961; and Ruse, M., *The Darwinian Revolution: Science Red in Tooth and Claw*. Chicago: University of Chicago Press, 1979, for details of these developments.

25. See, for example, Weismann, A., *Das Keimplasma: Eine Theorie der Vererbung*. Jena: Gustav Fischer, 1892.

26. Romanes, G. J., *Life and Letters*. London: Longmans, Green, 1896.

27. In particular the unfortunate historical association between neo-Lamarckism and Lysenkoist positions during the Lysenko affair needs to be carefully avoided.

28. For details of this history, see Sarkar, Reductionism and molecular biology, and Fischer, E. P., and C. Lipson. *Thinking About Science: Max Delbrück and the Origins of Molecular Biology*. New York: Knopf, 1988, and references therein.

29. Though the terms, "mutant" and "mutation" were routinely used for this transformation, it remained an open question whether these "mutations" were mutations of genes. In 1943, for example, Luria and Delbrück carefully observe: "Naming such hereditary changes 'mutations' of course does not imply a detailed similarity with

any of the classes of mutations that have been analyzed in terms of genes for higher organisms. The similarity may be merely a formal one" (Luria, S. E., and M. Delbrück. Mutations of bacteria from virus sensitivity to virus resistance. *Genetics* 28: 491–511, 1943, p. 492). Even in the late 1950s a few skeptics such as Hinshelwood (for example, in Dean, A. C. R., and C. N. Hinshelwood. Aspects of the problem of drug resistance in bacteria. In *Drug Resistance in Microorganisms*. G. E. W. Wolstenholme, and C. M. O'Connor, editors. London: J. & A. Churchill, 1957, pp. 4–24) would maintain that any reference to genes in bacteria would actually be a reference to complex chemical reaction networks.

30. d'Herelle, F., *The Bacteriophage and Its Behavior*. Baltimore: Williams and Wilkins, 1926; Gratia, A. Studies on the d'Herelle phenomenon. *Journal of Experimental Medicine* 34: 115–131, 1921; Burnet, F. M. "Smooth-rough" variation in bacteria in its relation to bacteriophage. *Journal of Pathology and Bacteriology* 32: 15–42, 1929.

31. Luria, S. E. and M. Delbrück, Mutations of bacteria. According to Luria, he was led to the idea of the fluctuation test while watching the operation of slot machines during a faculty dance at the Bloomington Country Club in Indiana. If these machines were programmed to return money at random, the returns would form a Poisson distribution clustered around a mean and there would be virtually no jackpots. If, however, the machines were programmed so that they return occasional jackpots with many very tiny returns, the returns would fluctuate much more widely. The average return would be the same in both cases. The way in which the machines were programmed could only be determined by actually observing the actual distribution of the returns and not from the mean alone. Luria immediately applied this insight to the question of the distribution of mutants in bacteria. For further details see Luria, S. E., *A Slot Machine, a Broken Test Tube: An Autobiography*. New York: Harper and Row, 1984, pp. 74–79.

32. Yang, Y. N., and P. Bruce Wright. Rough variation in *V. cholerae* and its relation to resistance to cholera-phage (Type A). *Journal of Pathology and Bacteriology* 38: 187–200, 1934.

33. Luria, S. E. and M. Delbrück (Mutations of bacteria) do not cite them. Neither do Newcombe, H. B. Origin of bacterial variants. *Nature* 164: 150–151, 1949, Lederberg, J., and E. Lederberg. Replica plating and indirect selection of bacterial mutants. *Journal of Bacteriology* 63: 399–406, 1952, or Cavalli-Sforza, L. L., and J. Lederberg. Isolation of preadaptive mutants by sib selection. *Genetics* 41: 367–381, 1956, which were critical papers in the establishment of the neo-Darwinian view. Cavalli-Sforza and Lederberg do list them in their bibliography in an earlier paper (Cavalli-Sforza, L. L., and J. Lederberg. Genetics of resistance to bacterial inhibitors. In *Symposium: Growth Inhibition and Chemotherapy*. Rome: Istituto Superiore di Sanita, 1953, pp. 108–142) but do not discuss them in the text. A recent article by Lederberg is responsible for drawing attention to the contribution of Yang and Bruce White (Lederberg, J. Replica

plating and indirect selection of bacterial mutants: isolation of preadaptive mutants in bacteria by sib selection. *Genetics* 121: 395–399, 1989).

34. Luria, S. E. and M. Delbrück, Mutations of bacteria, p. 493.

35. Ibid.

36. From this point, however, the distinction between "variation" and "mutation" and related terms will be ignored for the sake of convenience of expression.

37. Lea, D. E., and C. A. Coulson. The distribution of the number of mutants in bacterial populations. *Journal of Genetics* 49: 264–285, 1949. As a historical curiosity, it is worth noting that it is possible that Fisher might have solved the Luria–Delbrück distribution first. Crow (Crow, J. F., R. A. Fisher, A centenniel view. *Genetics* 124: 207–211, 1990) recalls that while he found the Luria–Delbrück argument convincing, he thought that the mathematical treatment "shoddy and confusing" (ibid., p. 210). Consequently, in 1946, he approached Fisher with the problem. Fisher "leaned back in his chair, thought for perhaps a minute, took a scrap of paper, and wrote a generating function" (ibid.). Crow, not understanding the formula yet, put that scrap of paper aside, intending to work on it later and then lost it!

38. Stewart, F., D. Gordon, and B. Levin. Fluctuation analysis: the probability distribution of the number of mutants under different conditions. *Genetics* 124: 175–185, 1990. There were many previous attempts to generalize the Luria–Delbrück distribution but none to this extent. For example, phenotypic lag is considered in Armitage, P. The statistical theory of bacterial populations subject to mutation. *Journal of the Royal Statistical Society B* 14: 1–40, 1952, and Koch, A. L. Mutation and growth rates from Luria–Delbrück fluctuation tests. *Mutation Research* 95: 129–143, 1982. The latter and Mandelbrot, B. A population birth-and-mutation process I: explicit distributions for the number of mutants in an old culture of bacteria. *Journal of Applied Probability* 11: 437–444, 1974 also consider differential fitnesses (growth rates), of the original and mutant strains. The effect of these factors on the distributions will be considered in section 12.5.

39. Newcombe, H. B., Delayed phenotypic expression of spontaneous mutations in Escherichia coli. *Genetics* 33: 447–476, 1948; Demerec, M., and U. Fano. Bacteriophage-resistant mutants in Escherichia coli. *Genetics* 30: 119–136, 1945.

40. Witkin, E. M., Genetics of resistance to radiation in *Escherichia coli*. *Genetics* 32: 221–248, 1947; Ryan, F. J., On the stability of nutritional mutants of bacteria. *Proceedings of the National Academy of Sciences (USA)* 34: 425–435, 1948.

41. Demerec, M., Production of staphylococcus strains resistant to various concentrations of penicillin. *Proceedings of the National Academy of Sciences (USA)* 31: 16–24, 1945; Oakberg, E. F., and S. E. Luria. Mutations to sulfonamide resistance in *Staphylococcus aureus*. *Genetics* 32: 249–261, 1947; Demerec, M. Origin of bacterial resistance to antibiotics. *Journal of Bacteriology* 56: 63–74, 1948.

42. Ryan, F. J., L. K. Schneider, and R. Ballentine. Mutations involving the requirement of uracil in Clostridium. *Proceedings of the National Academy of Sciences (USA)* 32: 261–271, 1946; Alexander, H. E., and J. Leidy. Mode of action of streptomycin on type b Hemophilus influenzae. *Journal of Experimental Medicine* 85: 607–621, 1947; Curcho, M. de la G. Mutation to tryptophan independence in *Erbethella typhosa*. *Journal of Bacteriology* 56: 374–375, 1948.

43. Newcombe, H. B., Delayed phenotypic expression; Ryan, F. J., Distribution of numbers of mutant bacteria in replicate cultures. *Nature* 169: 882–883, 1952.

44. Newcombe, H. B., Origin of bacterial variants; Witkin, E. M., Genetics of resistance.

45. Ryan, F. J. *et al.*, Mutations involving uracil.

46. Ryan, F. J., Distribution of numbers.

47. Ryan, F. J., Distribution of numbers; Ryan, F. J., Adaptation to use lactose in *Escherichia coli. Journal of General Microbiology* 7: 69–88, 1952. This mutation is particularly relevant here because it is one of those studied by Cairns, J. *et al.*, The origin of mutants. In fact, that work used Ryan's studies as one of its starting points.

48. Ryan, F. J., Distribution of numbers.

49. Armitage, P., The statistical theory. Note that Ryan does not explicitly invoke phenotypic lag or any other specific mechanism.

50. Dean, A. C. R. and C. N. Hinshelwood, Aspects of drug resistance.

51. Eriksen, K. R., Studies on the mode of origin of penicillin resistant staphylococci. *Acta Pathologica et Microbiologica Scandinavica* 26: 269–279, 1949. However, Eriksen's experimental results were not particularly convincing because small samples were taken from the cultures which made the statistical tests inefficient as Cavalli-Sforza and Lederberg (Genetics of resistance) point out. For Hinshelwood's position, see, for example, Hinshelwood, C. N. Chemistry and bacteria. *Nature* 166: 1089–1092, 1950; and Dean, A. C. R., and C. N. Hinshelwood. 1952. The resistance of *Bact. Lactis aerogenes* to proflavine (2: 8-diaminoacridine). I: the applicability of the statistical fluctuation test. *Proceedings of the Royal Society B* 139: 236–250, 1952.

52. Cavalli, L. L., Genetic analysis of drug-resistance. *Bulletin of the World Health Organization* 6: 185–206, 1952; Michison, D. A. The occurrence of independent mutations to different types of streptomycin resistance in *Bacterium coli. Journal of General Microbiology* 8: 168–185, 1953.

53. For example, Mitchison, D. A., The occurrence of independent mutations, studied three types of strains of *Bacterium coli* which had differences in growth rates while being selected for streptomycin resistance by fluctuation analysis.

54. Cavalli, L. L., Genetic analysis of drug-resistance.

55. Newcombe, H. B., Origin of bacterial variants.

56. Ibid., p. 150.

57. For replica plating, see Lederberg, J. and E. M. Lederberg, Replica plating; for sib-selection, see Cavalli-Sforza, L. L. and J. L. Lederberg, Isolation of preadaptive mutants.

58. In practice Cavalli-Sforza and Lederberg (Isolation of preadaptive mutants) had to increase cell concentration in each cycle because of slow selection.

59. Stahl, F. Bacterial genetics: a unicorn in the garden. *Nature* 335: 112–113, 1988. Lederberg was certainly aware of this but seems not to have regarded it as particularly relevant (see Cavalli-Sforza, L. L. and J. Lederberg, Genetics of resistance, p. 121).

60. Ryan, F. J., On the stability of nutritional mutants; Distribution of numbers; Adaptation to use lactose.

61. This does not, of course, endorse in any way criticisms of the sort made by Hinshelwood, C. N., Aspects of drug resistance, and chemistry and bacteria.

62. Luria, S. E. and M. Delbrück, Mutations of bacteria. Note that they do not explicitly state that one alternative is the negation of the other; they simply ignore intermediate possibilities.

63. Lea, D. E. and C. A. Coulson, The distribution of number of mutants.

64. Armitage, P., The statistical theory. The objection just noted in the text is independent of whether the last one, that of the selective media being lethal for non-mutants, is valid. Indeed, phenotypic lag had already been invoked by Newcombe, H. B., Origin of bacterial variants, even before Lea and Coulson had calculated the Luria–Delbrück distribution.

65. Cavalli-Sforza, L. L. and J. Lederberg, Isolation of preadaptive mutants.

66. Lederberg, J. and E. M. Ledergerg, Replica plating; Newcombe, H. B., Origin of bacterial variants. Both experiments rely on visual recognition and make no quantitative arguments whatsoever. What they do show, however, is that *some* mutations were spontaneous (random).

67. The discussion in this section is not intended to be complete. For details, see Foster, P., Directed Mutation in *Escherichia coli: Theory and Mechanisms*, in Tauber, A. I. (Ed.), *Organism and the Origins of Self*. Dordrecht: Kluwer, 1991, pp. 213–235; and Sarkar, On the possibility of directed mutations.

68. Shapiro, J., Observations on the formation of clones containing *araB-lacZ* cistron fusions. *Molecular and General Genetics* 194: 79–90, 1984.

69. Cairns, J. *et al.*, The origin of mutants.

70. See note 47.

71. Hall, B., Adaptive evolution that requires multiple spontaneous mutations I. *Genetics* 120: 887–897, 1988.

72. Lenski, R. E., Are some mutations directed? emphasizes the importance of this point.

73. Levin, B., D. Gordon, and F. Stewart. Is natural selection the composer as well as the editor of genetic variation? [unpublished]; Cairns, J. *et al.*, The origin of mutants.

74. Mittler, J. E. and R. E. Lenski, New data on excisions of Mu from E. coli MCS2 cast doubt on directed mutation hypothesis, *Nature* 334: 173–175, 1990; Shapiro, J., Observations; Cairns, J. *et al.*, The origin of mutants.

75. See the contribution by Foster, Directed mutation, for further detail on this point and, also, discussion of other experimental results of Hall.

76. Mittler, J. E. and Lenski, R. E., New data on excisions.

77. Cairns, J. *et al.*, The origin of mutants. Note that these authors interpret the shift as a Poisson component *added* on to a Luria–Delbrück distribution, but this is an *interpretation* and, therefore, subject to legitimate questioning.

78. Stewart, F. *et al.*, Fluctuation analysis.

79. The second and the last of these cannot be shown using the analysis of Stewart, F. *et al.*, Fluctuation analysis.

80. The importance of this case has been particularly emphasized by Charlesworth, D., B. Charlesworth, and J. J. Bull, Origin of mutants disputed, *Nature* 336: 525, 1988, and Tessman, I., Origin of mutants disputed, *Nature* 336: 527, 1988.

81. Cairns has replied to this objection by observing that there did not appear to be any difference in fitness between mutant and nonmutant types in his experiments (Cairns, J., Origin of mutants disputed, *Nature* 336: 527–528, 1988).

82. Stewart, F. *et al.*, Fluctuation analysis.

83. Lenski, R. E., M. Slatkin, and F. J. Ayala, Mutation and selection in bacterial populations: alternatives to the hypothesis of directed mutation. *Proceedings of the National Academy of Sciences (USA)* 86: 2775–2778.

84. Lenski, R. E., M. Slatkin, and F. J. Ayala, Mutation and selection; Lenski, R. E., M. Slatkin, and F. J. Ayala, Another alternative to directed mutation, *Nature* 337: 123–124, 1989.

85. Levin, B., D. Gordon, and F. Stewart, Is natural selection the composer?

86. An obvious exception is the hypothesis of differential fitness.

87. Cairns, J. *et al.*, The origin of mutants. These mechanisms will be discussed in detail in the text.

88. Stahl, F., Bacterial genetics; Davis, B. D., Transcriptional bias: a non-Lamarckian mechanism for substrate-induced mutations, *Proceedings of the National Academy of Sciences (USA)* 86: 5005–5009, 1989.

89. For example, the mechanism suggested by Davis, B. D., Transcriptional bias, can be ruled out (in part due to the experimental reports of Davis himself).

90. See the contribution by Foster, Directed mutation, for further discussion of these mechanisms and the current state of the experimental research.

91. The description of the mechanisms given here are simplified, and sometimes generalized, versions of those proposed by the original authors. Though they are intended to remain faithful to the spirit of those versions, they might disagree in detail.

92. Cairns, J. *et al.*, The origin of mutants.

93. Cairns, J. *et al.*, The origin of mutants, require the same organelle to contain the reverse transcriptase and monitor the performance of the RNA molecules. Thus the interaction from the feedback process to the organelle is automatically ensured. However, there is no conceptual reason to require the same organelle to do all this. Hence, the three requirements have been separated in the version presented in the text.

94. Stahl, F., Bacterial genetics.

95. Davis, B. D., Transcriptional bias.

96. It should perhaps be emphasized that probably none of the proponents of these mechanisms, especially Davis, B. D., Transcriptional bias, would choose to call the mechanisms "neo-Lamarckian." Moreover, all that is being suggested here is that they are only "neo-Lamarckian" in the sense that they satisfy the criteria of the explication of the term given in section 12.2.

97. Cairns, J. *et al.*, The origin of mutants. It is, perhaps, worth emphasis that there is no evidence, as yet, in favor of these mechanisms.

98. Davis, B. D., Transcriptional bias, has argued that such a mechanism, itself, would have adaptive value. This is quite likely and such a mechanism might well have arisen in a thoroughly neo-Darwinian fashion. However, the new mutations that such a mechanism would make possible would still show the characteristics of being neo-Lamarckian. Thus the basic question, whether there is any neo-Lamarckian mutagenesis, remains the same.

99. See Sarkar, S., Reductionism and molecular biology, for a detailed examination of this issue.

100. For details, see Sarkar, S., Reductionism and molecular biology, and Wimsatt, W. C. Reductive explanation: a functional account. *Boston Studies in the Philosophy of Science* 32: 671–710, 1976.

101. Cairns, J. *et al.*, The origin of mutants, p. 145.

102. Sarkar, S., Reductionism and molecular biology, and Wimsatt, W. C. Reduction and reductionism. In *Current Research in the Philosophy of Science*. P. D. Asquith and H. Kyburg, editors. Philosophy of Science Association. pp. 352–377, 1978.

103. Sarkar, S., Reductionism and molecular biology, and Models of reduction; Kauffman, S. A., Articulation of parts; Wimsatt, W. C., Reductive explanation.

104. The classic accounts of "theory reduction" are Nagel, E. *The Structure of Science: Problems in the Logic of Scientific Explanation*. New York: Harcourt, Brace & World, 1961, and Schaffner, K. Approaches to reduction. *Philosophy of Science* 34: 137–147, 1967. For arguments that attempt to show that such classic accounts of "theory reduction" cannot capture the flavor of research in molecular biology, see Sarkar, S. Reductionism and molecular biology, Wimsatt, W. C., Reductive explanation, and, especially, Hull, D. Reduction in genetics—biology or philosophy? *Philosophy of Science* 39: 491–499, 1972.

105. Luria, S. E. and M. Delbrück, Mutations of bacteria. The same point can be made about most of the research of the Phage Group which usually tried to infer informations about processes within the virus by looking at interactions of the virus, as a whole, with other entities. Though, the present article argues for reductionist research in the context of the problem being considered here, examples such as this clearly show the value of non-reductionist research in biology in other contexts. Of course it is almost trivial to assert the same value for non-reductionist research in most of evolutionary biology.

106. Cairns, J. *et al.*, The origin of mutants, p. 145.

107. Delbrück, M., A physicist looks at biology, *Transactions of the Connecticut Academy of Sciences* 38: 173–190, 1949; Stent, G. S., That was the molecular biology that was, *Science* 160: 390–395, 1968; Bohr, N., Light and life, *Nature* 131: 421–423, 457–459, 1933.

108. The important point is that the willingness to use such tests show that there was no prior commitment only to investigate biological problems in a reductionist fashion. It is not being suggested that Luria and Delbrück were deliberately attempting to keep their research strategy nonreductionist. In fact, Delbrück at least partly operated with an assumption that the only way in which his nonreductionist hopes might be realized was by pushing reductionist physical methods to the extreme and finding a paradox that defied such explanation. For details of this history, see Sarkar, S., Reductionism and molecular biology, and Fischer, E. P. and C. Lipson, *Thinking About Science*.

109. The only class of exceptions to the general reductionist explanations offered by molecular biology, at present, is that of functional explanations. For details of the problems that functional explanations pose for reductionist accounts, see Sarkar, S., Reductionism and molecular biology, and Sarkar, S., Natural selection, hypercycles and the origin of life. In *PSA 1988*, Vol. 1, A. Fine and J. Leplin, editors, pp. 196–206, 1988 (chapter 6 of this volume).

110. Wimsatt, W. C., Teleology and the logical structure of function statements. *Studies in the History and Philosophy of Science* 3: 1–80, 1972.

13   Directional Mutations: Fifteen Years Afterward

13.1

Some fifteen years after Cairns, Overbaugh, and Miller (1988) suggested the existence of what they called "directed" mutations in *Escherichia coli*, it is instructive—and humbling—to note that there is yet no agreement about: (i) the precise range of the phenomena; (ii) the specific models of mutagenesis responsible for this type of mutagenesis; or (iii) even what such mutations should be called. One fact has been established beyond reasonable controversy: that there is a class of stress-induced stationary phase mutations in bacteria and, at least, in several other unicellular organisms that are distinct from the much more common growth-dependent mutations. It is clear that such stress-induced stationary phase mutations routinely occur in *E. coli* and yeast (*Saccharomyces cerevisiae*), but the other unicellular organisms in which they occur have not been as systematically investigated. In *E. coli*, much of the work has focused on the FC40 strain, which has a prolific production of stress-induced stationary phase $lac^- \to lac^+$ mutations (at a rate of about 1 per 10^7 cells per day) (Foster 2000a,b; Hastings and Rosenberg 2002). This is a strain in which the *lac* operon is deleted from the chromosome, but a revertible lac^- allele is present on the conjugal plasmid F. This allele can be reverted to lac^+ by any mutation in the 130 base pair region that induces a required frameshift without creating a nonsense codon. However, not all aspects of stress-induced mutations are seen in *E. coli* FC40 (Foster 1992; 1999b). Besides E. coli, bacteria that are reported to show such mutations include *Salmonella typhimurium, Bacillus subtilis, Pseudomonas sp.*, and *Clostridium sp.*; eukaryotes include, besides *S. cerevisiae, Candida albicans* (Foster 2000a). Stress-induced stationary phase mutations are beginning to be recognized as a likely general feature of the microbial world, as was predicted in chapter 12 (Kivisaar 2003).

Since the early 1990s it has been clear that all these mutations do not constitute a single phenomenon (Foster 1999b; Kivisaar 2003). It now appears to be the case that there are at least two classes of stress-induced mutations: point mutations and adaptive gene amplification (Anderson, Slechta, and Roth 1998; Rosenberg 2001; Hastings and Rosenberg 2002; Hendrickson et al. 2002). The former class is well established; the existence of the latter remains partly controversial (Foster 2000b). Point mutation involves: (i) DNA breakage; (ii) recombinational break repair (which, surprisingly, is not required during growth-dependent mutation); (iii) possibly, a transient failure of mismatch repair (see, however, Foster 1999a and the response from Harris et al. 1999); (iv) possibly, a special mutator DNA polymerase; and (vi) very likely, transition to a hypermutable state. In contrast, during amplification, multiple copies of a leaky mutant allele may be generated through DNA duplication resulting in reversion at the phenotypic level. For instance, in *E. coli* FC40, a leaky mutant *lac*⁻ allele may be amplified to 20–50 copies providing sufficient gene expression to exhibit a Lac⁺ phenotype. Even if the existence of both classes is granted, it is controversial whether particular cases fall into one class or the other. Both types of process may also occur simultaneously; neither precludes the other.

By now it is clear that stress-induced stationary-phase mutations are not simply artifacts of poor experimental design as some skeptics suggested in the late 1980s and early 1990s—see chapter 12 and Sniegowski and Lenski (1995). Perhaps most important, these mutations have also been observed when natural populations of bacteria (beyond the laboratory strains) experience stress due to environmental factors (see Kivisaar 2003 for a review). Moreover, supporting a claim originally made by Cairns, Overbaugh, and Miller (1988), there is some recent evidence that this form of mutagenesis is under genetic control in some cases (Bjedov et al. 2003). The last-mentioned study examined 787 worldwide natural isolates of *E. coli* from diverse ecological niches. Consequently, there are likely to be nontrivial evolutionary implications of the existence of this form of mutagenesis, and of the nature of the mechanisms that are responsible for it.

However, the continued uncertainties noted above (in the first paragraph), that is, the failure to find general principles governing all forms

Directional Mutations 349

of stress-induced mutations, underscore the fact that the evolutionary significance of these mutations remains far from clear, and little better understood than it was in the early 1990s. Thus much of what was said in chapters 11 and 12 requires no revision. But none of the issues that were left unresolved in those chapters can yet be brought to successful closure. The critical unresolved question that remains is whether certain organisms have evolved mechanisms to control their mutation and, thus, the speed of their evolution. Whether any of these mechanisms are of any relevance to organisms with segregated germ-lines is also an open question. Foster (1999a) has suggested otherwise and is probably correct. However, some of these mechanisms may have analogues in the control of somatic hypermutation in the mammalian immune system (Kivisaar 2003). They may thus be critical to the evolution of immunity in "higher" animals, which remains largely an unsolved puzzle. Bjedov et al. (2003) have also argued that similar mechanisms may be involved in several forms of age-dependent mutagenesis in "higher" organisms including those inducing cancer and those in the germ-line itself. While the vehemence of the original controversy, recorded particularly in chapter 12 (see also Keller 1992), has partly dissipated, almost all the interpretive issues remain open.

13.2

However, some of the more spectacular initial claims of "directed" mutation that were discussed in chapter 12 have not survived further scrutiny. Surprisingly for research conducted in the 1990s, the classical techniques discussed in chapter 12 (fluctuation analysis, replica plating, and sib-selection) continued to be used in this work (for a review, see Sniegowski and Lenski 1995). At least some of the original mutations studied by Cairns, Overbaugh, and Miller (1988) in the SM195 strain of *E. coli* are not induced by the selective environment: they almost certainly arose before exposure to the that environment (Prival and Cebula 1996). Similarly, some of Shapiro's (1984) results (using a strain of *E. coli* carrying a *Mu* prophage with a temperature-sensitive repressor [*cts*62]) have also been shown to be accountable without recourse to "directed" mutation; these mutations can be induced by anaerobic starvation in the absence of lactose (Mittler and Lenski 1992; Foster and Cairns 1994; Maenhaut-Michel and Shapiro 1994; Sniegowski 1995). Finally, Hall's (1988) original interpretation of "directed" mutation to enable salicin utilization in *E. coli* K-12 has also turned out not to be tenable. The original interpretation required the simultaneous occurrence of two rare events: excision of the IS*103* element from the *bglF* allele (which encodes a phosphotransferase required for salicin utilization) and a mutation (at the *bglR* promoter site from $bglR^0$ to $bglR^+$). The most surprising result was that both rare events seemingly occurred in the presence of salicin—see chapter 12 (sec. 12.4). However, subsequent work has shown that excision alone allows cells to grow slowly on salicin (Mittler and Lenski 1992; Hall 1994). There is no anticipatory mutagenesis (Symonds 1989) in these experiments. However, whereas Mittler and Lenski (1992) claim that the required excision occurs even in the absence of salicin, Hall (1994) detected no such excisions unless salicin is present—this issue remains unresolved (Foster 1999b). As Foster (2000b, p. 21) summarizes the situation: "Some of the more dramatic cases of 'directed' mutation have now been shown to have other causes, so at this point we are left with rather few examples of apparent directedness that are still unexplained." The remaining examples nevertheless underscore the complexity of mutagenesis and its control at the molecular level, requiring, at the very least, some expansion of the received view of evolution beyond "random" or "blind" mutation.

13.3

The term "adaptive mutation" has become the most popular form of reference to what Cairns, Overbaugh, and Miller (1988) and the editors of *Nature* initially called "directed" mutation (Foster 1999b). The most uncontroversial characteristic feature of these mutations is that they occur in nondividing cells (and are, thus, distinct from the mutations that occur during cellular growth—see, for instance, Hall 1994 and Foster 1999b). The following discussion will be entirely restricted to mutations that satisfy this criterion. For some, "directed" mutation implied either (a) instruction (for instance, from a template), or (b) transfer of information; both these mechanisms were ruled out very early (see Foster and Cairns 1992). Moreover, if it is required that "directed" mutations occurred only in the presence of the selective environment (as Cairns, Overbaugh, and Miller 1988 originally seemed to want), there is no evidence for such mutations (see Foster 2000a). Lenski (1989) attempted a less extreme definition of "directed" mutation, arguing that a mutation is directed if and only if the mutation rate is enhanced in the environment in which it would be selected. Sarkar (1990; the same definition is defended in chapters 11 and 12) argued that the biconditional is too strong; a mutation should be viewed as directed if the mutation rate is enhanced in an environment in which it would be selected. Otherwise, a mutation becomes undirected simply because its rate is also enhanced by the presence of some nonspecific mutagen in its environment (besides being enhanced by exposure to a selective environment).

Foster (1999b, p. 74) argues that these mutations must satisfy two criteria: "(*a*) non-selected mutations did not occur during selection; and (*b*) the selected mutations did not arise under nonspecific stress, such as starvation." Replacing the requirement of the occurrence of the mutation with that of an enhanced mutation rate, Sarkar's definition accepts only the second of Foster's two criteria. Hall (1994, p. 110) offers yet two other criteria: (i) the mutations "occur during environmentally induced stress, but only when the mutation will relieve that stress"; and (ii) they "occur only in the genes that are under selection, i.e., they are not the results of a generally increased mutation rate under the selective conditions." These assumptions are equivalent to those used by Foster: Hall's first criterion is Foster's second, and vice versa. Once again, Sarkar's definition accepts only Hall's first criterion. In both cases, the reason for rejecting the other criterion is that, presumably, the directionality of mutagenesis should not be lost if the mechanisms of mutagenesis depended on some nonspecific increase in DNA mutability under stress. An example of this sort is the hypermutable state model (see sec. 13.4 below), which is taken to be a paradigmatic mechanism for "directed" mutations. Meanwhile, Sarkar's definition of "directed" captures at least one sense of "adaptive" presently being used (see Rosenberg 2001, p. 504, box 1). Presumably because of the unpalatable connotations of "directed" mentioned above (in the first paragraph of this section), that term faded from use.

Several creative alternative definitions have been considered in the literature: (i) SLAM, stressful-lifestyle associated mutation (Rosenberg, Thulin, and Harris 1998) (however, this terms apparently conjures up violent behavior or, at least, "a loud noise" [Foster 1999b, p. 59]); (ii) SAM, starvation-associated mutation (Bridges 1995) (however, "SAM" is already an abbreviation for S-adenosylmethionine); (iii) SPM, stationary-phase mutation

(Grigg and Stuckey 1966); (iv) anticipatory mutagenesis (Symonds 1989); and (v) selection-induced mutation (Hall 1994). Except for (v), none of these potential names captures what is puzzling about these mutations: that there is a positive correlation between the occurrence of a mutation and the fitness change that it enhances.

"Adaptive" mutation is undoubtedly an advance over these terms, and its use in the present sense goes back to Delbrück (1946). Even as the controversy over directed mutations first erupted, it was used in the same sense by Tlsty, Margolin, and Lum (1989). Cairns and Foster (1991) began using that term instead of "directed" mutation in the early 1990s. However, even this term is not perfect: care must be taken to distinguish between the products of the special forms of mutagenesis being considered here and "any useful mutation formed at any time" (Rosenberg 2001, p. 504, box 1). To avoid this problem, Sarkar (1993) suggested the term "directional" mutation, in analogy with directional selection in population genetics. Directional mutation is supposed to have the same definition as directed mutation in the sense of chapters 11 and 12 as explained above, but does not carry a connotation of instruction or information transfer. Directionality also connotes some degree of nonspecificity, in agreement with mechanisms such as the hypermutable state model. Moreover, in analogy with selection, directional mutation carries the connotation of being a type of process; this gives it some advantage over adaptive mutation. Because of these advantages, "directional" mutation will be used in this chapter though there is little chance that its use will be widespread. (A different, and possibly more important, reason for avoiding the term "adaptive" [because it implies an unproven claim of adaptation] will be given at the end of sec. 13.4.)

13.4

Of the various mechanisms suggested for directional mutation in the early 1990s (see chapter 12), one always appeared more promising than the others because it violated the fewest of the tenets of the received view of evolution: this is the hypermutable state model (Hall 1990). In particular, this model retained the standard evolutionary "random variation—natural selection" (sometimes called "trial and error") mechanism while moving it to a lower level of organization (from a population to the interior of a cell).

Directional Mutations

As chapter 12 details, random variants are produced while, under stress, the cell undergoes a decline of some of its self-repair mechanisms. A variant that allows escape from stress stabilizes as a mutation in what is essentially a selective process. Thus, what is "random" at one level of organization is directional at a higher level. Notice that the received view incorporates the same logic at the level of individuals and populations: "random" mutations occur at the level of individuals, selection occurs at the level of the population. The result is directional evolution in the direction of increased adaptation. Sarkar (1993) showed that the distribution of mutants arising from a transient hypermutable state is a Poisson distribution. (Essentially the derivation consists only of noticing the fact that the compound of a Poisson distribution [modeling entry to the hypermutable state] and a binomial distribution [modeling the exit as a mutant from the hypermutable state] is itself a Poisson distribution.) This result was consistent with the observations reported by Cairns, Overbaugh, and Miller (1988) and others.

By now, there is ample evidence for the existence of a transient hypermutable state, at least for some classes of directional mutations (Rosenberg, Thulin, and Harris 1998; Foster 1999b, 2000b; Rosenberg 2001; Slechta et al. 2002; Tompkins et al. 2003). In *E. coli*, the error-prone DNA polymerase IV (Pol IV) has been conclusively implicated in hypermutation (Tompkins et al. 2003; Layton and Foster 2003), though it is clearly not required for all directional mutations (Foster 1999b, 2000b). However, two puzzles remain:

(i) The hypermutable state model (like all "trial and error" models) predicts that cells with directional mutations will also contain nondirectional mutations at a higher than random frequency (Foster 1993). Initial results appeared to verify this prediction (Boe 1990; Hall 1990; Foster 1997; Torkelson et al. 1997; Rosche and Foster 1999; Godoy, Gizatullin, and Fox 2000). However, a semiquantitative analysis by Rosche and Foster (1999) seems to indicate that only about 10 percent of the directional mutants arise from transient hypermutator cells, whereas almost 97 percent of double mutants arise from these cells. The 10 percent figure agrees with a theoretical analysis by Ninio (1991; see also Cairns 1998). However, Ninio's calculation depends sensitively on the values of a wide variety of parameters that cannot be easily estimated, and it is unclear how robust that result is. Consequently, this intriguing agreement between the experimental results and the original calculation should be interpreted with considerable caution.

(ii) Roth et al. (2003) recently constructed three quantitative versions of the hypermutable state model. In contrast to previous treatments, they explicitly consider the accumulation of lethal or deleterious mutations, along with the directional mutation, in transient hypermutator cells. (Such an accumulation constitutes an inevitable genetic [mutational, rather than substitutional] load that is the cost of hypermutability.) Once these other mutations are taken into account, quantitative agreement disappears between the models and the known data (for instance, the known mutation rates, and the observation that only 0.1 percent of the cells enter the hypermutable state [Torkelson et al. 1997]). Instead, Roth et al. suggest that the production of directional mutants is described by an adaptive leaky gene amplification model that relies on the amplification of the leaky mutant *lac* allele during growth within clones undergoing selection (see also Anderson, Slechta, and Roth 1998; and Hendrickson et al. 2002). The matter remains unresolved. Critically, the argument that the required genetic loads are too high is an argument from plausibility, and not yet backed by experiment.

The results of Roth et al. (2003) also raise an important conceptual issue. Sniegowski and Lenski (1995) correctly pointed out that the term "adaptive" was being used far too loosely in these discussions. Mutation rates may well be enhanced under stress, and this may increase the production of that mutation which is "selected for" in these experiments. But, mutation rate enhancement may lead to an increased genetic load, as found in Roth et al.'s models. Sniegowski and Lenski argue that, in such a circumstance, mutation rate enhancement is not an adaptation. This argument raises a valid question about the past evolutionary history of mutation rate enhancement: why mechanisms leading to it happen to exist. However, it is far from conclusive. Whether or not a phenotype is beneficial depends on the environment. (Fitness is a relational property, capturing the relationship between an organism and its environment.) Moreover, whether variability in organisms, including mutability, phenotypic plasticity, and so on, is selected for requires an analysis of the range of environments it has experienced in its evolutionary history. Neither the theoretical nor the experimental work on directional mutations has seriously broached this issue, though several papers have recently pointed out that microbial populations in nature undergo long periods of starvation-induced stress (Kivisaar 2003).

13.5

The original results of Cairns, Overbaugh, and Miller (1988) spurred considerable mathematical interest in the Luria–Delbrück distribution (describing the mutants arising during the growth of clones prior to plating on the selective medium) and on the foundations of fluctuation analysis. As Cairns, Overbaugh, and Miller (1988) realized, Luria and Delbrück's (1943) original experimental protocol involved the use of a plating medium that was lethal to nonmutants and thus incapable of detecting any slow formation of directional mutants. Part of the original argument—though, in retrospect, not a particularly important part—lay in the observed deviations in the distribution of mutants from the Luria–Delbrück distribution. The trouble was that, in spite of many sporadic attempts to analyze the distribution, there was no fast, easy algorithm that could be used in a wide variety of contexts (in particular, to model different assumptions about cell growth and phenotype expression). Luria and Delbrück (1943) had provided no expression for the form of the distribution. Lea and Coulson (1949) provided a seminal early analysis that repays study even today. However, Lea and Coulson's distribution had an infinite variance, which raised valid questions about its reliability. (Some other models, for instance, one due to Haldane, remained unpublished at the time—see Sarkar 1991 for an account of Haldane's solution.) Critics of Cairns, Overbaugh, and Miller (1988) argued that a wide variety of mechanisms could account for the observed deviations from the Luria–Delbrück distribution, as discussed in some detail in chapter 12.

Stewart, Gordon, and Levin (1990) initiated the new work on the Luria–Delbrück distribution. Soon afterwards, Ma, Sandri, and Sarkar (1991, 1992) provided a simple computationally efficient algorithm for computing the expected distribution of mutants (which Stewart 1990 dubbed the "MSS" algorithm). Sarkar, Ma, and Sandri (1992) generalized these results, showing that the infinite variance (and other moments) of the original calculation was an artifact of unnecessary approximations introduced by Lea and Coulson. Lea and Coulson had, in effect, assumed that the final microbial populations, after clonal growth, were infinite in size. Once Sarkar, Ma, and Sandri (1992) removed this counterfactual approximation, all moments of the distribution became finite. With the publication of these papers, computation of the Luria–Delbrück distribution was no longer an issue. Jaeger and Sarkar (1995) have provided further minor generalizations, allowing differential fitness of mutants and nonmutants, while Pakes (1993), Kemp (1994), Goldie (1995), and Prodinger (1996) have since refined the mathematical analysis of the asymptotics of the distribution. Zheng (1999) provides a fairly comprehensive recent review.

The expected distribution of mutants can now be theoretically modeled for almost any biologically plausible experimental protocol and is no longer restricted to the one followed in Luria and Delbrück's original experiment. Chapter 12 gave a qualitative argument that almost all deviations from the Luria–Delbrück distribution due to other factors would be in the direction of the Poisson distribution (in the sense of having a lower variance). That claim is now quantitatively established.

Fluctuation analysis is not limited to the study of microbial mutagenesis. The model on which it is based is that of clonal growth accompanied by occasional mutation. Thus, the growth of cell lines by mitosis fits the model. Mutations during such growth—for instance, those that potentially induce cancer in cell lines—can be analyzed using the Luria–Delbrück distribution. Moreover, since this distribution depends critically on mutation rates, the experimentally observed distribution of mutants can be used to estimate mutation rates in cell lineages. (Luria and Delbrück 1943 were the first to note this important fact.) Before the flurry of recent theoretical work on the distribution, one limiting factor for such a use of fluctuation analysis was the difficulty of computing the Luria–Delbrück distribution. With that limitation now removed, there are now much better methods to estimate mutation rates than were previously available. Asteris and Sarkar (1996) provided Bayesian methods for the estimation of mutation rates using fluctuation analysis. Because of the use of a Bayesian methodology, results of multiple experiments can be combined to yield more accurate estimates of the mutation rate than can be obtained from individual experiments. Zheng (2002) has since produced estimation protocols based on classical statistics, but there is little philosophical or practical reason to prefer these over the more robust Bayesian methods.

13.6

What short-term lessons (fifteen years is not much time in the history of science) can be drawn from the controversy over directional mutations in microbial species? There are at least six:

(i) The vehemence of the original controversy cannot be explained by ordinary cognitive factors alone. (Keller 1992 correctly emphasized this point.) Biologists, particularly in the United States, were not easily willing to have the basic framework of the received view of evolution questioned, even with respect to relatively minor components, *and only in the domain of some unicellular organisms*. Partly, the strength of this defense of the received view can only be explained by the peculiar sociopolitical context of the United States in which, for instance, the teaching of evolution in schools is continually under threat from a tiny but vocal (and politically influential) group of religious fanatics, that is, the so-called scientific creationists/intelligent design theorists. Any criticism of evolutionary theory provides ammunition to these miscreants—hence arises a proscription of criticism of the received view of evolution within biology.

(ii) However, the religious fervor of the United States alone does not provide a complete explanation. The long shadow of the Lysenko affair also plays a role, even after half a century. For many biologists, any criticism of the received view of evolution that in effect opens the door to the possibility of the inheritance of acquired characteristics is reminiscent of the Lysenko era and the violent suppression of genetics in Stalin's Soviet Union. Moreover, for those with long memories, any mention of Lamarck also conjures up the many instances of fraud perpetrated in the name of Lamarckism (interpreted as the inheritance of acquired characteristics) during the first several decades of the twentieth century. Talk of directional mutation fell afoul of both these ghosts from the past lurking at the back of the collective memory of the biological community.

(iii) Leaving aside the question of vehemence, it is instructive to note that most of those who were fervently opposed to the very possibility of directional mutation were evolutionary biologists (see, for instance, Sniegowski and Lenski 1995 as well as the literature discussed in chapter 12) who, by and large, did not believe that the received view of evolution inherited from the 1920s and 1930s requires any essential modification in the light of molecular biology. The future will show whether this conservatism turns out to have been entirely justified. So far, the theoretical arguments advanced by the conservatives have contributed only marginally to a resolution of the dispute compared with the explosion of experimental results during the last decade. However, those analyzing the future history of this controversy will be faced with the troublesome issue that the received view of evolution is not a well-defined entity. If the received view is committed to the idea that mutation occurs at the level of the cell, then the hypermutable state model is in conflict with that view. (This is the position taken in chapter 12, where the received view is called "neo-Darwinism.") However, if it is committed only to blind variation at some level of organization, no matter which, then it is not in conflict. (This is the position advanced by Sniegowski and Lenski 1995.) An expansion of evolutionary theory to allow for multiple levels of variation and selection is, nevertheless, an expansion of the view of evolution inherited from the 1930s (see chapter 1).

(iv) It is, however, already clear that the conservatism shown by Lenski (1989) and others has had a healthy impact on the original controversy. As mentioned in section 13.2, several of the original spectacular claims of directional mutation did not survive the intense scrutiny they received from skeptics. Skepticism about radical new claims has typically been part of science and one of the major reasons why science is believed to be self-correcting. The controversy over directional mutation underscores the remarkable extent to which it is methodologically justified: any claim that science makes as being capable of providing more reliable knowledge compared to other epistemological enterprises at least implicitly assumes that the results of science have been subjected to rigorous scrutiny before general acceptance.

(v) Yet another traditional—and conservative—theme that emerges from these developments is that both theoretical innovation and reliable theory is driven by experiment (rather than speculative freedom). Fluctuation analysis has been available since Luria and Delbrück's (1943) introduction of the technique. Yet, the subtleties of the technique—in particular, the behavior of the distribution of mutants under various ancillary experimental conditions that violated parts of the original protocol—were rarely systematically theoretically modeled until the experiments of Cairns, Overbaugh, and Miller (1988) and those that followed forced the issue. (Armitage [1952] is an exception.) The clarification was not difficult; it was just that, until the issues became experimentally relevant, there was no tangible impulse to explore these rather than other techniques. Similarly, the elaboration of the hypermutable state model and the adaptive gene amplification model have both been driven by experiment. It is likely that future theoretical developments will similarly be driven by experiment, rather than questions about evolutionary theory.

Directional Mutations

(vi) When the controversy over directional mutations began, fluctuation analysis based on the Luria–Delbrück distribution and some other techniques (mainly replica plating and sib-selection) initially emerged as the preferred analytic tool that was supposed to resolve the controversy. These methodologies were essentially nonreductionist; Sarkar (1990; see also chapter 12) argued that such nonreductionist methodologies were not powerful enough to decide between the competing claims, and that, in the era of molecular biology, only an essentially reductionist strategy of searching for mechanisms could resolve the controversy. While the controversy is yet to be resolved, reductionist strategies have almost entirely replaced fluctuation analysis and other nonreductionist methodologies. (Some exceptions were mentioned earlier, in section 13.2; see also Sniegowski and Lenski 1995—however, even these were from the early 1990s.) At least to this extent, the initial advocacy of reductionism has been borne out.

References

Anderson, D. I., Slechta, E. S., and Roth, J. R. 1998. "Evidence that Gene Amplification Underlies Adaptive Mutability of the Bacterial *lac* Operon." *Science* 282: 1133–1135.

Armitage, P. 1952. "The Statistical Theory of Bacterial Populations Subject to Mutation." *Journal of the Royal Statistical Society B* 14: 1–40.

Asteris, G., and Sarkar, S. 1996. "Bayesian Procedures for the Estimation of Mutation Rates from Fluctuation Experiments." *Genetics* 142: 313–326.

Bjedov, I., Tenaillon, O., Gérard, B., Souza, V., Denamur, E., Radman, M., Taddei, F., and Matic, I. 2003. "Stress-induced Mutagenesis in Bacteria." *Science* 300: 1404–1409.

Boe, L. 1990. "Mechanism for Induction of Adaptive Mutations in Escherichia coli." *Molecular Microbiology* 4: 597–601.

Bridges, B. A. 1995. "Starvation-associated Mutation in *Escherichia coli* Strains Defective in Transcription Repair Coupling Factor." *Mutation Research* 329: 49–56.

Cairns, J. 1998. "Mutation and Cancer: The Antecedents to Our Studies of Adaptive Mutation." *Genetics* 148: 1433–1440.

Cairns, J., and Foster, P. L. 1991. "Adaptive Reversal of a Frameshift Mutation in *Escherichia coli*." *Genetics* 128: 695–701.

Cairns, J., Overbaugh, J., and Miller, S. 1988. "The Origin of Mutants." *Nature* 335: 142–145.

Delbrück, M. 1946. "[Discussion to article by A Lwoff]." *Cold Spring Harbor Symposia in Quantitative Biology* 11: 154.

Foster, P. 1992. "Directed Mutation: Between Unicorns and Goats." *Journal of Bacteriology* 174: 1711–1716.

Foster, P. L. 1993. "Adaptive Mutation: The Uses of Adversity." *Annual Review of Microbiology* 47: 467–504.

Foster, P. L. 1997. "Nonadaptive Mutations Occur on the F′ Episome during Adaptive Mutation Conditions in *Escherichia coli*." *Journal of Bacteriology* 179: 1550–1554.

Foster, P. L. 1999a. "Are Adaptive Mutations Due to a Decline in Mismatch Repair? The Evidence Is Lacking." *Mutation Research* 436: 179–184.

Foster, P. L. 1999b. "Mechanisms of Stationary Phase Mutation: A Decade of Adaptive Mutation." *Annual Review of Genetics* 33: 57–88.

Foster, P. L. 2000a. "Adaptive Mutation: Implications for Evolution." *BioEssays* 22: 1067–1074.

Foster, P. L. 2000b. "Adaptive Mutation in *Escherichia coli*." *Cold Spring Harbor Symposia on Quantitative Biology* 45: 21–29.

Foster, P. L., and Cairns, J. 1992. "Mechanisms of Directed Mutation." *Genetics* 131: 783–789.

Foster, P. L., and Cairns, J. 1994. "The Occurrence of Heritable *Mu* Excisions in Starving Cells of *Escherichia coli*." *The European Molecular Biology Organization Journal* 13: 101–105.

Godoy, V. G., Gizatullin, F. S., and Fox, M. S. 2000. "Some Features of the Mutability of Bacteria during Nonlethal Selection." *Genetics* 154: 49–59.

Goldie, C. 1995. "Asymptotics of the Luria–Delbrück Distribution." *Journal of Applied Probability* 32: 840–841.

Grigg, G. W., and Stuckey, J. 1966. "The Reversible Suppression of Stationary Phase Mutation in *Escherichia coli* by Caffeine." *Journal of Bacteriology* 179: 4620–4622.

Hall, B. G. 1988. "Adaptive Evolution That Requires Multiple Spontaneous Mutations. I. Mutations Involving an Insertion Sequence." *Genetics* 120: 887–897.

Hall, B. G. 1990. "Spontaneous Point Mutations That Occur More Often When Advantageous Than When Neutral." *Genetics* 126: 5–16.

Hall, B. G. 1994. "On Alternatives to Selection-induced Mutation in the *bgl* Operon of *Escherichia coli*." *Molecular Biology and Evolution* 11: 159–168.

Harris, R. S., Feng, G., Ross, K. J., Sidhu, R., Thulin, C., Longerich, S., Szigety, S. K., Hastings, P. J., Winkler, M. E., and Rosenberg, S. M. 1999. "Mismatch Repair Is Diminished during Stationary-phase Mutation." *Mutation Research* 437: 51–60.

Hastings, P. J., and Rosenberg, S. M. 2002. "In Pursuit of a Molecular Mechanism for Adaptive Gene Amplification." *DNA Repair* 1: 111–123.

Hendrickson, H., Slechta, E. S., Bergthorsson, U., Andersson, D. I., and Roth, J. R. 2002. "Amplification-mutagenesis: Evidence That 'Directed' Adaptive Mutation and General Hypermutability Result from Growth with a Selected Gene Amplification." *Proceedings of the National Academy of Sciences (USA)* 99: 2164–2169.

Jaeger, G., and Sarkar, S. 1995. "On the Distribution of Bacterial Mutants: The Effects of Differential Fitness of Mutants and Non-mutants." *Genetica* 96: 217–223.

Keller, E. F. 1992. "Between Language and Science: The Question of Directed Mutation in Molecular Genetics." *Perspectives in Biology and Medicine* 35: 292–306.

Kemp, A. W. 1994. "Comments of the Luria–Delbrück Distribution." *Journal of Applied Probability* 30: 822–828.

Kivisaar, M. 2003. "Stationary Phase Mutagenesis: Mechanisms That Accelerate Adaptation of Microbial Populations under Environmental Stress." *Environmental Microbiology* 5: 814–827.

Layton, J. C., and Foster, P. L. 2003. "Error-prone DNA Polymerase IV Is Controlled by the Stress-response Sigma Factor, RpoS, in *Escherichia coli*." *Molecular Microbiology* 50: 549–561.

Lea, D. E., and Coulson, C. A. 1949. "The Distribution of the Number of Mutants in Bacterial Populations." *Journal of Genetics* 49: 264–285.

Lenski, R. E. 1989. "Are Some Mutations Directed?" *Trends in Ecology and Evolution* 4: 148–151.

Luria, S. E., and Delbrück, M. 1943. "Mutations of Bacteria from Virus Sensitivity to Virus Resistance." *Genetics* 28: 491–511.

Ma, W. T., Sandri, G. vH., and Sarkar, S. 1991. "Novel Representation of Exponential Functions of Power Series Which Arise in Statistical Mechanics and Population Genetics." *Physics Letters A* 155: 103–106.

Ma, W. T., Sandri, G. vH., and Sarkar, S. 1992. "Analysis of the Luria–Delbrück Distribution Using Discrete Convolution Powers." *Journal of Applied Probability* 29: 255–267.

Maenhaut-Michel, G., and Shapiro, J. A. 1994. "The Roles of Starvation and Selection in the Emergence of *araB-lacZ* Fusion Clones." *European Molecular Biology Organization Journal* 13: 5229–5244.

Mittler, J. E., and Lenski, R. E. 1992. "Experimental Evidence for an Alternative to Directed Mutation in the *bgl* Operon." *Nature* 356: 446–448.

Ninio, J. 1991. "Transient Mutators: A Semiquantitative Analysis of the Influence of Translation and Transcription Errors on Mutation Rates." *Genetics* 129: 957–962.

Pakes, A. G. 1993. "Remarks on the Luria–Delbrück Distribution." *Journal of Applied Probability* 30: 991–994.

Prival, M., and Cebula, T. A. 1996. "Adaptive Mutation and Slow-growing Revertants of an *Escherichia coli* lacZ amber Mutant." *Genetics* 144: 1337–1341.

Prodinger, H. 1996. "Asymptotics of the Luria–Delbrück Distribution via Singularity Analysis." *Journal of Applied Probability* 33: 282–283.

Rosche, W. A., and Foster, P. L. 1999. "The Role of Transient Hypermutators in Adaptive Mutation in *Escherichia coli*." *Proceedings of the National Academy of Sciences (USA)* 96: 6862–6867.

Rosenberg, S. M. 2001. "Evolving Responsively: Adaptive Mutation." *Nature Review Genetics* 2: 504–515.

Rosenberg, S. M., Thulin, C., and Harris, R. S. 1998. "Transient and Heritable Mutators in Adaptive Mutation in the Lab and in Nature." *Genetics* 148: 1559–1566.

Roth, J. R., Kofoid, E., Roth, F. P., Berg, O. G., Seger, J., and Andersson, D. I. 2003. "Regulating General Mutation Rates: Examination of the Hypermutable State Model for Cairnsian Adaptive Mutation." *Genetics* 163: 1483–1496.

Sarkar, S. 1990. "On the Possibility of Directed Mutagenesis in Bacteria: Statistical Analyses and Reductionist Strategies." In A. Fine, M. Forbes, and L. Wessels, eds., *PSA 1990: Proceedings of the 1990 Biennial Meeting of the Philosophy of Science Association*, vol. 1. East Lansing: Philosophy of Science Association, pp. 111–124.

Sarkar, S. 1991. "Haldane's Solution of the Luria–Delbrück Distribution." *Genetics* 127: 257–261.

Sarkar, S. 1993. "Beyond Neo-Darwinism: The Challenge of Directed Mutations." *Philosophical Studies from the University of Tampere* 50: 69–84.

Sarkar, S., Ma, W. T., and Sandri, G. vH. 1992. "On Fluctuation Analysis: A New, Simple, and Efficient Method for Computing the Expected Number of Mutants." *Genetica* 85: 173–179.

Shapiro, J. A. 1984. "Observations on the Formation of Clones Containing *araB-lacZ* Cistron Fusions." *Molecular and General Genetics* 194: 79–90.

Slechta, E. S., Liu, J., Andersson, D. L., and Roth, J. R. 2002. "Evidence That Selected Amplification of a Bacterial *lac* Frameshift Allele Can Stimulate Lac$^+$ Reversion

(Adaptive Mutation) with or without General Hypermutability." *Genetics* 191: 945–956.

Sniegowski, P. D. 1995. "A Test of the Directed Mutation Hypothesis in *Escherichia coli* MCS2 Using Replica Plating." *Journal of Bacteriology* 177: 1119–1120.

Sniegowski, P. D., and Lenski, R. E. 1995. "Mutation and Adaptation: The Directed Mutation Controversy in Evolutionary Perspective." *Annual Review of Ecology and Systematics* 26: 553–578.

Stewart, F. M. 1990. "Fluctuation Analysis: The Effect of Plating Deficiency." *Genetica* 84: 51–55.

Stewart, F. M., Gordon, D. M., and Levin, B. R. 1990. "Fluctuation Analysis: The Probability Distribution of the Number of Mutants under Different Conditions." *Genetics* 124: 175–185.

Symonds, N. D. 1989. "Anticipatory Mutagenesis?" *Nature* 337: 119–120.

Tlsty, T. D., Margolin, B. H., and Lum, K. 1989. "Differences in the Rates of Gene Amplification in Nontumorigenic and Tumorigenic Cell Lines as Measured by Luria–Delbrück Fluctuation Analysis." *Proceedings of the National Academy of Sciences (USA)* 86: 9441–9445.

Tompkins, J. D., Nelson, J. L., Hazel, J. C., Leugers, S. L., Stumpf, J. D., and Foster, P. L. 2003. "Error-prone Polymerase, DNA Polymerase IV, Is Responsible for Transient Hypermutation during Adaptive Mutation in *Escherichia coli*." *Journal of Bacteriology* 185: 3469–3472.

Torkelson, J., Harris, R. S., Lombardo, M.-J., Nagendran, J., Thulin, C., and Rosenberg, S. M. 1997. "Genome-wide Hypermutation in a Subpopulation of Stationary Phase Cells Underlies Recombination-dependent Adaptive Mutation." *European Molecular Biology Journal*: 185–189.

Zheng, Q. 1999. "Progress of a Half Century in the Study of the Luria–Delbrück Distribution." *Mathematical Biosciences* 162: 1–32.

Zheng, Q. 2002. "Statistical and Algorithmic Methods for Fluctuation Analysis with SALVADOR as an Implementation." *Mathematical Biosciences* 176: 237–252.

14 From Genes as Determinants to DNA as Resource: Historical Notes on Development, Genetics, and Evolution

14.1 Introduction

H. J. Muller's 1926 address to a symposium on the gene at the International Congress of Plant Sciences was entitled "The Gene as the Basis of Life."[1] For Muller, genes were capable of self-reproduction; consequently, they must have autocatalytic properties. On this basis, Muller argued, the "gene ... arose coincidentally with growth and 'life' itself."[2] Not only were genes thus constitutive of life, Muller went on, but all of evolution must be explained from a genetic basis: "in all probability all specific, generic, and phyletic differences, of every order, between the highest and lowest organisms, the most diverse metaphyta and metazoa, are ultimately referable to changes in ... genes."[3] The same year, Muller's mentor, T. H. Morgan, published *The Theory of the Gene*. The book summarized fifteen years of research, primarily on the fruit-fly (*Drosophila melanogaster*), that established the hegemony of genetics in twentieth-century biology. The purpose of this concluding chapter is to reflect on the history of genetics during that century, sketch how it came to dominate discussions of both development and evolution (helping to maintain their long divorce), and finally speculate on how the emergence of genomics and proteomics may be leading to a radically different agenda for biology.

Trained as a turn-of-the-century embryologist, Morgan had denied the full significance of both Darwinism and Mendelism at least until 1910 when he discovered sex-limited Mendelian inheritance of a trait (the mutant white eye in *D. melanogaster*).[4] That discovery spawned a pathbreaking research program in genetics. By 1926, Morgan and his laboratory had investigated over 400 mutant characters of *D. melanogaster*. Through the systematic use of linkage mapping (invented by Morgan's student,

A. H. Sturtevant, in 1913) these characters were partitioned into four linkage groups corresponding to the four chromosome pairs of Drosophila. The publication of *The Theory of the Gene* marked the completion of one of the most innovative research programs of twentieth-century biology.

That nothing was known about the developmental genesis of these traits at the level of cell, tissue, or organ had not in any way impeded these investigations. Thus, by 1926, Morgan had not only come to accept and insist on Mendelism as the theory of heredity, he was ready to demand a sharp divorce of genetics from development:

> Between the characters, that furnish the data for the [Mendelian] theory and the postulated genes, to which the characters are referred, lies the whole field of embryonic development. The theory of the gene, as here formulated, states nothing with respect to the way in which the genes are connected with the end-product or character. The absence of information relating to this interval does not mean that the process of embryonic development is not of interest for genetics ... but the fact remains that the sorting out of the characters in successive generations can be explained at present without reference to the way in which the gene affects the developmental process.[5]

Morgan was not the first to suggest such a strategy of genetic analysis; in 1914, William Bateson, in his presidential address to the British Association for the Advancement of Science, had also noted that the possibility of this separation is the characteristic feature of the new Mendelian genetics.[6]

Meanwhile, genes were slowly acquiring the material reality that most skeptics of Mendelism had long demanded of them. Muller's successes at inducing mutations through physical processes, particularly X-rays, added confidence to the position that genes were associated with definite material objects.[7] The physical interpretation of Mendelism helped establish what came to be called classical (transmission) genetics. In the 1930s and early 1940s, genes were thought to be composed of protein; nucleic acids composed of only four nucleotide bases (A: adenine; C: cytosine; G: guanine; and T: thymine) were believed not to be complex enough to provide the variability required to specify the several hundred known genes. However, in 1944, Avery, MacLeod, and McCarty demonstrated experimentally that, at least in bacteria, genes were composed of DNA.[8] The same year, the physicist, Erwin Schrödinger, in a book called *What Is Life?*, produced an ingenious combinatorial argument showing that even composites from a small number of building blocks can have more than the amount of variety required of genes.[9] While the significance of this argument was largely unrecognized in the 1940s, Schrödinger's book played a key role in encouraging physical scientists to tackle biological problems, leading to the rapid expansion of molecular biology in the 1950s.

The most crucial development in physical studies of the gene was the decipherment of the structure of DNA by Watson and Crick in 1953.[10] Although what typically gets emphasized is the double-helical structure of the model, what is critical to its eventual role in biology is the model of genetic specificity it incorporates. The term "specificity" was introduced in a genetic context by H. A. Timoféeff-Ressovsky and N. W. Timoféeff-Ressovsky only in 1926.[11] However, the specificity of gene action was a presumption of genetics from its inception. Originally proposed as a one-to-one correspondence between gene and trait, the idea survived in an increasingly mitigated form throughout the twentieth century. The double helix provided a model of specificity entirely new in biology: specificity was achieved by the order of arrangement of nucleotide bases, on the possibility of which only Schrödinger had speculated. This model ushered in the age of biological information: information interpreted as a sequence or arrangement of bases became the model of specificity for genetics.[12] Most important, it led to the view that genes were the sole purveyors of biological information. Crick summarized the view in what he called the "central dogma" of molecular biology: "This states that once 'information' has passed into protein *it cannot get out again*. In more detail, the transfer of information from nucleic acid to nucleic acid, or from nucleic acid to protein may be possible, but transfer from protein to protein, or from protein to nucleic acid is impossible."[13]

The contrast here is with the older physical model of specificity, stereospecificity, dating back to the immunologist Paul Ehrlich's side-chain theory from the 1880s, which had begun to dominate structural studies in biology in the 1920s and 1930s.[14] The rise of the informational perspective also reified the view, articulated by Muller, that genes as determinants of biological features were special, different from the other resources used by organisms during development. The specificity of the gene–gene product (nucleic acid or protein) relationship was informational and thus different from specificity at every other level of biological organization, which remained physical (or stereospecific). Thus arose the view of DNA as the master molecule in charge of development—see section 14.3. Section 14.2, meanwhile, discusses the displacement of other views of development during the process of establishing the hegemony of genetics.

14.2 Evocators of Development

Around 1900, for biologists in the field and in the laboratory, it was far from obvious that organismic traits could be inherited through discrete units like Mendel's factors. There were two problems:

(i) Discrete Mendelizing traits were rare. Most traits varied continuously (or were "quantitative") and were often normally distributed around a population mean, as hypothesized by the biometricians.[15] Moreover, their inheritance seemed to follow rules such as the biometricians' law of ancestral inheritance.

(ii) Developmentally, not only was there no one-to-one correspondence between traits and hereditary factors, there was also ample evidence that the relation between them was not even determinate.

For instance, the German zoologist, R. Woltereck, studied morphologically distinct strains of *Daphnia* and *Hyalodaphnia* species from different lakes. These were pure lines that maintained their form through several generations of parthenogenesis. Woltereck focused on continuous traits such as head-height at varying nutrient levels. For both genera, the phenotypes varied between different pure lines, were affected by some environmental factors such as nutrient levels, were almost independent of others such as the ambient temperature, and showed cyclical variation with factors such as seasonality. Moreover, the response of a phenotype to the same environmental change was not identical in different pure lines. Woltereck drew "phenotype curves" to depict this phenomenon. These curves changed for every new variable that was considered. There was thus potentially an almost infinite number of them and Woltereck coined the term *"Reaktionsnorm"* to indicate the totality of the relationships embodied in them.[16] (It was only later that Woltereck's individual phenotype curves came to be called norms of reaction, or reaction norms.)

Woltereck argued that what was inherited was this *Reaktionsnorm* and that hereditary change consisted of a modification of that norm. Even W. Johannsen, who first explicitly made a sharp distinction between genotype and phenotype, endorsed the concept of the reaction norm, which he thought to be "nearly synonymous" with "genotype."[17] Only slightly later, H. Nilsson-Ehle coined the term "plasticity" to describe the non-unique relation of the genotype to the phenotype and argued that this relation has general adaptive significance.[18] This view found resonance in the Soviet Union, where the norm of reaction (understood as what Woltereck had described as individual phenotypic curve) emerged as a concept of central importance in genetics. An avoidance of genetic determinism was clearly concordant with the Soviet program of producing an interpretation of science based on dialectical materialism; phenotypic plasticity, as modeled by variable reaction norms, furthered that project. However, in the West (that is, the United States and Europe outside the Soviet Union), where Johannsen's sharp genotype–phenotype distinction became part of the standard picture of genetics, the subsequent decades witnessed a general trend to emphasize the constancy and causal efficacy of the genotype at the expense of the complexity of its interactions. The norm of reaction remained a relatively ignored concept during this period.[19]

Ironically, the conceptual reticulation of classical genetics that helped maintain the primacy of the gene also emerged from developments in the Soviet Union. There, in the 1920s, an active genetical research group formed around the pioneering population geneticist, S. Chetverikov.[20] In 1922, one member of the group, D. D. Romashoff, discovered the *Abdomen abnormalis* mutation in *Drosophila funebris* that resulted in the degeneration of abdominal stripes.[21] There was individual variability in the mutant phenotype, which Romashoff interpreted as a difference in the strength of the mutation's effect. The manifestation of the mutation depended on environmental factors—in particular, on the dryness and liquid content of food—but Romashoff could not rule out the possible influence of other loci. Another member of that group, N. W. Timoféeff-Ressowsky, studied the recessive *Radius incompletus* mutation of *D. funebris*.[22] In mutant flies, the second longitudinal vein did not reach the end of the wing. Timoféeff created different pure lines, each homozygous for this mutation. Descendants included phenotypically normal flies. The proportion of normals was fixed for each pure line but varied between lines. External factors had little influence; the differences between the lines were apparently under the control of genotypic factors. Some lines gave a large proportion of mutants but manifested the mutation weakly; in others, the converse was realized. There were many intermediate lines.

The Soviet work was carefully followed by the German neuroanatomist, O. Vogt, who was a frequent visitor to Moscow because of a project to dissect Lenin's brain to demonstrate his genius.[23] Vogt, long committed to a genetic interpretation of psychoses, introduced two new concepts to describe Timoféeff's results: a mutation's "expressivity" was the extent of its manifestation; its "penetrance" was the proportion of individuals carrying it that manifested any effect at all. The differences between distinct lines were entirely ignored in Vogt's definitions. Expressivity and penetrance became properties of the gene rather than a property of a mutation relative to a constant genetic background. Yet, for historical reasons that are not entirely clear, Timoféeff enthusiastically endorsed the new concepts.[24] What the original results of Romashoff and Timoféeff had shown was a *predictable complexity* in the genotype–environment interaction. Both datasets permitted the construction of norms or reaction, even though Vogt's reinterpretation made such a move moot. Two related aspects of that reinterpretation deserve emphasis: (i) Vogt ignored the *systematic* differences between pure lines; and (ii) he explicitly introduced expressivity and penetrance as properties of genes on par with, though allegedly different from, dominance.

The introduction of expressivity and penetrance constituted a convoluted reticulation of the structure of Mendelian genetics by an ad hoc complication introduced to the concept of the gene. Besides having their standard transmission properties, genes were no longer only recessive or dominant (or displaying varying degrees of dominance); they also had degrees of expressivity and penetrance. There was no clear distinction between expressivity and dominance: expressivity, as defined by Vogt, is indistinguishable from the degree of dominance. In retrospect, the purpose that the new concepts served was to maintain a genetic etiology in the face of recalcitrant phenotypic plasticity induced by the complexity of genotype–environment interactions. Variability in the phenotypic manifestation of a trait became a result of a gene's expressivity and (indirectly) its penetrance. If the presence of a gene *for a trait* nevertheless failed to produce the trait, a genetic etiology for the trait was still maintained by simply positing that the gene had incomplete penetrance. If the presence of that gene led to the presence of the trait but only to some variable degree, the gene was still responsible for the trait but had variable expressivity. The terms "penetrance" and "expressivity" were introduced into the English literature by C. H. Waddington in his *Introduction to Modern Genetics*, where they were incorrectly attributed to Timoféeff.[25] Waddington's book, along with Timoféeff's growing prominence within Western genetics, made the terms common currency by the 1950s. Phenotypic plasticity—an almost inevitable outcome if development is the result of a suite of different factors, rather than only of the genotype—was relegated to irrelevance by mystifying the concept of the gene.

Waddington's role in this story is curious. Though trained primarily as an embryologist, Waddington came to recognize the significance of the new genetics very early. In 1924, Spemann and Mangold had discovered the "organizer," a region of the early embryo (at the gastrula stage) that seemed to direct subsequent development.[26] This led to an active research agenda by many embryologists to identify the "active principle" of the organizer. Committed reductionists believed this to be a chemical; Spemann, himself, had more holistic leanings. Waddington was among those to demonstrate that dead "organizers" could induce cell differentiation. By 1938 he had come to view organizers as "evocators" of development: "[t]he factor which, in the development of vertebrates, decides which of the alternative modes of development shall be followed is the organiser, or, more specifically, the active chemical substance of the organiser which has been called the evocator."[27] Waddington argued that changes of this sort are discrete, that is, there are definite developmental pathways with no intermediates between them. Because genes were also discrete, Waddington argued that "genes ... act in a way formally like ... evocators, in that they *control* the choice of alternative."[28]

For Waddington, the *aristopedia* class of alleles (*aristopedia, aristopedia-Spencer*, and *aristopedia-Bridges*) at the spineless locus of the third chromosome of *Drosophila melanogaster* provided an apposite example. The presence of the first two alleles from this class (*aristopedia* and *aristopedia-Spencer*) led to the transformation of the arista into a tarsus. In the case of the third (*aristopedia-Bridges*), the change was less marked but, even in this case, there was no true intermediate. Rather, a smaller number of segments were altered, thus showing that a discrete change had taken place. Waddington's invocation of the language of "control" would be of critical significance after the advent of molecular biology—see section 14.3. What is critical here is that his work marks the first serious attempt to synthesize genetics and development, and it presumes, without argument, the primacy of the gene. Following through on this assessment of the importance of genes, in the 1940s, Waddington shifted the focus of his research from classical embryology to the genetic control of tissue differentiation in *Drosophila*.[29]

If Morgan had merely argued for a divorce of genetics from development, Waddington, in effect, demanded the subjugation of the second to the first. A quote, though from a later period, emphasizes this point: "we know that genes determine the specific nature of many chemical substances, cell types, and organ configurations; and we have every reason to believe that they ultimately control all of them."[30] Given the dominance of developmental genetics in developmental biology since the 1960s, Waddington's choice of "control" hardly seems unusual today. But, in the embryology of the 1920s and 1930s (and earlier periods), reproduction was recognized as an important component of development: a full developmental cycle included reproduction. From a developmental perspective, the one from which Waddington emerged, it makes just as much, if not more, sense to explicate and emphasize the developmental determination of genetics through the control of reproduction, rather than to stipulate the genetic control of development. Nevertheless, Waddington made that fateful move with far-reaching consequences for the study of development in the twentieth century.

14.3 The Age of the Master Molecule

As noted in section 14.1, the construction of the double helix model for DNA and the informational model of biological specificity in 1953 radically altered the conceptual terrain of biology, at least at the organismic and lower levels of organization. E. Schrödinger had already speculated on the existence of a "hereditary code-script" in 1943; starting in 1954, another physicist, G. Gamow, began an explicit program of deciphering the "genetic code."[31] The hope was to discover substantive properties of the code from simple formal rules incorporating functional assumptions about the efficiency and fidelity of information storage and transmission. As the mathematician, S. W. Golomb, put it: "[i]t will be interesting to see how much of the final solution [of the coding problem] will be proposed by the mathematicians before the experimentalists find it, and how much the experimenters will be ahead of the mathematicians."[32]

As is often the case, biology was not kind to the mathematicians: the theoretical program of deciphering the code was an unmitigated failure. The code that was experimentally deciphered in the early 1960s had none of the elegance envisioned by the theorists. In spite of this failure, this theoretical research program had one lasting consequence: it helped bring

Historical Notes on Development, Genetics, Evolution 373

to prominence the idea that the genome should be construed as a computer program. The emergence of this idea was encouraged by the context in which it occurred: this was the period that saw the beginning of large-scale digital computation.[33]

Two papers from 1961, with radically different agendas, explicitly introduced the idea of the genome as a blueprint and a program to be interpreted during development. In their classic paper laying out the details of the operon model for gene regulation, F. Jacob and J. Monod concluded: "The discovery of regulator and operator genes, and of repressive regulation of the activity of structural genes, reveals that the genome contains not only a series of blue-prints, but a co-ordinated program of protein synthesis and the means of controlling its execution."[34] What is critical about this passage is that agency resides in the genome: it controls the execution of the instructions in it. The fact that these instructions were already being interpreted as information gave credence to the metaphor of a genomic program. The operon model solved the decade-old problem of enzymatic adaptation through gene regulation. Later it became the standard model of gene regulation for most prokaryotic genes, as discussed below.

A much more extended and careful discussion of programming and computation came in a paper the major purpose of which was to delimit the domain of molecular biology, that is, prevent its intrusion into organismic biology. In "Cause and Effect in Biology," E. Mayr notoriously distinguished "proximate" causes investigated by molecular biology from "ultimate" causes that are only provided by evolutionary biology. Evolution is the programmer producing a code that plays itself out in an individual, allowing individual behavior to be purposive:

An individual who—to use the language of the computer—has been "programmed" can act purposefully.... Natural selection does its best to favor the production of codes guaranteeing behavior that increases fitness.... The purposive action of an individual, insofar as it is based on the properties of its genetic code, therefore is no more nor less purposive than the actions of a computer that has been programmed to respond appropriately to various inputs.[35]

Once again, agency resides in the genome, but because of natural selection and, in contrast to Jacob and Monod's interpretation of the operon, not because of physical or chemical mechanisms.

The critical feature of the operon model was that the regulation of gene activity apparently occurred at the genetic level. This was an unexpected development: while the problem of gene regulation was recognized as being critical to understanding development since a pioneering paper by J. B. S. Haldane in 1932, it was generally believed that the mechanism of control would operate from the cellular level.[36] (Developmental holists believed that the mechanism would operate from even higher levels, for instance, from that of the tissue or organ.) In 1962, in *New Patterns in Genetics and Development*, Waddington seized on the operon model to argue that regulation at the genetic level provides an explanation of tissue differentiation.[37] Differentiation was thus a matter of switching genes on or off. Even more controversially, Waddington interpreted other spatial developmental phenomena—histogenesis, morphogenesis, pattern formation, and so on—as special cases of differentiation.[38] Thus begun the program of a developmental genetics, of explaining development from a genetic basis, which took over the study of development in the 1970s.

A detailed history of developmental genetics is yet to be constructed. From the perspective of that subdomain, the crucial developments were the discoveries in the 1980s of the homeobox sequence and *HOX* genes, which were supposed to control much of morphogenesis.[39] *HOX* and similar genes do have significant regulatory roles in many species. Nevertheless, the confidence of geneticists in "master control genes" for development went far beyond what the data justified. This confidence was reflected in the initiation of the Human Genome Project (HGP) (and, later, other sequencing projects) in the 1990s. Blind sequencing of genomes was supposed to reveal the mechanisms by which biological processes operated at all levels of organization.

The trouble is that, by the late 1980s, there was ample reason to believe that DNA sequences alone would reveal little about biology even at the cellular level, let alone at higher levels. The informational model for DNA sequences as functional genes worked well provided that two conditions were satisfied:

(i) a *sufficiency* condition—inspection of the presence of a DNA sequence in a cell is sufficient to infer a capacity to produce the encoded protein; and

(ii) a *uniqueness* condition—a single DNA sequence produces exactly the encoded protein.

If these two conditions are satisfied, the genetic code can be used—as a look-up table—to predict the amino acid sequence of the encoded protein. For prokaryotes, these conditions are satisfied: all DNA sequences, besides regulatory ones and sequences specifying transfer or ribosomal RNA (tRNA and rRNA), code for proteins and do so uniquely.

However, for eukaryotes, this picture begins to unravel.[40] Besides the standard genetic code, mitochondrial DNA and even nuclear DNA in some taxa use variant codes. The extent of such variation is at present unknown. Coding and regulatory regions of DNA are interspersed with long strands of DNA with no identifiable function.[41] These nonfunctional regions, when occurring within structural (or coding) genes, are transcribed into mRNA only to be spliced out before translation at the ribosome. (Such noncoding regions are called "introns"; coding regions are "exons."[42]) mRNA is also routinely edited through a variety of other mechanisms; bases are added and removed, sometimes in the hundreds. Perhaps the most surprising— and, in retrospect, the most important (see section 14.4)—discovery was that of alternative splicing: the same mRNA transcript can be spliced in a variety of ways, leading to a set of different proteins. There is no evidence to suppose that the control of alternative splicing can be brought under the aegis of any simple genetic model such as the operon.

Of late it has even become controversial that, without significant modification of the concept of biological information, any informational model of biological specificity can survive. The few attempts to rescue that model deny any claim that genes are the sole purveyors of biological information (see, for instance, chapter 10). But if they are not, developmental genetics, by itself, has no prospect of providing an adequate model of development. There is more to the phenotype than what can be specified by the genotype.

14.4 After the Human Genome Project

The initiation of the Human Genome Project (HGP) was perhaps the most contentious episode in the history of science policy in biology to date. If the HGP is judged by the explicit promises that its proponents made in the late 1980s and 1990s to secure public support (and funding), it has been an unmitigated failure, the most colossal misuse ever of scarce resources for biological research. In 1992, W. Gilbert claimed: "I think there will be a change in our philosophical understanding of ourselves.... Three billion bases [of a human DNA sequence] can be put on a single compact disc (CD), and one will be able to pull a CD out of one's pocket and say, 'Here's a human being; it's me!'"[43] Today the claim seems laughable. None of the promises of Gilbert's radical genetic reductionism has been borne out. Proponents of the HGP promised enormous immediate medical benefits. Arguably, at least, there have not been any. Gilbert routinely promised the birth of a new theoretical biology. Instead, emphasis now is on informatics: the design of computational tools to store and retrieve sequence information efficiently and reliably, with little expectation that any great theoretical insight is forthcoming. Commenting on the complexity of sporulation choice by an organism no more complex than *Bacillus subtilis*, C. Stephens recently pointed out:

Despite the explosive rate at which sequence databases are growing, and the concomitant increase in computing power available for sifting through them, sequence gazing alone cannot predict with confidence the precise functions of the multitude of coding regions in even a simple genome! Experimental analysis of gene function is still critical, a thought that brings with it the realization that the era of genomic analysis represents a new beginning, not the beginning of the end, for experimental biology.[44]

In one sense, from a perspective that takes the social responsibility of science seriously, to the extent that basic research should provide tangible immediate social benefits, this failure of the HGP is no doubt unfortunate. However, it is not unexpected: in the late 1980s and early 1990s, scientific skeptics of the HGP routinely pointed out that it would not deliver on its promises.[45] More important, social skeptics worried about the use of DNA sequences for discrimination in health care and employment as well as social stigmatization. The failure of the HGP to deliver on its explicit promises provides an argument against the rationale for such uses of DNA and thus assuages some of these social worries, *provided that the failure is publicly recognized.* It must even have been abundantly clear to the proponents of the HGP that their original promises were unrealistic, leaving them vulnerable to charges of fraud in their presentations to public funding bodies.

Nevertheless, no biologist, including those who were initially skeptical of the project on scientific grounds, should any longer denounce the scientific results of the HGP. At the very least, the HGP has killed the facile genetic reductionism of the heyday of developmental genetics. There is little reason any more to suspect that claims of straightforward and irrevocable genetic determination of complex human traits will ever again be credible. It may even spell the extinction of molecular genetics itself, first transforming it into genomics, and then replacing it with proteomics.

The reason the HGP may have such radical implications for biology is because of the startling properties discovered of the human genome sequence when compared to other species' sequences:

(i) The most important surprise from the HGP was that there are probably only about 31,000 genes in the human genome compared to an estimate of 140,000 as late as 1994.[46] Among the eukaryote genomes that have been fully sequenced, the human estimate remains the highest (in 2001), but not by much. Plant genomes are expected to contain many more genes than in the human sequence. It is already known that the mustard weed, *Arabidopsis thaliana*, has 26,000 genes, almost as many as the human. Morphological or behavioral complexity is not correlated with the number of genes that an organism has. This has been called the G-value paradox.[47]

(ii) The number of genes is also not correlated with the size of the genome, as measured by the number of base pairs. *D. melanogaster* has 120 million base pairs but only 14,000 genes; the worm *Caenorhabditis elegans* has 97 million base pairs but 19,000 genes; *Arabidopsis thaliana* has only 125 million base pairs, while humans have 2,900 million base pairs.[48]

(iii) At least in humans, the distribution of genes on chromosomes is highly uneven. Most of the genes occur in highly clustered sites.[49] Most genes that occur in such clusters are those that are expressed in many tissues—the so-called housekeeping genes.[50] However, the spatial distribution of cluster sites appears to be random across the chromosomes. (Cluster sites tend to be rich in C and G, whereas gene-poor regions are rich in A and T.) In contrast, the genomes of arguably less complex organisms, including *D. melanogaster*, *C. elegans*, and *A. thaliana*, do not have such pronounced clustering.

(iv) Only 2 percent of the human genome codes for proteins, whereas 50 percent of the genome is composed of repeated units. Coding and other functional regions (including regulatory regions) are interspersed by large areas of "junk" DNA of no known function. However, some functional regions, such as *HOX* gene clusters, do not contain such junk sequences.

(v) Hundreds of genes appear to have been transferred horizontally from bacteria to humans and other vertebrates, though apparently not to other eukaryotes.[51]

(vi) Once attention shifts from the genome to the proteome, a strikingly different pattern emerges. The human proteome is far more complex than the proteomes of the other organisms for which the genomes have so far been sequenced. According to some estimates, about 59 percent of the human genes undergo alternative splicing, and there are at least 69,000 distinct protein sequences in the human proteome. In contrast, the proteome of *C. elegans* has at most 25,000 sequences.[52]

At the very least, except in rare cases, the presence of a particular DNA sequence allows very little to be inferred about what happens in the proteome, let alone at higher levels of organization. At most, that piece of DNA is a potential resource for use during development. Dethroned DNA must find its place among other developmental resources. Some of these other resources are transferred intergenerationally through the material continuity of reproduction (for instance, through the maternal cytoplasm in most "higher" animals). Others are acquired from the environment (for instance, by accretion by some marine animals). Nevertheless, DNA may be special in many ways; as will be argued in section 14.5, there is a strong case to be made for disparity between DNA and other molecular constituents of cells. All the same, DNA, and ipso facto the gene, can no longer be the locus of agency responsible for the structural and behavioral repertoire of living forms, including their remarkable diversity.

14.5 Concluding Remarks

In the proteomics age, the most important problem in the philosophy of biology is to conceptualize the functional role of DNA within the cell so as to explain the organization and other properties of the genome. This chapter will end with a preliminary attempt to do so by explicating one speculative model that makes some novel predictions, though these have yet to be fully operationally disambiguated from predictions of other more traditional models. The new model will tentatively be called the sequestered modular template (SMT) model of the cell. The construction of this model begins with the observation that the cell is probably the first spatially delimited living structure to have evolved. As biochemists realized in the 1910s and 1920s, the cell's functions are primarily carried out by proteins, mainly enzymes. There are two types of such functions: those that maintain structural and behavioral integrity, and those that encourage reproductive proliferation. Evolutionary biology puts an emphasis on the latter type of function. But the former are as (perhaps even more) important for at least two reasons:[53] (i) without the maintenance of structural and behavioral integrity at least up to reproductive age, there is no question of reproduction; and (ii) in many organisms, especially sexually reproducing organisms, cellular functions continue beyond reproductive age.

Maintenance of integrity, as well as reproduction, requires the production of replacement parts. Enzymes wear out (in spite of being catalysts); transport-enabling molecular moieties on cell membranes get damaged, as do the membranes themselves. They must be replaced. There are two obvious ways to carry out replacement part production: (a) directly, by growth and fission of the relevant type; and (b) indirectly, using a template. Whether or not, during evolution, the second strategy originally arose and got fixed entirely by accident rather than selection, it has at least two advantages:[54]

(i) Suppose that cellular processes are based on a small repertoire of basic chemical mechanisms (as is true in contemporary organisms). Then the direct process of growth and fission would be catholic: the conditions under which one molecular type gets produced will very likely lead to the production of many other molecular types. Indirect reproduction permits preferential control.

(ii) Templates can be sequestered from environmental insults in a way that the active molecules cannot. The latter must necessarily interact with the environment to maintain cellular functions.

For the cell, it makes sense to have templates and, then, to sequester them. There is thus a critical disparity between the templates and the product molecules: DNA and genes are thus special compared to the other molecular constituents of the cell. It makes even more sense to make these templates as physically stable as possible. It is again probably entirely an accident that the first templates were structurally simple molecules: most likely, RNA, the variation in which was entirely combinatorial (that is, in sequence). But template integrity was better protected by a switch to a more stable form: DNA. (For instance, the base, uracil, U, is easily transformed to C by deamination; DNA uses the more stable T instead of U.) Enclosing templates by a membrane helps protection: eukaryotes achieve it by producing nuclei (and also enclosing some genes in mitochondria and plastids). After enclosure, further tinkering to increase template protection would be evolutionarily advantageous. Thus, it makes sense to cluster genes when possible: protecting clustered sites is easier than protecting widely dispersed sites. Clustering happens in humans, as noted in section 14.4. The puzzle is why it does not seem to occur, or occur as much, in the other genomes that have so far been sequenced. A possible resolution of this puzzle is that these genomes are smaller in size, resulting in less scope for clustering.

It also makes sense that genes used as templates for many functions, and those that are critical resources in early development, should receive the most protection. *HOX* genes deserve and get such attention. From the perspective of the SMT model, resources should thus be preferentially deployed to protect such genes from mutation. Here, the SMT model makes a prediction partly in variance with the received gene-based evolutionary model. That model would explain the evolutionary conservation of such genes by the deleterious selective effects of such mutations. The SMT model claims that, in addition, repair mechanisms preferentially target such sites. Turning this qualitative claim of differential prediction of the two models into an exact claim will require a quantitative analysis of the SMT model.

Modularity enters this model at two levels: (i) modularity of the genes themselves; and (ii) modularity of functional subregions (exons) of genes. At the genetic level, modularity is achieved because, by and large, genes are nonoverlapping and, much more important, they are separated from each other by long strands of nonfunctional DNA, which helps prevent gene disruption during recombination, a presumably physical inevitability of chromosome duplication. There is obviously a trade-off between this benefit and that of clustering. At the subgenic level, the benefits of modularity were clearly articulated by Gilbert in 1978.[55] Recombination in introns allows the combinatorial production of new proteins that are still likely to be at least partly functional because component parts have not lost their structural integrity. Gilbert also argued that point mutations at intron–exon boundaries can potentially alter splicing patterns and generate radically different proteins. According to the SMT model, this would be undesirable. The SMT model is inherently conservative: it predicts that such mutations are rare. Moreover, with respect to this mechanism of generating diversity, the SMT model is consistent with the strategy used by the immune systems of mammals, in which recombination rather than somatic mutation (anywhere, and not just at intron–exon boundaries) is the preferred mode of the generation of diversity (though both processes are known to occur).

If modules are being carefully protected—and, therefore, evolutionarily conserved—it makes sense to use the same modules for a variety of purposes. From this perspective, alternative splicing makes sense as a way to utilize templates efficiently. Two predictions of the SMT model about alternative splicing are:

(i) that there is an inverse correlation between genome-wide mutation rates over evolutionary time scales and the degree of alternative splicing in taxa; and

(ii) that this is a result of mechanisms at the cellular level.

A low number of genes and a high level of alternative splicing implies that organisms so constructed rely on the use of a large segment of the available combinatorial space at the level of modules within proteins. That organisms of this sort appear to be more structurally and behaviorally complex than others suggests a strong correlation between modularity at the proteomic level and evolvability. These arguments show the fallacy of any attempt at reading an organism off from its DNA sequence alone.

There are several other arguments for the SMT model:

(i) The same gene is often "co-opted" for different functions during the course of evolution. Typically, co-option follows duplication. For instance, the aggregation of the cellular slime mold, *Dictyostelium discoideum*, during times of stress uses a 495-residue long cellular adhesion molecule (CAM). The evolution of multicellularity is believed to have involved use of multiple copies of the corresponding DNA. Eventually, these copies diverged and were then co-opted to produce variants such as N-CAM for neuronal aggregation and H-CAM for hepatic aggregation.[56] From the perspective of the SMT model, it makes sense to transform and use redundant copies of a template.

(ii) Transfer RNA (tRNA) and ribosomal RNA (rRNA) are obviously critical to the function of a cell. DNA specific to such RNA (and not to proteins) forms a tiny fraction of the total genome. The number of complete sets of such DNA sequences is correlated with the size of the genome. For instance, for tRNA genes, the human mitochondrial genome has 1 complete set; the bacterium, *Escherichia coli*, has 100 such sets; and the human nuclear genome has 1,000 complete sets. The SMT model predicts that the correlation of genome size and the number of copies reflects the number of such sequences that are likely to be simultaneously necessary: it makes sense to have exactly the optimal amount of some resource. It is unclear whether the received gene-based evolutionary model would make the same prediction. Most important, if the last argument is correct, then these different copies should not evolve independently of each other. Rather, their evolution should be concerted, as, indeed, appears to be the case.[57]

(iii) The various types of RNA-editing systems that have so far been observed, usually classified as insertional or substitutional, rely on a highly heterogeneous class of mechanisms. Consequently they must have arisen independently in different lineages. Because RNA editing often corrects errors in transcription, the SMT model predicts that any available mechanism should be recruited for this purpose, and this would be encouraged throughout evolutionary history.[58] Thus, it makes sense that it evolved independently several times, resulting in a heterogeneous class of editing mechanisms.

These and other similar arguments suggest that the SMT model merits further exploration in future work.

There is, however, one central unresolved issue: the model, as sketched above, assumes that the cell is the locus of agency, that is, the level at which it is appropriate to model and tabulate benefits, costs, and accidents. But, is what is good, bad, or neutral, for the cell also the same at higher levels of organization in multicellular organisms? Cancer trivially shows that it is not always so. Buss and many others have noted the possibility of conflicts of interest between different levels of biological organization.[59] It is far from clear that all the arguments given above will carry over to higher levels of organization than the cell. Finally, there is no reason to suppose that agency resides at exactly one level of organization. If the SMT model is to be successful, it must be able to cope with the possibility and likelihood of distributed agency.

Acknowledgments

This work was supported by Grant No. SES-0090036, 2002–2003, from the United States National Science Foundation.

Notes

1. Muller (1962), pp. 188–204.

2. Ibid., p. 200.

3. Ibid., p. 195. A decade later, Dobzhansky (1937, p. 11) would echo the same sentiment, defining evolution to be a "change in the genetic [allelic] composition of populations."

4. Morgan (1910).

5. Morgan (1926), p. 26.

6. Bateson (1914).

7. Muller (1927).

8. Avery, MacLeod, and McCarty (1944).

9. Schrödinger (1944). Sarkar (1991) provides an assessment of Schrödinger's conceptual achievements.

10. Watson and Crick (1953a). See chapter 1 (sec. 1.3.1) for a discussion of the significance of the double-helix model.

11. See Timoféeff-Ressovsky and Timoféeff-Ressovsky (1926).

12. This started with Watson and Crick (1953a,b); see chapters 8 and 9 for details.

13. Crick (1958), p. 153; emphasis in the original. Thiéffry and Sarkar (1998) provide a critical history of the central dogma.

14. See Silverstein (1989).

15. See Sarkar (1998) for a discussion of biometry. Provine (1971) provides a detailed history of this dispute.

16. See Woltereck (1909), p. 135. For historical detail, see Sarkar (1999).

17. See Johannsen (1911), p. 133.

18. Nilsson-Ehle (1914).

19. See Sarkar (1999) for further details.

20. Adams (1980).

21. Romaschoff (1925).

22. Timoféeff-Ressowsky (1925).

23. For this curious history, see Laubichler and Sarkar (2002).

24. See Timoféeff-Ressovsky and Timoféeff-Ressovsky (1926).

25. See Waddington (1938).

26. See Spemann and Mangold (1924).

27. See Waddington (1939), p. 37, elaborated in Waddington (1940).

28. Waddington (1939), p. 37; emphasis added.

29. Waddington (1962), p. 14.

30. Ibid., p. 4; this aspect of new patterns in genetics and development will be further discussed in sec. 14.3.

31. Gamow (1954a,b). For a history, see chapter 9.

32. Golomb (1962), p. 100.

33. For more on this history, see Kay (2000).

34. Jacob and Monod (1961), p. 354.

35. Mayr (1961), pp. 1503–1504.

36. See Haldane (1932).

37. Waddington (1962), pp. 20, 23.

38. Ibid., pp. 1–3.

39. For a balanced analysis, noting both the importance of, and the interpretive excesses about, these discoveries, see McGinnis (1994).

40. For details of the processes mentioned in this paragraph, as well as a guide to the literature, see chapters 8 and 9.

41. The existence of such DNA initially came as a relief since it resolved the C-value paradox, that genome size was not correlated with organismic complexity (see Cavalier-Smith 1985).

42. This terminology was introduced by Gilbert (1978).

43. Gilbert (1992), p. 96.

44. Stephens (1998), p. R47.

45. See, for instance, Sarkar and Tauber (1991), and Tauber and Sarkar (1992, 1993).

46. See Hahn and Wray (2002) and the references therein.

47. Hahn and Wray (2002).

48. Ibid.

49. These have often been likened to "urban centers" while the gene-poor regions have been likened to "deserts." See ⟨http://www.ornl.gov/sci/techresources/Human_Genome/project/journals/insights.html⟩.

50. Lercher, Urrutia, and Hurst (2002).

51. International Human Genome Sequencing Consortium (2001). (This remains controversial.)

52. Hahn and Wray (2002).

53. See, in this context, chapter 1 (sec. 1.2.1).

54. Haldane (1942) was the first to note the value of template-based reproduction and presciently suggested it as a model for gene duplication.

55. See Gilbert (1978).

56. See Ohno and Holmquist (2001).

57. See Wagner (2002).

58. See Covello and Gray (1993) for a traditional evolutionary interpretation of RNA editing; see also Gray (2000) for a recent review.

59. Buss (1987), in particular, has argued for the relevance of such conflicts for the evolution of patterns of development, especially germ-line sequestration. See also Falk and Sarkar (1992).

References

Adams, M. B. 1980. "Sergei Chetverikov, the Kol'tsov Institute, and the Evolutionary Synthesis." In E. Mayr and W. B. Provine, eds., *The Evolutionary Synthesis: Perspectives on the Unification of Biology*. Cambridge, Mass.: Harvard University Press, pp. 242–278.

Avery, O. T., MacLeod, C. M., and McCarty, M. 1944. "Studies on the Chemical Nature of the Substance Inducing Transformation of Pneumococcal Types: Induction of Transformation by a Deoxyribonucleic Acid Fraction Isolated from Pneumococcus III." *Journal of Experimental Medicine* 79: 137–157.

Bateson, W. 1914. "Address of the President of the British Association for the Advancement of Science." *Science* 40: 287–302.

Buss, L. 1987. *The Evolution of Individuality*. Princeton: Princeton University Press.

Cavalier-Smith, T., ed., 1985. *The Evolution of Genome Size*. Chichester: Wiley.

Covello, P. S., and Gray, M. W. 1993. "On the Evolution of RNA Editing." *Trends in Genetics* 9: 265–268.

Crick, F. H. C. 1958. "On Protein Synthesis." *Symposia of the Society for Experimental Biology* 12: 138–163.

Dobzhansky, T. 1937. *Genetics and the Origin of Species*. New York: Columbia University Press.

Falk, R., and Sarkar, S. 1992. "Harmony from Discord." *Biology and Philosophy* 7: 463–472.

Gamow, G. 1954a. "Possible Mathematical Relation between Deoxyribonucleic Acid and Proteins." *Biologiske Meddelelser udviket af Det Kongelige Danske Videnskabernes Selskab* 22(3): 1–11.

Gamow, G. 1954b. "Possible Relation between Deoxyribonucleic Acid and Protein Structures." *Nature* 173: 318.

Gilbert, W. 1978. "Why Genes in Pieces?" *Nature* 271: 501.

Gilbert, W. 1992. "A Vision of the Grail." In D. J. Kevles and L. Hood, eds., *The Code of Codes*. Cambridge, Mass.: Harvard University Press, pp. 83–97.

Golomb, S. W. 1962. "Efficient Coding for the Desoxyribonucleic Acid Channel." *Proceedings of the Symposium for Applied Mathematics* 14: 87–100.

Gray, M. W. 2000. "RNA Editing: Evolutionary Implications." In *Nature Encyclopedia of Life Sciences*. London: Nature Publishing Group. Available at ⟨http://www.els.net/⟩.

Hahn, M. W., and Wray, G. A. 2002. "The G-Value Paradox." *Evolution and Development* 4: 73–75.

Haldane, J. B. S. 1932. "The Time of Action of Genes, and Its Bearing on Some Evolutionary Problems." *American Naturalist* 66: 5–24.

Haldane, J. B. S. 1942. *New Paths in Genetics*. New York: Random House.

International Human Genome Sequencing Consortium. 2001. "Initial Sequencing and Analysis of the Human Genome." *Nature* 409: 860–921.

Jacob, F., and Monod, J. 1961. "Genetic Regulatory Mechanisms in the Synthesis of Proteins." *Journal of Molecular Biology* 3: 318–356.

Johannsen, W. 1911. "The Genotype Conception of Heredity." *American Naturalist* 45: 129–159.

Kay, L. E. 2000. *Who Wrote the Book of Life? A History of the Genetic Code*. Stanford: Stanford University Press.

Laubichler, M., and Sarkar, S. 2002. "Flies, Genes, and Brains: Oskar Vogt, Nikolai Timoféeff-Ressovsky, and the Origin of the Concepts of Penetrance and Expressivity." In L. S. Parker and R. Ankeny, eds., *Medical Genetics, Conceptual Foundations and Classic Questions*. Dordrecht: Kluwer, pp. 63–85.

Lercher, M. J., Urrutia, A. O., and Hurst, L. D. 2002. "Clustering of Housekeeping Genes Provides a Unified Model of Gene Order in the Human Genome." *Nature Genetics* 31: 180–183.

Mayr, E. 1961. "Cause and Effect in Biology." *Science* 134: 1501–1506.

McGinnis, W. 1994. "A Century of Homeosis, a Decade of Homeoboxes." *Genetics* 137: 607–611.

Morgan, T. H. 1910. "Sex Limited Inheritance in *Drosophila*." *Science* 32: 120–122.

Morgan, T. H. 1926. *The Theory of the Gene*. New Haven: Yale University Press.

Muller, H. J. 1927. "Artificial Transmutation of the Gene." *Science* 66: 84–87.

Muller, H. J. 1962. *Studies in Genetics*. Bloomington: Indiana University Press.

Nilsson-Ehle, H. 1914. "Vilka erfarenheter hava hittills vunnits rörande möjligheten av växters acklimatisering?" *Kunglig Landtbruksakademiens Handlingar och Tidskrift* 53: 537–572.

Ohno, S., and Holmquist, G. P. 2001. "Evolutionary Developmental Biology: Gene Duplication, Divergence and Co-option." In *Nature Encyclopedia of Life Sciences*. London: Nature. Available at ⟨http://www.els.net/⟩.

Provine, W. B. 1971. *The Origins of Theoretical Population Genetics*. Chicago: University of Chicago Press.

Romaschoff, D. D. 1925. "Über die Variabilität in der Manifestierung eines erblichen Merkmales (Abdomen abnormalis) bei *Drosophila funebris* F." *Journal für Psychologie und Neurologie* 31: 323–325.

Sarkar, S. 1991. "*What Is Life?* Revisited." *BioScience* 41(9): 631–634.

Sarkar, S. 1998. *Genetics and Reductionism*. New York: Cambridge University Press.

Sarkar, S. 1999. "From the *Reaktionsnorm* to the Adaptive Norm: The Norm of Reaction, 1909–1960." *Biology and Philosophy* 14: 235–252.

Sarkar, S., and Tauber, A. I. 1991. "Fallacious Claims for HGP." *Nature* 353: 691.

Schrödinger, E. 1944. *What Is Life? The Physical Aspect of the Living Cell*. Cambridge: Cambridge University Press.

Silverstein, A. M. 1989. *A History of Immunology*. San Diego: Academic Press.

Spemann, H., and Mangold, H. 1924. "Über induktion von embryonalanlagen durch implantation artfremder organisatoren." *Wilhelm Roux' Archiv für Entwicklungsmechanik der Organismen* 100: 599–638.

Stephens, C. 1998. "Bacterial Sporulation: A Question of Commitment?" *Current Biology* 8: R45–R48.

Tauber, A. I., and Sarkar, S. 1992. "The Human Genome Project: Has Blind Reductionism Gone Too Far?" *Perspectives on Biology and Medicine* 35(2): 220–235.

Tauber, A. I., and Sarkar, S. 1993. "The Ideology of the Human Genome Project." *Journal of the Royal Society of Medicine* 86: 537–540.

Thiéffry, D., and Sarkar, S. 1998. "Forty Years under the Central Dogma." *Trends in Biochemical Sciences* 32: 312–316.

Timoféeff-Ressovsky, H. A., and Timoféeff-Ressovsky, N. W. 1926. "Über das phänotypische Manifestieren des Genotyps. II. Über idio-somatische Variationsgruppen bei Drosophila funebris." *Wilhelm Roux' Archiv für Entwicklungsmechanik der Organismen* 108: 146–170.

Timoféeff-Ressowsky, N. W. 1925. "Über den Einfluss des Genotypus auf das phänotypen Auftreten eines einzelnes Gens." *Journal für Psychologie und Neurologie* 31: 305–310.

Waddington, C. H. 1938. *An Introduction to Modern Genetics*. London: George Allen and Unwin.

Waddington, C. H. 1939. "Genes as Evocators in Development." *Growth* 1: S37–S44.

Waddington, C. H. 1940. *Organisers and Genes*. Cambridge: Cambridge University Press.

Waddington, C. H. 1962. *New Patterns in Genetics and Development*. New York: Columbia University Press.

Wagner, A. 2002. "Gene Duplication and Redundancy." In *Nature Encyclopedia of Life Sciences*. London: Nature. Available at ⟨http://www.els.net/⟩.

Watson, J. D., and Crick, F. H. C. 1953a. "Molecular Structure of Nucleic Acids—A Structure for Deoxyribose Nucleic Acid." *Nature* 171: 737–738.

Watson, J. D., and Crick, F. H. C. 1953b. "Genetical Implications of the Structure of Deoxyribonucleic Acid." *Nature* 171: 964–967.

Woltereck, R. 1909. "Weitere experimentelle Untersuchungen über Artveränderung, speziell über das Wesen quantitativer Artunterschiede bei Daphnien." *Verhandlungen der deutschen zoologischen Gesellschaft* 19: 110–173.

Index

Acquired characters, 197, 205, 245–246, 288–290, 296–297, 308, 309, 311, 337n18, 337n20, 357
Adaptation, 18, 19, 133, 134, 216, 225, 233, 290, 354, 369
Adaptor hypothesis, 188–189, 215
Aesthetics, 4, 20, 161–175
Agency, 89, 94, 99n19, 378, 382
Alcoholism, 13, 89
Allele-sharing, 92
Allelic association, 92
Allostery, 10, 86, 110, 131, 194, 227, 240, 253n27
α-helix model (of protein structure), 35n2, 93
Amphibian decline, 97
Antibodies. *See* Immunity
Approximations, 61–62, 66, 107, 108, 239, 241, 253n26
Arabidopsis thaliana, 377
Architecture, 161–165, 174–175
Aristotle, 5, 15
Armitage, P., 314, 318, 358
Autocatalysis, 136, 146–149
Autism, 13, 89
Avery, O. T., 185, 366
Ayala, F. J., 323, 325

Bacillus subtilis, 347, 376
Balzer, W., 62–63, 75, 118, 119

Base-pairing. *See* Complementarity
Bateson, W., 88, 366
Bayesian methods, 356
Bergson, H., 71
Bernal, J. D., 7
Bernard, C., 7
Bipolar affective disorder, 13, 89, 90
Biometry, 99n10, 239, 306, 368
Blending inheritance, 287–288
Blind variation. *See* Randomness
Bohr, C., 7
Bohr, N., 71, 73, 332
Bohr effect, 7, 10, 11, 131
Bonner, J. T., 24
Boyle, R., 105
Bradypus tridactylus, 14–15
Brenner, S., 34, 213, 214, 215
Britten, R. J., 230
Burnet, F. M., 310
Buss, L., 382

Caenorhabditis elegans, 377, 378
Cairns, J., 293, 294, 303, 320, 321, 323, 326–328, 330, 332, 347–350, 352, 353, 355, 358
Candida albicans, 347
Carnap, R., 176n1, 176n4, 249n2, 264
Causality, 39n53, 65, 125–129, 155–156, 227, 268, 369, 373
Cavalli-Sforza, L., 317

Central Dogma (of molecular biology), 170, 183, 186, 193, 195, 197, 205, 210, 214, 216, 234, 245–246, 270, 367
Chance, B., 226–227
Chaucer, G., 261
Chargaff, E., 7, 219
Chemostat, 292
Chetverikov, S., 369
Clostridium septicum, 314
Co-option, 96
Code, comma-free, 170–172, 187–190, 214–218, 236, 241, 242, 248
Code, genetic, 22, 30, 73, 93, 94, 95, 120, 124, 125, 130, 134, 145, 153, 154, 169–174, 183–199, 209–221, 236, 241–248, 264, 274–278, 279, 372–375, 377
Code, Morse, 185, 209
Color, 161–165
Communication theory. *See* Information theory
Complementarity (base-pair), 1, 8, 86, 123, 136, 146, 149–151, 156–157, 185, 186, 191, 209–211, 244, 247
Complementarity, principle of, 71, 99n12, 120, 332
Complex traits, 87, 89, 110, 376
Computers, 24, 228, 267, 274, 373, 376
Conservation biology, 97
Contingency, 25
Control. *See* Gene regulation
Cooperativity, 7, 10
Coulson, C. A., 292, 312, 318, 355
Creationism, 357
Crick, F. H. C., 1, 34, 170, 172, 183, 185–188, 191, 194, 196, 211, 214–218, 241, 249n4, 263, 367
Cultural inheritance, 26
Cummings, R., 140n10
C-value paradox, 43n100, 383n41
Cybernetics, 65, 185, 194–196, 210, 222–224, 226–229, 234, 236, 268
Cytology, 86

Daphnia, 90, 368
Darwin, C., 1, 16, 26, 32, 287, 288, 303, 308, 309
Darwinism. *See* Neo-Darwinism
Davidson, D., 69
Davis, B., 326
Dawe, C. M., 62–63, 118, 119
d'Buffon, C., 15
Delbrück, M., 71, 73, 89, 120, 172, 185, 189, 209, 211, 217, 291–293, 309–313, 318, 332, 344n108, 352, 355, 356, 358
Demerec, M., 313
Descartes, R., 6, 105
Determinism, 25, 227, 265, 274, 281, 369, 376
Development, 14, 22, 24, 25, 30, 31, 85, 86, 89, 90, 94, 97, 111, 185, 197, 243, 245, 280, 288, 295, 365–368, 371–374
Developmental evolution, 4, 31–35, 94, 99n21
d'Herelle, F., 310
Dictyostelium discoideum, 381
Differentiation (cell), 230–231. *See also* Development
DNA. *See* Double-helix model; Linguistic metaphor
Dobzhansky, T., 4, 91, 175, 383n3
Dominance, 37n23, 88, 98
Double-helix model (of DNA), 1, 4, 7, 9, 21–23, 86, 93, 123, 139n7, 169, 187, 210, 367, 372
Drift, genetic, 30, 233, 235, 379
Drosophila funebris, 91, 369
Drosophila melanogaster, 88, 365, 371, 377
Drosophila pseudoobscura, 42n87

Eberthella typhosa, 314
Ecology, vii, viii, 92, 96, 97, 111–112, 280
Editing (RNA), 24, 94, 193, 221, 270, 277, 375, 382

Index

Ehrlich, P., 367
Eigen, M., 122, 135, 145, 146, 151
Electromagnetism, 23, 55, 75, 105
Empiricism, logical, 5, 12, 13, 55, 176n1, 176n4, 176n5
Endocrine system, 97
Enzyme induction. *See* Operon
Ephrussi, B., 185, 194, 209
Epigenetics, 28, 197, 245
Escherichia coli, 11, 131, 191, 193, 224, 235, 243, 250n9, 273, 292, 293, 296, 313–315, 319–322, 347, 349, 350, 354, 382
Evolution, vii, viii, 4–5, 16–20, 26–35, 86, 96, 100n26, 111, 122, 127–129, 133, 135–138, 145–157, 175, 225, 229, 233–236, 274, 275–277, 287–290, 303–309, 334, 348, 349, 353, 354, 357, 358, 365, 378–382
Evolution, developmental. *See* Developmental evolution
Evolution, received view, 21, 26–34, 303, 304, 353, 357, 358, 380–382
Evolutionary ethics, 164
Evolutionary developmental biology, 32
Exaptation, 133, 134
Explanation, 3, 5–14, 60–75, 87, 90, 106–108, 117, 119, 120–129, 146, 154–157, 183, 205–207, 222, 229, 236–242, 280, 330–333
 functional, 3, 16–20, 55–59, 106, 121–138, 146, 154–157, 345n109
 reductionist (*see* Reductionism)
Explanatory weight, 9, 37n22, 90
Expressivity, 85, 91, 92, 370

Feedback, 11, 110, 194, 195, 223, 224, 226, 227, 230, 231, 327, 328, 343n93
Feyerabend, P., 56, 76n3
Fisher, R. A., 27, 30, 239, 261, 306, 339n37

Fitness, 16, 17, 26, 126, 127, 235, 305, 307, 310, 321, 323–324, 334, 354
Fluctuation test, 291–296, 310–314, 319–325, 329, 332, 338n31, 349, 356, 358, 359
Fodor, J., 68–69
Foster, P. L., 349, 350–353
Function, 3–4, 14–20, 106, 125–127, 134, 135, 155–156, 161–165, 174–175, 378–379

G-value paradox, 33, 377
Galen, 6
Gamow, G., 170, 172, 187, 196, 210–215, 372
Gatlin, L., 234
Gene, cryptic. *See* Editing (RNA)
Gene conversion, 29
Gene regulation, 222, 223–231, 248, 373. *See also* Operon
Gene–environment interaction, 91, 269
Gene–enzyme relationship, 22, 184, 208, 264, 367
Genericity, 95–96
Genetic program, 23–25
Genetics
 classical, 12, 13, 21, 22, 24, 28, 40n61, 57, 70, 71, 72, 85–92, 109–111, 121, 365–367
 developmental, 13, 31, 95, 372, 374–376
 molecular, vii, 12, 71, 72, 86, 87, 109, 121, 169, 377
 population, 92, 111, 288–289
 reverse, 87
Genomics, 12, 25, 33–34, 377
Genotype–phenotype relation, 26, 27, 31, 32, 85, 368, 369
Gilbert, W., 25, 375–376, 380
Golomb, S. W., 172, 189, 217
Gould, S. J., 133, 134, 372
Gratuity, 93, 228–229
Gravitation, 105, 237

Haldane, J. B. S., 27, 30, 223, 238, 259n13, 281n6, 292, 306, 355, 374
Haldane, J. S., 7
Hall, B. G., 294, 295, 321, 349–351
Hardy–Weinberg rules, 27
Hartley, R. V. L., 261
Harvey, W., 6, 105
Helmholtz, H. v., 105, 106
Hemophilus influenza, 314
Hempel, C. G., 13, 77n15
Heritability, 108–109
Heuristics, 4, 55, 88
Hinshelwood, C. N., 315
Hogben, L., 7, 91
Holism, 7, 10–11, 280, 371, 374
Holliday, R., 29
Homeobox, 374
Homeostasis, 11
Hooker, C. A., 75, 77n7
Hopkins, F. G., 7
HOX genes, 374, 377, 380
Hull, D. L., 72, 73, 117, 118, 121
Human Genome Project, viii, 3, 20, 33, 40n58, 87, 173–175, 198, 246, 374–377
Huntington's disease, 279
Huygens, C., 105
Hyalodaphnia, 90, 368
Hypercycles, 145–146, 150–153, 157

Immunity, 86, 173, 184, 208, 247, 263, 311, 349, 381
Incommensurability, 76n3
Individualism, methodological, 96, 111, 252n24
Informatics, 376
Information
  biological, 4, 20–25, 29, 86, 110, 120, 149, 169–174, 183–199, 205, 206, 208–219, 222–223, 231–236, 241–242, 244–249, 261–281, 294, 331, 350, 367, 372, 374, 375
  semiotic, 264, 270–280
  theory, 194–196, 231–236, 261, 264, 265–270, 273, 274
Intelligence, 13
Intelligent design. *See* Creationism

Jablonka, E., 28
Jacob, F., 10, 11, 24, 226, 373
Johannsen, W., 26, 88, 368–369
Jukes, T. H., 30

Kalmus, H., 224
Kammerer, P., 289
Kant, I., 7, 71
Kauffman, S. A., 64–65, 67–68, 75, 118, 119, 331
Kemeny, J., 55
Kimura, M., 30, 196, 233, 234, 259n17, 267
King, J. L., 30
Kitcher, P., 72, 73, 118, 121
Kleene, S. C., 227

Lamarck, J.-B., 26, 28, 106, 290, 303, 308, 309, 357
Lamarckism, 28, 287, 291, 295, 303, 304, 306–310, 319, 320, 326, 328, 329, 333, 336n17, 337n27, 343n96, 357
Lamb, M. J., 28
Lea, D. E., 292, 312, 318, 355
Lederberg, J., 86, 89, 185, 187, 209, 224, 225, 238, 250n4, 309, 316–318
Legett, A. J., 253n26
Lenin, V. I., 369
Lenski, R. E., 321–323, 325, 348, 350, 354, 358
Levin, B., 313, 321, 323, 325
Lewontin, R. C., 141n19, 145, 147, 150, 259n17, 305
Lindegren, C. C., 29
Linguistic metaphor, 23–25, 40n64, 190, 191, 193, 198–199, 206, 264, 270

Linkage, 85, 88, 92, 109, 237
Linum usitatissimum, 296
Load, genetic (mutational), 354
Load, genetic (substitutional), 233
Lock-and-key fit, 10, 109, 186, 209, 240, 263
Loeb, J., 7
Logical positivism. *See* Empiricism, logical
Luria, S. E., 185, 209, 291–293, 310–314, 318, 332, 344n108, 356, 358
Luria–Delbrück distribution, 292–294, 314, 318, 320, 321, 323–325, 355–356, 359
Lwoff, A. 172
Lysenko, T. D., 289, 307, 333, 336n14, 337n27, 357

Manic depression. *See* Bipolar affective disorder
Materialism, 106
Matthaei, J. H., 190, 217
Maxwell, J. C., 23
Maynard Smith, J., 270
Mayr, E., 16, 18, 24, 77n10, 118, 119, 337n20, 373
Mazia, D., 167, 249n4
McClintock, B., 28
Mechanical philosophy, 5, 7, 8, 16, 36n12, 55, 70, 71, 97, 105, 106
Mechanics, 78n23, 105, 237
Meiotic drive, 28–29
Mendel, G., 26, 88, 332
Mendelism, 12, 21, 26–31, 85, 88, 99n10, 109, 239, 288, 289, 306, 365, 366, 368
Metabolism, 2
Metaphor. *See* Information, biological; Linguistic metaphor
Model organisms, 34, 91, 279
Models, 21–23, 68, 85, 166
Modern Synthesis, 26. *See also* Evolution, received view

Modularity, 380–382
Molecular shape, 9, 93, 110, 186, 209, 211, 212, 241, 263, 367
Monod, J., 10, 11, 24, 191, 194, 224–229, 231, 253n27, 373
Morgan, T. H., 31, 86, 88, 89, 365, 366, 372
Morphogenesis, 4, 374. *See also* Development
Muller, H. J., 89, 365–367
Multicellularity, 381
Mutagenesis, 4, 21, 27, 94, 185, 190, 192, 209, 218, 277, 288–298, 305–329, 333–335, 337n29, 347–359
adaptive (*see* Mutagenesis, directional)
directional, 5, 290–298, 307–329, 331–335, 347–359

Nagel, E., 6, 55, 56, 60–63, 68, 72, 75, 106, 109
Naturalism, 164, 177n5
Natural selection, 16–20, 26–31, 122, 127–129, 133, 135–138, 145–153, 155–157, 196, 229, 233, 235, 253n28, 267, 275–277, 287–289, 305, 306, 325, 328–329, 352–354, 373, 379
Nature–nurture dispute, 13–14
Neo-Darwinism, vii–viii, 26, 185, 216, 287–291, 295–298, 303–310, 316, 319, 325–329, 333, 365. *See also* Evolution, received view
Neo-Lamarckism. *See* Lamarckism
Neurobiology, 111, 237
Neuroticism, 13
Neutral theory (of molecular evolution), 18, 29–31, 233
Neutrality paradox, 30, 32
Newcombe, H. B., 313, 315–316
Newton, I., 164
Nickles, T., 61, 75, 252n21
Nilsson-Ehle, H., 368
Ninio, J., 353
Nirenberg, M. W., 190, 217

Norm-of-reaction, 42n91, 90, 91, 369–370
Nyquist, H., 281n2

Operon model, 11, 24, 86, 93, 110, 194, 224–229, 240, 241, 253n27, 373, 374
Oppenheim, P., 55
Optics, 55, 75, 237
Organizer, 371
Orgel, L. E., 170
Origin of life, 135, 136, 145, 150–153, 198, 246

Pardee, A. B., 224, 225
Pauling, L, 1–2, 8, 23, 86, 93, 184, 208, 263
Pearson, K., 99n10
Penetrance, 85, 91, 92, 370
Persistence, principle of, 18–19
Phage, 120, 235, 292, 309–319
Phenogenesis. *See* Development
Phenotypic lag, 294, 314, 324, 325, 339n38
Phenotypic plasticity, 14, 42n91, 279, 354, 368–371
Photography, 161–162, 174–175
Physicalism, 10, 56, 67, 68, 154, 156–157, 176n5, 207, 219, 229, 236–238, 242, 247, 249n2, 252n20, 296. *See also* Physical warrants
Physical warrants, 67, 154–157, 237, 238, 252n20, 331
Physics, fundamental, 166–169
Poisson distribution, 291–295, 311, 312, 320, 323–325, 338n31, 353, 356
Polydactyly, 110, 261
Polygenic traits. *See* Complex traits
Polymerase chain reaction, 244
Project K, 34–35
Protein folding, 93, 173, 185
Proteomics, 14, 34–35, 377, 378
Proximate-ultimate distinction, 16, 18, 19

Quantitative trait locus (QTL), 92
Quantum mechanics, 66, 67, 167–169, 251n19
Quastler, H., 195, 232

Randomness, 27, 41n76, 287, 290, 292, 293, 305, 306, 328–329, 350, 352, 353
Reading disability, 13
Redistributive plating, 292, 315–316, 318
Reduction. *See also* Reductionism
  abstract, 108
  strong, 108, 111, 239 (*see also* Reductionism, physical)
  weak, 108, 111, 237
Reductionism, viii, 2, 5–14, 16, 32, 33, 25, 55–80, 87–98, 106–113, 117–122, 128–129, 135–138, 172, 206, 229, 236–244, 247, 251n19, 252n22, 280, 296, 303, 304, 329–334, 335n3, 344n104, 344n105, 344n108, 345n109, 359, 371
  constitutive, 57–59, 68–70, 118, 119, 139n3
  explanatory, 57–59, 63–68, 118–122, 129, 135–138, 139n3, 146, 158n3, 168, 174, 251n19
  genetic, 3, 6, 12–14, 31, 33, 35, 87–92, 109, 110–111, 237, 238, 265, 280–281, 365, 370, 371, 376
  physical, viii, 2, 3, 6–12, 35, 87, 88, 280
  theory of, 57–63, 74–76, 118–121, 137, 251n19, 344n104
Redundancy, 95–96
Relativity, 78n23, 167–169, 237, 251n19
Replica plating, 292, 316–318, 349, 359
Research programs, 21–22, 71–75, 106, 111, 366
Research strategies, 4, 55, 64, 70–74, 112, 117, 119, 120, 330
Romanes, G. J., 42n78, 288, 309
Romashoff, D. D., 91, 369, 370
Rosenberg, A., 69, 72, 73, 118–122

Roth, J. R., 354
Roux, W., 86
Ruse, M., 117, 118
Ryan, F. J., 292, 293, 314, 318

Saccharomyces cerevisiae, 29, 95, 224, 296, 347
Salmon, W., 64
Salmonella typhimurium, 347
Sanger, F., 212
Sarkar, S., 65–68, 73–75, 252n21, 291, 294, 295, 331, 351–353, 355–356, 359
Schaffner, K., 56, 57, 60–63, 66, 68, 71–73, 75, 117, 119, 120, 121
Schizophrenia, 13, 89
Schneider, T. D., 196, 234–236, 267
Schrödinger, E., 22, 170, 184–184, 209, 210, 218, 264, 366–367, 372
Screening-off, 64
Segregation (germline), 197, 246, 288, 297
Segregation (Mendelian), 88, 92, 109
Segregation distortion, 29
Selection, levels of, vii, 26, 111, 137, 234, 335n3, 335n4, 358, 382
Selection, units of. *See* Selection, levels of
Semantics, 36n8, 263–264, 268
Sequestered modular template (SMT) model, 378–382
Servomechanics, 223
Sexual orientation, 13
Sexual reproduction, 298n1
Shannon, C. E., 231–232, 235, 261–262, 265–267, 273, 274
Shapiro, J. A., 247–248, 319, 320, 321, 349
Shimony, A., 74, 77n9, 78n23, 253n26
Sib-selection, 292, 316–318, 349, 359
Sickle-cell trait, 18, 110, 261, 279
Signal. *See also* Gene regulation
Slatkin, M., 323, 325
Smuts, J. C., 7

Sneed, J., 62, 75, 118
Sociobiology, 33, 164, 173
Specificity, 10, 184, 197, 208–211, 219, 225, 232, 261, 262, 264, 269, 275, 278, 367
Spemann, H., 371
Spiegelman, S., 187, 249n4
Splicing (RNA), 24, 94, 192, 193, 221, 277, 375, 381
Stahl, F. W., 318, 326, 327
Staphylococcus aureus, 314
Statistical mechanics, 55, 267
Statistical relevance, 64
Statistics, 261, 292, 296, 303, 304, 320, 329, 332. *See also* Bayesian methods
Stegmüller, W., 62
Stent, G., 139n5
Stephens, C., 376
Stewart, F. M., 313, 321, 323, 325, 355
Stereospecificity. *See* Molecular shape; Structure–function relationship
Stoichiometry, 275–277
Structure–function relationship, 9, 184, 241
Sturtevant, A. D., 366
Supervenience, 69–70, 109, 111, 118, 119, 121
Suppes, P., 55, 56, 62, 76n4
Symmetry, 162, 167–168
Synthetic identities, 61, 68, 106

Telegraph, 261
Teleology, 7, 16, 71, 140n10
Teleomechanism, 7, 71
Theories, 6, 12, 13, 21, 23, 55, 60–63, 73, 75, 106–108, 118, 119, 140n12, 166, 205, 251n19
Thermodynamics, 55, 74
Timoféeff-Ressovsky, N. W., 91, 367, 369–370
Tolerance, 95
Traits, 278–280, 367
Trypanasoma brucei, 193, 259n12

Umbarger, H. E., 224
Universality (of cellular mechanisms), 31, 34

Vitalism, 7, 71, 109, 119
Vogt, O., 91, 369
Vrba, E. S., 133, 134

Waddington, C. H., 259n17, 370–372, 374
Wallace, A. R., 1, 16, 26
Waterton, C., 14–15
Watson, J. D., 1, 21, 33, 170, 185, 191, 211, 367
Weak interactions rule, 9
Weaver, W., 1
Weismann, A., 288, 296, 309
Wiener, N., 194, 223
Williams, G. C., 233, 259n17, 267
Wilson, E. O., 33
Wimsatt, W. C., 57, 61, 63–64, 73, 117–119, 126–127, 155, 252n21, 331
Woese, C. R., 172
Woltereck, R., 90, 368–369
Woodger, J. H., 6, 76n2
Wright, S., 27, 30, 306

Yates, R. A., 224
Ycas, M., 213
Yockey, H. P., 192

Zheng, Q., 356